CW00518281

Sustainable Finance and Banking

The Financial Sector and the Future of the Planet

Marcel Jeucken

Earthscan Publications Ltd
London • Sterling, VA

First published in the UK and USA in 2001
by Earthscan Publications Ltd

ISBN: 1 85383 766 0

Typesetting by PCS Mapping & DTP
Printed and bound in the UK by Biddles Ltd, Guildford
Cover design by Declan Buckley

For a full list of publications please contact:

Earthscan Publications Ltd
120 Pentonville Road, London, N1 9JN, UK
Tel: +44 (0)20 7278 0433
Fax: +44 (0)20 7278 1142
Email: earthinfo@earthscan.co.uk
http://www.earthscan.co.uk

22883 Quicksilver Drive, Sterling, VA 20166-2012, USA

Earthscan is an editorially independent subsidiary of Kogan Page Ltd and publishes in association with WWF-UK and the International Institute for Environment and Development

A catalogue record for this book is available from the British Library

Library of Congress Cataloging-in-Publication Data

Jeucken, Marcel
[Duurzaam bankieren. English]
Sustainable finance and banking : the financial sector and the future of the planet / Marcel Jeucken.
p. cm.
Rev. translation of: Duurzaam bankieren.
Includes bibliographical references and index.
ISBN 1-85383-766-0
1. Sustainable development—Finance. 2. Banks and banking—Netherlands. I. Title.

HC79.E5 J4813 2001
332.1'068—dc21

2001002398

This book is printed on elemental chlorine-free paper

Contents

Part II Banking and Sustainability

Part III In Reflection

Appendices

List of Boxes, Figures and Tables

Boxes

Figures

TABLES

Foreword

Jan Pronk

Sustainable development has featured more and more prominently on the international agenda for governments and business over the last few decades. Environmental policy in The Netherlands originally focused on business, the main source of environmental damage. The financial sector was explicitly brought on board for the first time in the late 1990s. The 1997 policy document on the environment and the economy was the first to describe a role for the financial sector. It states that capital and financiers will need to be involved in integrating the environment into business practice and the economy.

Three initiatives that have strengthened the role of the financial sector in environmental policy are:

1. The Green Investment Incentive Scheme, launched in 1995, which offers individuals, companies and public authorities loans at 2 or more per cent below market rates for green or environmentally friendly projects.
2. The Netherlands Bankers' Federation and the government's multi-year energy conservation agreement, concluded in 1996, which aims to increase energy efficiency by 25 per cent over 1995 levels by 2006.
3. An Environmental Council for the Banking Sector, set up in 1999, which enables the banking world and the government to consult on how the financial sector can do more to promote sustainable development. Topics debated include how banks can get involved in the clean development mechanism and joint implementation, and what banks are contributing to funding for soil remediation.

There have also been sustainable banking initiatives at the international level, including UNEP's important declaration on banking and the environment. Financial institutions which sign up to this declaration recognize their duty to stimulate sustainable development and commit themselves to integrating environmental considerations into their internal management and commercial decisions. So far some 175 financial institutions around the world have signed the declaration. It is vital that sustainable banking should not be confined to just a few countries, as banking activities and capital flows are not confined by national borders.

The financial sector is involved in the environment in many ways:

- as investors, providing businesses with capital for investment;
- as developers of financial products that can strengthen sustainable development, such as energy-saving or green investment funds;
- as stakeholders who have an interest in preventing businesses running environmental risks; and
- as polluters.

The financial sector used to view the environment as a source of financial risk, for example when companies discover their site is polluted. Nowadays, the opportunities presented by the environment are also being discovered. Banks in The Netherlands and elsewhere are increasingly developing sustainable banking products, such as environment-related investment funds, green mortgages and sustainable insurance policies. Trailblazers began on a small scale in the 1960s and 1970s. New challenges lie in climate issues and in coming up with financial innovations and ideas to help businesses manage carbon dioxide in a cost-effective way. The financial sector can give sustainable development a tremendous boost with activities of this kind.

This book describes such activities in detail, along with other sustainable banking and finance issues. It gives an interesting view of sustainable banking from inside the sector. It combines countless examples, blueprints and analyses to provide a detailed overview of the trend towards sustainable banking. The book is written from the angle of the sector as a whole, not from the viewpoint of a particular bank or institution. It contains a wealth of background information and examples of the products and activities of various banks. The author is the first to tackle this subject in such broad and accessible terms, making his book of interest not only to experts in the financial sector or government but also to people with no specialist banking or financial expertise. I have no doubt that this book will inspire many people to take up the challenge of sustainable banking.

Jan Pronk
Minister of Housing, Spatial Planning and the Environment
The Netherlands
April 2001

Foreword

Hans Smits

Much of the 20th century's focus has been on economic progress, in which mankind has made giant steps. Increasingly, side effects such as loss of biodiversity, climate change and various forms of environmental pollution are manifest and need attention. The same is true for social issues such as poverty alleviation and equal development chances for all. Therefore, the issue of sustainability has become more and more important. In my opinion it will be one of the key issues for the 21st century. As such, sustainable development is about the welfare of human beings and a natural environment that does not reduce possibilities for future generations, without losing sight of the economic continuity of the current generation.

The path towards sustainability will involve citizens, non-governmental organizations (NGOs), governments, companies and, obviously, the financial sector as well. In addition to our historical socioeconomic objectives, our mission statement puts the ecological dimension well: 'The Rabobank Group believes sustainable growth in prosperity and wellbeing requires careful nurturing of natural resources and the living environment. Our activities will contribute to this development.' These activities are numerous, and include standardized environmental risk assessments, and products such as our Green Bank for green savings, investments and mortgages; our RG Sustainable Equity Fund; and our environmental loans, leases, mortgages and insurance products. Obviously, we also continuously try to improve our internal environmental care as well and report on all sustainability issues in a transparent way. Moreover, we are a signatory of the UNEP Statement by Financial Institutions on the Environment and Sustainable Development, and a member of the World Business Council for Sustainable Development. Furthermore, we think climate change is a major issue and therefore, among others, participate in the Prototype Carbon Fund of the World Bank.

As a large, international, all-finance and cooperative bank and a major global player in agri-business finance, we are quite naturally involved in the concept of sustainable development. In our policies we focus on the sustainability leaders of today and also help companies which are having difficulties in integrating sustainability into their activities. In my opinion sustainable banking is about both supporting the innovative and proactive companies, and stimulating the reactive companies. Alliances and cooperation between NGOs, governments, companies, consumers and the financial sector are a natural outcome of this 'double strategy' and will pave the way towards sustainability. As the major financial player in The Netherlands with a large local network, our bank is well placed to do so and is involved in various such alliances. In my opinion, this joining of forces is the best strategy for achieving more sustainable development.

This book is impressive in both its scope and depth and is, to my knowledge, the first major publication which deals integrally with all aspects of sustainability and banking. My compliments go to the author for making the book's structure and content accessible and of interest to readers from the financial community, the environmental movement, business, government and society in general. I sincerely think this book can improve the dialogue within the financial sector and with their stakeholders, and propel many towards a sustainable future.

Hans Smits
Chairman, Rabobank Group
The Netherlands
April 2001

Acknowledgements

'Mistakes exist
in the proximity of the truth
and so mislead us.'
Tagore

Although the process of writing this book started as a translation of a Dutch book called *Duurzaam Bankieren* (*Sustainable Banking*, Jeucken, 1998b), the process of rewriting turned out to be so inspiring that in fact more of a new book has evolved. *Duurzaam Bankieren* appeared in 1998 in celebration of the Rabobank organization's centennial. The book focused strongly on experiences and issues contained within the Rabobank. However, some prescience in 1997 had already resulted in the Dutch version being split up into a general section (the basic text) and 20 articles contributed by colleagues of the Rabobank organization, which specifically sketched the developments and issues within their bank. A broader translation which is intrinsically separate from the Rabobank was straightforwardly accomplished by eliminating the articles overly specific to Rabobank from this book. Furthermore, this new book has been a fully private undertaking.

Writing this book has been inspiring in that I took a closer look at international developments and more systematically assessed developments in other major banks in the world. Moreover, much has happened in this sector since the mid-1990s! This amended approach eventually resulted in an entirely new chapter (9), where in a comprehensive framework differences between leading international banks are analysed. Other chapters have been revised considerably to represent the – in my view – current state-of-the-art of banking and sustainability and to represent international developments and issues as well. Two chapters (2 and 6) focus slightly on the situation in The Netherlands, explaining current law and politics relevant to sustainability in banking. Dutch environmental policy is considered to be a worldwide leader and by taking one country as an example, a more in-depth analysis is possible. However, references are made to international deviations from the Dutch standards. Finally, *Sustainable Finance and Banking* has made an attempt to give introductions to environmental issues to people with no background in this area and, vice versa, has made an attempt to give an introduction to finance in general to people with no background in banking. In short, the book has been written with an eye not just to people working within finance or the environmental community but also to governments, non-governmental organizations (NGOs), companies and individuals that have an interest in the role that the financial sector can play in achieving sustainable development.

Many people have assisted me in writing this book, in some form or the other. Firstly, I would like to thank my parents and Marjan for their continued support and simply for being there. Secondly, I would like to thank the Dutch Ministry of Environment, Interpolis Re, Rabobank International and Rabobank Nederland for their financial support, for without their support you would not be reading these words. Thirdly, I would like to thank the translation team of CPLS Text & Copy and the publisher Earthscan for their excellent work and cooperation. Although there is much reason to spend a few words on every individual, there is simply not enough scope. So in all gratitude I would like to thank all of you: Glen Armstrong, Duncan Austin, Onno Frank van Bekkum, Theo van Bellegem, Hans van den Berg, Jesper Berggren, Marieke van Berkel, Peter Blom, Wim Boonstra, Jan Jaap Bouma, Toon Bullens, Jacqueline Cramer, Richard Cooper, Breda Cummins, Eric van Dam, Rowan Davies, Albert Dierick, Daan Dijk, Theo Dijkstra, Wim van Dinten, Alois Flatz, Claude Fussler, Olav-Jan van Gerwen, Marina de Gier, Robert Goodland, Michaël de Groot, Wim Hafkamp, Bregje Hamelynck, Sybren de Hoo, Chris Horgan, Harry Hummels, Frances Innocent, Jeroen Jansen, Karin Jansens, Walter Kahlenborn, Brigitta van Kanten, Gerard Keijzers, Bert Kersten, Gijs Kloek, Andreas Knörzer, Jan van der Kolk, Henk Korte, Bart-Jan Krouwel, Hugo Kuijer, Willem Lageweg, Peter van Lamoen, Wim Lange, Jacqueline Aloisi de Larderel, Marischka Leenaers, Marc Leistner, Frances MacDermott, Hans Megens, Folkert van der Molen, Jan Willem de Moor, Michel Negenman, Kenneth Newcombe, Ikuo Nishimura, Sandra Odendahl, Richard Paardenkoper, Jan Pronk, Rik van Reekum, Robert Rubinstein, Rikkert Scholten, Lucas Simons, Jonathan Sinclair Wilson, Jan-Pieter Six, August Sjauw-Koen-Fa, Hans Smits, Lucas van Spengler, Piet Sprengers, Thomas Steiner, Hans Stielstra, Allerd Stikker, Susie Day, Bouwe Taverne, Rob van Tilburg, Dorine Tinga, Peter van der Veer, Roel van Veggel, Pier Vellinga, Paul van de Ven, Jan Vinkenvleugel, Jaap van der Vlies, Herman Vollebergh, Michiel van Voorst, Jan van Winkel and Dana Younger. Many other people have helped shape my thoughts and knowledge. Though I cannot mention each of you here, I do owe each of you thanks as well. Obviously, no one mentioned here but me takes responsibility for the contents of this book.

I would like to finish by expressing my hope that by writing this book I have furthered awareness of the need for sustainability and the ways in which the financial sector may contribute to its achievement.

Marcel Jeucken
Utrecht, The Netherlands
June 2001

List of Acronyms and Abbreviations

AAU	assigned amount units
ABA	American Bankers Association
ADB	Asian Development Bank
AFDB	African Development Bank
AFN	alternative fulfilment of needs
ALARA	as low as reasonably achievable
ATM	automatic teller machine
BBA	British Bankers' Association
BBVA	Banco Bilbao Vizcaya Argentaria
BIS	Bank for International Settlements
BLB	Bayerisches Landesbank
BSB	soil remediation for Dutch industrial sites
BSCH	Banco Santander Central Hispano
BSE	bovine spongiform encephelitis
BTM	Bank of Tokyo-Mitsubishi
CBA	Canadian Bankers Association
CBS	Dutch Central Bureau for Statistics
CDB	Caribbean Development Bank
CDM	clean development mechanism
CEO	chief executive officer
CER	certified emissions reductions
CERCLA	Comprehensive Environmental Response, Compensation and Liability Act (US)
CERES	Coalition for Environmentally Responsible Economies
CFC	chlorofluorocarbon
CIBC	Canadian Imperial Bank of Commerce
CO_2	carbon dioxide
CoP6	Sixth Climate Conference
CPB	Dutch Central Planning Bureau
CSFI	Centre for the Study of Financial Innovation
CSG	Credit Suisse Group
DGXI	*former abbreviation for the* Environment Directorate-General (EC)
DJSGI	Dow Jones Sustainability Group Index
DNS	debt-for-nature swaps
DSI 400	Domini 400 Social Index (US)
DTTI	Deloitte Touche Tohmatsu International
EA	environmental assessment
EBRD	European Bank for Reconstruction and Development

EC	European Commission
ECA	export credit agencies
ECI	environmental condition indicators
EDC	Export Development Corporation
EDI	environmental damage insurance
EEA	European Environment Agency
EIA	environmental impact assessment
EIB	European Investment Bank
EIF	European Investment Fund
EIRIS	Ethical Investment Research and Information Service
EMAS	European Eco-Management and Audit Scheme
EMS	environmental management system
EMU	European Monetary Union (EU)
EPE	environmental performance evaluation
EPI	environmental performance indicators
ERU	emission reduction units
EtAc	ethical accounting
ETEI	Emissions Trading Education Initiative
EU	European Union
FEM	*Financieel Economisch Magazine* (Dutch journal)
FEMAS	European Eco-Management and Audit Scheme for the financial sector
FDI	foreign direct investment
FI	financial intermediary
FORGE	Financial Organisations Review and Guidance on the Environment
GAAP	generally accepted accounting principles
GAP	global action plan
GDP	gross domestic product
GEF	Global Environment Facility
GEMI	global environmental management initiative
GMO	genetically modified organism
GNP	gross national product
GRI	Global Reporting Initiative
GWh	gigawatt-hour
HDI	human development indicators
HSBC	Hong Kong and Shanghai Bank Corporation
IADB	Inter-American Development Bank
IBRD	International Bank for Reconstruction and Development
ICC	International Chamber of Commerce
ICSID	International Centre for Settlement of Investment Disputes
IDA	International Development Association
IEDI	integrated environmental damage insurance
IET	international emissions trading
IFC	International Finance Corporation
IISD	International Institute for Sustainable Development
ILO	International Labour Organization

IMF	International Monetary Fund
IPCC	Intergovernmental Panel on Climate Change
IRN	International Rivers Network
IRRC	Investor Responsibility Research Centre
ISA	international accounting standards
ISO	International Organization for Standardization
IUCN	World Conservation Union (*formerly* International Union for Conservation of Nature and Natural Resources)
JEXIM	Japanese Export–Import Bank
JI	joint implementation
KfW	Kreditanstalt für Wiederaufbau
KWh	kilowatt-hour
LASER	Dutch National Service Schemes
LBB	Landesbank Berlin
LCA	lifecycle assessment
LDCs	least developed countries
LETS	local exchange trading system
LG	Landesgirokasse
MDB	multilateral development bank
MEA	Millennium Ecosystem Assessment
MIGA	Multilateral Investment Guarantee Agency
MITI	Ministry of Trade and Industry (Japan)
MJA	Dutch multi-year energy-savings agreement
MNE	multinational enterprise
MPI	management performance indicators
MSCI	Morgan Stanley Capital International
NAB	National Australia Bank
NEPP	National Environmental Policy Plan (The Netherlands)
NGO	non-governmental organization
NL-EIA	Dutch tax relief on energy investments
NOVEM	Dutch Organization for Energy and Environment
NOVIB	Dutch Organization for International Development and Cooperation
NOx	nitrogen oxides
NPI	National Provident Institution
NPRI	national pollutant release inventory
NVB	Dutch Association of Banks
ODA	official development assistance
OECD	Organisation for Economic Co-operation and Development
OPI	operational performance indicators
PAGE	pilot analysis of global ecosystems
PCBs	polychlorinated biphenyls
PCF	Prototype Carbon Fund
PPA	prospective purchase agreements
PR	public relations
PV	photovoltaic
REEF	Renewable Energy and Energy Efficiency Fund

RIVM	Dutch institute for health and environment
ROaE	return on average equity
SBA	Swiss Bankers' Association
SDG	Sloar Development Group
SEDS	sustainable economic development scenarios
SEP	Dutch electricity-producing companies
SERM	Safety and Environmental Risk Management
SFE	Sydney Futures Exchange
SHS	solar home systems
SME	small- to medium-sized enterprise
SO_2	sulphur dioxide
S&P 500	Standard & Poor's 500
SPC	special purpose company
SRI	social responsible investing
SVN	social venture network
TRI	toxic release inventory (US)
UBS	Union Bank of Switzerland
UN	United Nations
UNCED	United Nations Conference on Environment and Development
UNDP	United Nations Development Programme
UNEP	United Nations Environment Programme
UNEP-FSI	Financial Services Initiative of the United Nations Environment Programme
UNFCC	United Nations Framework Convention on Climate Change
UNRISD	United Nations Research Institute for Social Development
US EPA	US Environmental Protection Agency
US FDIC	US Federal Deposit Insurance Corporation
UV	ultraviolet
VAMIL	Voluntary Depreciation of Environmental Investments Scheme (The Netherlands)
VBDO	Dutch Association for Investors in Sustainable Development
VfU	German Banking Association for Environmental Management
VNO-NCW	Dutch confederation of Industry and Employers
WB	World Bank
Wbb	Dutch Soil Protection Act
WBCSD	World Business Council for Sustainable Development
WBG	World Bank Group
WCED	World Commission on Environment and Development
Wm	Environmental Management Act (The Netherlands)
WRI	World Resources Institute
WRR	Dutch Scientific Council for Government Policy
WTO	World Trade Organization
WWF	*formerly known as* World Wide Fund For Nature *and* World Wildlife Foundation

1

Introduction

Trends

Humankind's awareness of its dependence on the environment goes back to the very beginning of human history. Through the centuries, the scale, degree and location of environmental problems and awareness have evolved correspondingly.[1] One can speak of *structured* environmental awareness since the Industrial Revolution. Less than 30 years ago, the world of science also acknowledged the severity of environmental problems. It became clear that it was no longer just a matter of incidents – such as an oil tanker accident off the coast of La Coruña in Spain – but that the existence of all of humanity was threatened by a silent global environmental crisis. In the 1980s especially, it became evident that economic development, which had brought significant prosperity, also caused not only social but environmental abuses as well. Concern for the environment has now been translated into laws; most countries are trying to pay back the environmental debts that have been incurred and stimulate preventive actions.

In view of the North–South problem, the concept of sustainable development was introduced into the political lexicon. Emphasis was laid on the interrelationship between environment and economy. Initially, protection of the environment was interpreted as a burden, an increase in business costs. As time passed, businesses nevertheless began seeing a positive relationship between the environment and economics and began opening up to the idea of environmental concern. A growing awareness among consumers, producers, employees and competitors is prompting an increasing number of businesses to go on the offensive in terms of sustainable development. The realization that pursuing sustainable development is an integral component of doing business is starting to hit home with many people in the business world; prospects of additional revenues exist in addition to considerable cost savings.

Besides being a sign of social accountability, an offensive stance in terms of protecting the environment is often necessary because of business continuity. Consumers and producers are making demands on end products and semi-manufactured goods respectively. Competitors distinguish themselves through new environmentally friendly or sustainable products. Some businesses are entirely dependent on finite natural resources. In order to secure permanent business continuity, care in handling these resources is essential; soft drinks manufacturers must think about where they are going to get clean drinking water in the future. Another example is the cooperation between Unilever and WWF concerning the conservation and management of fishing grounds. These initiatives, which go beyond short-term interests and are based on the principal of sustainable development, are examples of sustainable business.[2]

A similar move towards sustainability is perceivable in banks and other financial institutions, although they are still somewhat behind the times. This is related to the perception that banking is a relatively clean industry[3] and to the fact that concern for environmental aspects is equated with meddling in the affairs of their business relations. A 1990 survey revealed that financiers had little interest in the environmental concerns of their business partners (*Tomorrow*, 1993).[4] The environment and sustainable development are nevertheless full of risks for banks (a customer faced with having to decontaminate his soil, for instance) as well as opportunities (particular investment products or internal environmental care, for instance). In the US particularly, the risks for banks rose substantially in the 1980s due to a number of lawsuits (direct liability). US banks therefore began paying attention to environmental aspects before their European counterparts. Certain European banks are now running ahead of those in the US, particularly concerning product development and financing the environmental industry directly, with Swiss banks having been active for some time. Dutch, British and German banks are also ahead of their counterparts in the south of Europe in this respect. The range of activities is also continually evolving. Interesting developments are the foundation of the Dow Jones Groups Sustainability Index, in which a very reputable party set a benchmark at the end of 1998 for sustainable investment in the market and the collaboration between governments and businesses (including BP, Deutsche Bank and Rabobank) in the World Bank's Prototype Carbon Fund (PCF) in 1999. The PCF is being regarded as a vehicle that will enable the participating parties to gain knowledge and experience so that economic solutions can be generated to fight the problem of climate change. The interest banks show in sustainable development has evolved rapidly over the last few years.[5] Various banks perceive the importance and opportunities of sustainability (whether implicitly or otherwise) and have signed declarations, such as those by the International Chamber of Commerce (ICC) and the United Nations Environment Programme (UNEP), in which they endorse common and individual responsibility for bringing about sustainable development.

The role of banks in the achievement of sustainable development is significant considering the intermediary role that they play in society. This last point explains the concern that governments, the European Union (EU), the United Nations (UN) and non-governmental organizations (NGOs) are showing over the effect of banking activities on sustainability.[6] If sustainable business is to succeed at the macro level, the attitude taken by banks will be critical. A bank transforms money into place, term, size and risk in an economy and, as such, it affects economic development. This influence is not only quantitative but can be qualitative, since banks can influence the nature of economic growth. Its financing policy is one way for a bank to create opportunities for sustainable business. An example is funds that are specifically designed for investment in environmentally friendly ways, such as green funds. But banks can go a step further by applying premium differentiation (not based on financial values), for instance, in which a certain investment or credit application must satisfy return or risk-management requirements from a sustainability perspective.

PARADIGM SHIFT

Should banks use financial instruments to allow sustainable development in their own 'sustainable' dealings? That is, base their credit and investment policy on sustainability ratios instead of exclusively financial ratios? Defined like this, there seems to be little place for sustainable banking in the current economic/social paradigm, in which so much is determined by financial ratios. Generally speaking, sustainable development can be achieved through incremental improvements in the production process and in the durability of products.[7] These steps must be taken, and the private sector and a variety of banks have taken up the challenge. In theoretical terms, they are referred to as first-order change processes (see Voogt, 1995, or Watzlawick et al, 1973). Many things are possible in this way and considerable steps can be made towards sustainability.

We may ask ourselves whether sustainable development could be achieved without having to revise current norms and values or the current worldview. Modern society is dominated by economic materialism. There is nothing fundamentally wrong with this – in fact it has resulted in considerable prosperity. But this orientation and fixation on material economic growth has brought with it undesirable side effects, such as dire poverty in developing countries, environmental problems, declining social cohesion, wars and the threat of war. Not only are the problems directly related to the single-minded pursuit of wealth, some problems are even statistically recorded as economic growth (and therefore as increases in wealth). The drive to define, and recognize the importance of, sustainable development in fact implies that the current modes of wealth pursuit are too narrowly focused. A place must be found for immaterial aspects, or even better, a balance between the material and immaterial in the growth of prosperity.[8] The question is whether this can be achieved within the current economic system or economic orientation. Other methods of organization at the meta level are probably needed to take the place of the market mechanism now predominating, or complement it. Change processes of this nature are referred to as being 'second order'. This approach to the question of sustainability will be discussed in the last part of this book, with the final chapter attempting to combine this approach with a vision of banking and sustainability in the future.

GENERAL STRUCTURE OF THE BOOK

This book's objective is to contribute to expanding awareness with respect to sustainability and the steps that banks can take. It does not attempt to provide definitive answers, but does aim to stimulate thinking in terms of solutions. It also tries to stimulate people to go beyond their preconceptions by posing and discussing a number of essential questions. The primary target group is managers, directors and people working in all layers of the financial sector. However, the book will also be explicitly of interest to bank customers, governments, environmental organizations and scientists. The analyses and descriptions are primarily written from a Western perspective. The emphasis is therefore more on environmental pollution as a social problem than on erosion and poverty related to natural resources.[9] Box 1.1 shows the geographical scope of the book.

The book is organized into three parts. Part I offers a general introduction of environmental awareness, sustainable development, 'pure' banking, and sustainability and banking. It thus forms the framework of the discussions in

Box 1.1 Delineation of countries in this book

This book focuses on 'developed' countries, a category that classifies developed or Western countries on the basis of the following three criteria. Firstly, there are the 'high income' countries identified by the World Bank, 49 in all with a per capita gross national product (GNP) (1999, based on the World Atlas method) exceeding US$9,266.[10] Some 20 of these countries attribute the high per capita GNP to a combination of a relatively modest population (less than 100,000 people) and the fact of being a tourist paradise or tax haven. But prosperity also relates to aspects like life expectancy and schooling. So, the second criterion is the top 30 countries from the UN's Human Development Index (UNDP, 2000, p186). Within this group, two countries (Malta and Barbados) do not come into the 'high income' category of the World Bank and the top 21 are all OECD countries. Of the remaining seven countries, three have been involved in the past decade in war, or the threat of it (Israel, Cyprus and Slovenia) while two countries are city states (Singapore and Hong Kong); these have been omitted from consideration. The remaining two countries are members of both the OECD and the EU (Portugal and Greece). The final criterion is related to accessibility of data, which has led to the omission of the five countries above and inclusion of all EU countries.

The definitive selection comprises, therefore, OECD countries that are in both the 'high income' category of the World Bank and the top HDI countries of the UNDP (see Table 1.1).[11]

Table 1.1 *Geographical scope of this book*

Europe, EU	Luxembourg (1, 17)	**North America**
Austria (10, 16)	The Netherlands (14, 8)	Canada (12, 1)
Belgium (7, 7)	Portugal (35, 28)	US (2, 3)
Denmark (8, 15)	Spain (31, 21)	
Finland (21, 11)	Sweden (22, 6)	**Asia and Oceania**
France (13, 12)	United Kingdom (19, 10)	Japan (11, 9)
Germany (16, 14)	**Europe, non-EU**	Australia (18, 4)
Greece (36, 25)	Iceland (9, 5)	New Zealand (30, 20)
Ireland (23, 18)	Norway (6, 2)	
Italy (20, 19)	Switzerland (4, 13)	

Note: Numbers in brackets represent the ranking of each country by GDP per capita (on the basis of 'purchasing power parity' and in USD) and HDI respectively.

The 23 countries selected have an overall population of some 846 million, or about 14 per cent of the world's total population (23 per cent if China and India are excluded).[12] In respect of total world gross domestic product (GDP), these 23 countries account for a 75 per cent share (World Bank, 2000, pp274–275). This book concentrates specifically on these 23 countries, as a whole and individually. All references to 'developed' or 'Western' countries relate to this group. As for banking, only banks with headquarters or origins in developed countries are considered. The activities of these banks in developing countries are considered as well. Some background on most of the banks mentioned in this book is given in Appendix VIII.

Part II which examine the first-order change processes and the various approaches that banks and other actors in the financial sector can take in terms of solutions. These considerations are formulated with both negative and positive aspects; that is to say, both the opportunities and the threats are examined. Part II is the core of this book and looks at things from a pragmatic and descriptive angle. Part III reflects on Part II and explores second-order change processes that break through the existing economic paradigm. The potential role of banks (financial sector) in a new social order will be discussed in the last chapter.

In theory, any of the parts or chapters can be read separately though they form a concise whole. Part II is, for instance, accessible to readers with a practical interest in the interfaces between banks and sustainable development, while Chapter 10 is for readers with an interest in a more philosophical approach to the sustainability issue.

STRUCTURE BY CHAPTER

Chapter 2 outlines the development of environmental consciousness and the concept of sustainable development, beginning with the Ancient Greeks and moving on to the development of environmental policy in the 19th and 20th centuries. It reveals humans to be dependent on their ecological environment. The awareness of that dependence and the influence of human actions on that environment and themselves has grown through the centuries, and since the 1960s – albeit in fits and starts – has accelerated at a powerful rate. It is not only awareness of the environment that has evolved; government policy on the environment reveals trends that attempt to bring about a symbiosis of the environment and the economy. Sustainable development is thus often seen as an evolution of environmental consciousness, but it is really a revolution because the narrow economic thinking has to be broken. In this book sustainable development is predominantly summarized as the merging of the environment and the economy, since it is in this area that many improvements are possible. As a counterbalance to the domination of economic thinking, the environmental dimension of sustainable development weighs rather heavily in this book.

Chapter 3 considers what sustainable development means for businesses, and which phases or stances must be distinguished in this. This issue is also addressed in the light of the 'corporate governance' problem – how external verifiability is applied, and how far businesses are responsible for sustainable development. This is why attention is also paid to environmental reporting.

Chapter 4 discusses the roles of banks in sustainable development. An introduction to banking itself and the development of banks in recent decades is a prelude to the link with sustainability. This link is made by analysing the environmental pressure on banks, the stance of their stakeholders towards sustainability, and the potentially special role that banks can fulfil in sustainable development. Questions that arise are: Can banks promote sustainable development through their intermediation function? Would sustainable banking mean that environmental return is regarded as more important than financial return? Can we progress beyond 'offensive' banking, which tackles environmental consciousness from the standpoint of costs and revenues without considering the sustainability of business activities and credit provision?

In Part II, Chapter 5 looks at the opportunities that emanate from the integration of sustainable development in financial products. Can we conceive of new products that respond to the need for an emerging 'sustainability segment' – that is, businesses and consumers that want to invest or save in a sustainable manner? What opportunities for financing in general are offered by new markets created by the drive for sustainable development? Which new financial products or constructions would stimulate more sustainable investments through financing businesses or private consumers?

Chapter 6 explores the environmental risks associated with credit provision, participation and investments. A crucial aspect of this is that the risks for the bank are mainly determined by the risks of the business which is receiving credit facilities. Can separate methods be developed to analyse such risks in an efficient and systematic manner? In addition, is there an achievable synergy between the methodologies used in the products discussed in Chapter 5?

Chapters 7 and 8 deal with the supporting and organizational activities within banks. These chapters are strongly interrelated. Chapter 7 tackles what a bank can contribute to sustainable development within its own production process. That is, the possibilities for internal environmental care, for example the reduction of the environmental burden through the reduction of energy consumption and waste, and the implementation of sustainable building.

Chapter 8 then explores the organization of internal processes whose objective is to link sustainability to the way a bank operates: that is, the issue of how to instigate and facilitate the necessary first-order change processes. For the successful integration of sustainable development in the activities of banks, there has to be broad internal support for the environmental policy and the sustainability concept. A key to this is consciousness, and therefore much emphasis is laid on internal communication. Chapter 8 also considers the role of external communication and the issues concerning external verification.

In Part III, Chapter 9 reflects on the activities of 34 major international banks and provides insights into the differences between them. Chapter 10 then returns to the question of whether the incremental improvements discussed in Part II and Chapter 9 are sufficient to get sustainable development off the ground. The next question is, what second-order change processes have to be implemented to actually achieve a sustainable society? What are the conditions for this? A picture is given of the potential elements of such a new paradigm. This chapter therefore has a more philosophical angle.

Finally, Chapter 11 forms a link between the more philosophical chapter and the practical chapters in Part II. It takes a look into the future and investigates which developments, some of them already visible, may facilitate the achievement of sustainable development, and what role the banks may play in this.

Part I

Sustainability: A General Introduction

'A journey of 1000 kilometres begins with the first step.'
Tao

2

Environmental Consciousness and Sustainable Development

INTRODUCTION

Environmental problems are well described by the metaphor of the 'tragedy of the commons' (Hardin, 1968), in which an insufficient appreciation of the scarcity of resources leads to soil exhaustion, erosion and, with even wider implications, an interference with nature's ability to regenerate itself. This metaphor is frequently applied to the fishing sector. It is obvious to any fisher when overfishing is taking place, endangering future stocks.

Continuing the metaphor of the 'tragedy of the commons', every fisher knows that the problem is developing but has no incentive to tackle it. After all, by voluntarily reducing the size of his or her catch, the fisher would only enable others to catch more, with the net result that the level of overfishing would remain unchanged. The fact that the seas are common property ('common good') enables over-exploitation to develop, with the consequence that a classic market failure would result, one caused chiefly by a lack of information about the level of resources and the failure of individuals to transcend individual interests. This situation is sometimes referred to as the 'prisoner's dilemma' (see Appendix XII), which occurs when a person does not have enough information to predict another's behaviour and, as a result, works to achieve his or her own short-term interests, which are at odds with both parties' long-term interests.

Although insufficient appreciation of the scarcity of resources as the underlying factor in environmental problems, the problem is accentuated by a crippled system of property rights from which the negative external effects of production and consumption, such as environmental deterioration, derive.[1] If production factors are to be optimized, these external effects should be integrated into market prices. The environmental factor, however, is usually not included in any market decisions. Price incentives – and perhaps psychological, legal and social incentives as well – will ultimately ensure that the environmental factor is sufficiently included in production or consumption decisions. But the obvious question is whether this will be done in time for certain elements within the environment.[2]

Broadly speaking, there are four ways in which such market failures can be corrected. Firstly, by direct regulation, such as standards and licences. Secondly, by the use of economic policy instruments, such as regulating levies, subsidies, tax reductions and emissions trading (of CO_2 emissions, for example). Thirdly, by negotiations over property rights. Finally, via a broad category of pressure and persuasion assisted by jurisprudence and covenants, as well as by interested parties such as industrial customers, consumers, NGOs and employees. Psychological and social incentives can be based on altruism (environmental protection based on a commitment to future generations or a spiritual perspective) or self-interest (environmental requirements set by the government or the buyer).

The foundation for today's environmental policy, and also for sustainable development, can therefore be traced back to the realization that the environment and the economy are closely interrelated. An economic system cannot exist without an ecological system; the ecological system is the principal system from which the economic system is derived. After all, it is the ecological system that makes life itself possible and is both a source – in terms of natural resources – as well as a waste pit or sink – the recipient and processing system of waste materials and other emissions – for an economic system. This insight leads to the conclusion that maintaining the environment is a necessary condition for any economic activity and ultimately for mankind's survival.

The relationship between ecological and economic systems is explained further in Box 2.1.

The economic–ecological connections and the reactions of the public, scientists and politicians are further explored in this chapter. The following section outlines some compelling regional and international environmental questions while the next offers a historical overview of the development of the environmental consciousness of individuals and society. Growing insight has lead to the introduction of the concept of sustainable development, which is discussed on pages 21–7. This concept has been integrated into government policy in many countries. The section starting on page 27 examines in greater depth the issue of the governments' policies with regard to local, regional and international environmental questions. Changes in environmental consciousness, the idea of striving for sustainable development and the adoption of environmental policy are important for banks and business. Chapter 3 concentrates on the activities of business, after which Chapter 4 presents a framework for the reactions, activities and role banks may play.

ENVIRONMENTAL ISSUES

Environmental problems exist in diverse forms, dimensions and scales. In terms of scales, there are, roughly speaking, three identifiable levels. In the 1960s, developed countries saw an increase in the amount of public attention paid to local issues and to air and water pollution. Initially, the measures taken to control air pollution were typically 'end-of-pipe', which as a rule resulted in factories acquiring taller chimneys. This indeed reduced the local nuisance, while having no effect on the environmental impact. It simply transported the environmental nuisance over greater distances, sometimes to other countries. As a result of these developments, greater attention was paid to regional and international

Box 2.1 The coherence of economic and ecological systems

Within the Earth's biosphere, life is possible and various distinct ecological systems function. Ecosystems consist of communities of interacting organisms and the physical environment in which they live. Ecosystems are not just collections of species; they are systems consisting of organic and inorganic matter and natural forces that interact and change; they are intricately woven together by food chains and nutrient cycles (WRI, 2000, p3). Scale and extent can vary significantly, from a sand dune on the coast of France to a tropical rainforest, which is in itself a collection of smaller symbiotic ecosystems. References to 'the ecological system' in this book refer to these smaller and larger ecosystems in their entirety.

The extent and quality of the ecological system on Earth is determined by the extent and nature of naturally occurring raw materials, the penetration of solar energy and nature's capacity to absorb waste.

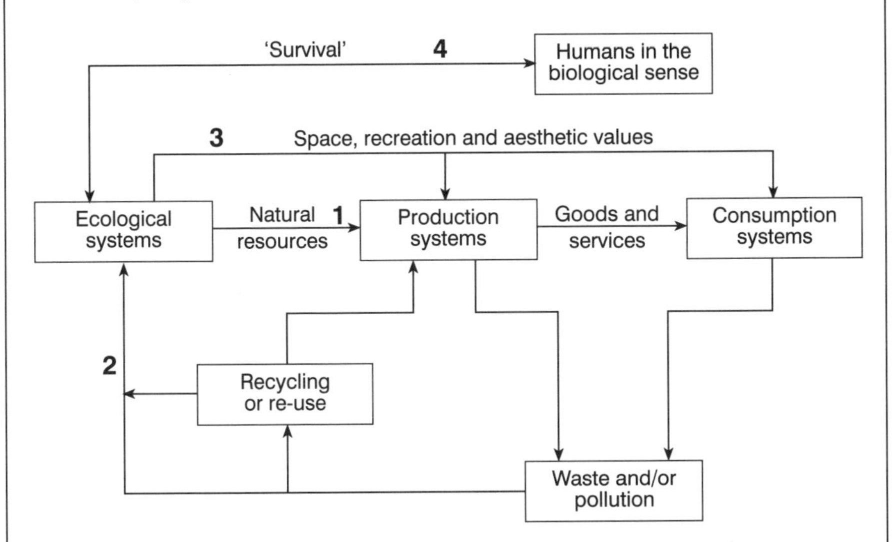

Figure 2.1 *The relationship between environment, people and the economy*

Figure 2.1 represents the relationship between economic and ecological systems (in physical terms); this is a closed system.[3] The economic system is in fact a subsystem of the ecological system and is bound to this 'umbrella' system by the use of resources (*1*), by the pollution and discharge of waste (*2*), and by the aesthetic value of nature and the opportunities it affords for recreation and space (*3*). People are not only bound to the ecological system as initiators of human actions but also in purely biological terms (*4*).[4] Human actions (*1* and *2*) influence the chances of human survival via the ecological system (4). The hole in the ozone layer and the increased greenhouse effect are the ultimate examples of this.[5]

Environmental concern focuses on the parts of the natural cycle that are harmed by economic activity (*2*) and/or where natural resources (*1*) and other aspects of the ecological system are exhausted or destroyed (*3*). With regard to natural resources (*1*), the exhaustibility of the supply of raw materials and their regenerative capacity are crucial. With regard to pollution (*2*), the ecological system's capacity for absorption is essential. For human life itself (*4*), the continuation of the ecological system is a requirement. Table 2.1 summarizes the numerous goods and services provided for us by just five ecosystems (WRI, 2000, p9). Cautious estimates put the worth of these goods and services worldwide at around US$33 trillion per year – about 180 per cent of worldwide GDP (Constanza et al, 1997, p256).

Table 2.1 *Goods and services from ecosystems*

Ecosystem	*Goods*	*Services*
Agro-ecosystems	Food crops; fibre crops; crop genetic resources	Maintain limited watershed functions (infiltration, flow control, partial soil protection); provide habitat for birds and soil organisms important to agriculture; build up organic soil matter; sequester atmospheric carbon; provide employment
Forest ecosystems	Timber and fuel wood; drinking and irrigation water; non-timber products (bamboo, vines, leaves, etc); food (mushrooms, honey, fruit, game); genetic resources	Remove air pollutants; emit oxygen; cycle nutrients; maintain array of watershed functions (infiltration, purification, flow control, soil stabilization); maintain biodiversity; generate soil; sequester atmospheric carbon; moderate weather extremes and impacts; provide employment and human and wildlife habitats; contribute aesthetic beauty and provide recreation
Freshwater systems	Drinking and irrigation water; fish; hydro-electricity; genetic resources	Buffer water flow (control timing and volume); dilute and carry away wastes; cycle nutrients; maintain biodiversity; provide aquatic habitat; provide transportation corridor; provide employment; contribute aesthetic beauty and provide recreation
Grassland ecosystems	Livestock (food, game, hides, fibre, etc); drinking and irrigation water; genetic resources	Maintain array of watershed functions (infiltration, purification, flow control, soil stabilization); cycle nutrients; remove air pollutants; emit oxygen; maintain biodiversity; generate soil; sequester atmospheric carbon; provide human and wildlife habitats; provide employment; contribute aesthetic beauty and provide recreation
Coastal ecosystems	Fish and shellfish; fish meal (animal feed); seaweed (for food and industrial use); salt; genetic resources	Moderate storm impacts (mangroves, barrier islands); provide marine and terrestrial habitats; maintain biodiversity; dilute and treat wastes; provide harbours and transportation routes; provide human and wildlife habitats; provide employment; contribute aesthetic beauty and provide recreation

Current collaborative research entitled 'pilot analysis of global ecosystems' (PAGE; WRI, 2000) shows that the quality of this basis for human life is deteriorating worldwide. Human activities are starting to significantly alter the Earth's basic chemical cycles – water, carbon and nitrogen – on which all ecosystems depend (WRI, 2000, p51). Throughout the world there is insufficient understanding of the effects of such changes and how they can be controlled. For this reason the UN, along with various governments and research institutes, started the Millennium Ecosystem Assessment (MEA) at the end of 2000. Studies of two ecosystems (grasslands and freshwater) are already under way (White et al, 2000, and Revenga et al, 2000, respectively). The aim of the MEA is to adopt an ecosystem-friendly approach in which society evaluates its decisions on land and resource use in terms of how they affect the capacity of ecosystems to sustain life for humans, animals, plants and other natural systems. Table 2.2 details the results from PAGE (WRI, 2000, p47).

Table 2.2 *The worldwide position and development of ecosystems*

	Agro	*Coastal*	*Forest*	*Freshwater*	*Grasslands*
Food/fibre production	4 ↓	3 ↓	4 ↑	4 ↓↑	3 ↓
Water quality	2 ↓	3 ↓↑	3 ↓	2 ↓	n/a
Water quantity	3 ↓	n/a	3 ↓	3 ↓	n/a
Biodiversity	2 ↓	3 ↓	2 ↓	1 ↓	3 ↓
Carbon storage	3 ↓↑	n/a	3 ↓	n/a	4 ↓

Source: WRI (2000, p11).
Note: A classification of the current state of ecosystems has been made on the basis of expert research worldwide: 1 = bad; 2 = inadequate; 3 = reasonable; 4 = good and 5 = excellent. An upwards arrow indicates a positive development, a downwards arrow shows the opposite. The use of two arrows shows that the underlying factors indicate a mixed picture; n/a = not assessed.

environmental questions, such as international air pollution and the contamination of rivers which flowed across borders. Increasing knowledge and awareness have finally focused attention on worldwide environmental issues, such as the hole in the ozone layer, biodiversity, the felling of rainforests and the intensified greenhouse effect. Two main concerns can be identified: the consumption of natural resources and environmental pollution.

One vital natural resource is water. Ensuring a sufficient supply of good quality water will be the problem of the 21st century. As far as other resources are concerned, the rapid growth of the human population makes their better management necessary. Ecosystems too are under pressure. Some facts (WRI, 2000): half of the world's wetlands were lost during the last century; logging and conversion have shrunk the world's forests by as much as half over 8000 years; tropical deforestation exceeds 130,000 square kilometres per year; fishing fleets are 40 per cent larger than the oceans can sustain; nearly 70 per cent of the world's major marine fish stocks are overfished or are being fished at their biological limit; one-third of the world's population lives in countries already experiencing moderate to high water scarcity, and that number could rise to two-thirds in the next 30 years; soil degradation has affected two-thirds of the world's agricultural lands in the last 50 years; between 1960 and 1990 some 30 per cent of tropical forest has been cleared in Asia (the worldwide average is 20 per cent); most of the world's original forests have been converted to agriculture; global food production is generally adequate to meet human nutritional needs, but problems with distribution mean that some 800 million people remain under-nourished; consumption of natural resources by modern industrial economies remains very high (in the range of 45 to 85 metric tons per person annually when all materials – including soil erosion, mining wastes, etc – are counted). The issue of natural resources boils down to good management, which entails the distinction between renewable and recyclable resources (eg forests and fisheries as well as minerals) and finite resources (eg gas and oil) being made explicit.

The second category of environmental problem is pollution. Here too a division into two main types is appropriate: pollution that accumulates in the ecological system (stock pollutants such as heavy metals and nuclear waste), and pollution that, to a certain degree, can be broken down by the ecological system (fund pollutants, like CO_2 and diverse organic materials). Sustainability would entail stock pollution being reduced to zero while fund pollution ought not to exceed nature's capacity to absorb it. In environmental policy there is a further distinction: that between stationary (eg factories) and mobile (eg cars) sources of emissions. In many countries the management of stationary sources appears to be easier to integrate into policy. Well known types of pollution are: the discharge of sulphur dioxide (SO_2) and nitrate oxides (NO_x), which can cause acid rain that attacks forests, buildings and human health; the discharge of carbon dioxide (CO_2), which can lead to an increase in the greenhouse effect and hence global climate changes; the discharge of chlorofluorocarbons (CFCs), which in the stratosphere above the North and South poles has led to increasingly large holes and has thereby reduced our protection against the sun's harmful radiation; the pollution of rivers with nitrates and phosphorus from industry and agriculture; and waste from production and consumption, soil pollution and noise pollution.

The above list is far from complete, but does give a broad picture of environmental problems. Appendix I illustrates the achievements and environmental impact of developed countries as regards protected natural areas, forests, water use, energy intensity in the economy, waste, and the discharge of SO_2 and CO_2. The discharge of SO_2 has diminished considerably in most developed countries. Developing countries, by contrast, have seen a sharp increase in SO_2 emissions. If the current trend continues, it is expected that SO_2 emissions in Asia will triple by 2010. Box 2.2 presents a little background to worldwide environmental issues.

ENVIRONMENTAL CONSCIOUSNESS

Interest in the environment has been around a long time. At the time of the Ancient Greek civilization and in Biblical times, the emphasis lay on nature.

Box 2.2 Some major international environmental issues examined

Water

Clean water (in addition to biodiversity) is a prerequisite for guaranteeing a sustainable food supply, public health, high quality living conditions and the development of nature. The world's freshwater systems are currently so degraded that their ability to support human, plant and animal life is in peril. Much of this degradation stems from habitat destruction, the introduction of non-native species, pollution, over-exploitation, and the construction of dams and canals. The water supply in many densely populated areas and the major conurbations of the world is already being used to its full capacity. Existing water sources are drying up through exploitation and wastage (in some developing countries 60–75 per cent of irrigation water does not reach crops). The agricultural sector throughout the world places by far the greatest demand on water consumption (around 70 per cent). It is estimated that in 2025, at least 3.5 billion people will be affected by the scarcity of water, equivalent to around 50 per cent of the total human population. In the US, 37 per cent of freshwater fish species, 67 per cent of mussels, 51 per cent of crayfish and 40 per cent of amphibians are threatened or have become extinct (WRI, 2000, pp188–191).

Food safety

Water consumption is closely linked to agricultural production. The provision of a food supply sufficient for a growing population has long been a political subject. It now appears that food production is sufficient to meet the needs of all people on Earth; the shortage of food in some developing countries is simply a distribution issue. Food safety (that is, food quality rather than quantity) is an issue of increasing priority in developed countries. 'Mad cows' (the BSE crisis), dioxins in chickens in several European countries, and genetically modified organisms (GMOs) such as those used in soya beans in Europe and the US, are problems that exercise many minds. The crises appear to be symptomatic of a growing aversion among consumers in developed countries to 'unnatural' products and production methods. In some countries there is, unsurprisingly, a move towards more organic agricultural methods.

Energy

Energy supplies are of vital importance. The industrial world would be unimaginable if it were devoid of reliable and affordable energy supplies. At the current time, the energy supply is almost entirely dependent on fossil fuels. Oil, natural gas and coal play crucial roles. The diminishing availability of fossil fuels is, however, inevitable. The increasing scarcity, price increases and possible far-reaching consequences of the use of fossil fuels make it essential that their use be drastically reduced. There are two ways in which this can be achieved: through the use of sustainable energy and by increasing the efficiency of their use. Renewable or sustainable energy sources are the sun, wind, biomass, hydropower, geothermic sources, currents such as tides, and energy deriving from temperature differences in the oceans.

Biodiversity

Biodiversity refers to the diverse living organisms in terrestrial and marine ecosystems, and the ecological complexes to which they belong. This includes diversity within species, between species, and within ecosystems (Brink, 2000, p38). Biodiversity provides mankind with many immediate advantages, such as ingredients for medicines and aesthetically pleasing elements. In addition, biodiversity is the basis for the continued provision by ecosystems of goods and services to people. Threats to biodiversity from all human sources are quickly reaching a critical level which may precipitate widespread changes in the number and distribution of species, as well as the functioning of ecosystems. Current extinction rates are 100 to 1000 times higher than pre-human levels. Competition from non-native plant and animal species – 'bioinvasions' – represents a relentless and growing threat to natural ecosystems, with some 20 per cent of the world's endangered vertebrate species threatened by exotic invaders. Bioinvasions are now considered to be the second greatest threat to global biodiversity after loss of habitat. The loss of biodiversity is especially tangible in aquatic environments such as coral reefs and freshwater habitats in rivers, lakes, and wetlands, and 58 per cent of the world's reefs and 34 per cent of all fish species are currently at risk from human activities (WRI, 2000 and Revenga, 2000).

Trade

In a growing global economy there is a progressive increase in transport, which goes hand in hand with environmental consequences that, until now, have barely been incorporated into pricing structures. Tighter environmental policies in developed countries could, in an increasingly global economy, lead to pollution havens as companies move their production plants to countries with less stringent environmental policies. In the EU, this could (once again) play an emphatic role in the forthcoming moves to incorporate some countries in Eastern Europe (Hager, 2000). However, research suggests that this is not necessarily the case (Jeucken, 1998a). Also, it seems that the World Trade Organization (WTO), which ensures that goods and services are traded as freely as possible, adheres so closely to the rules of free trade that countries are restricted in their freedom to impose environmental requirements on products and, especially, processes. Meanwhile, international treaties have been signed for the explicit purpose of restricting international trade – for example, in the case of contaminated waste and endangered species. Trade and environmental care remain a controversial issue.

Ozone layer

The ozone layer shields us from damaging ultraviolet (UV) radiation. Caused chiefly by the discharge of CFCs and halons, the growing hole in the ozone layer could lead to serious health hazards like skin cancer. The policy response to this included the Montreal Protocol, in which developed countries agreed to the total phasing out of these substances by the year 2010. Since 1987 most countries have reduced their consumption of ozone-depleting substances by 70 per cent (WRI, 2000, p179). The Protocol, and the results it has achieved, have become the model for international cooperation in the field of worldwide environmental problems. However, the ozone layer is still not safe. Firstly, there is the effect of time to contend with, as discharges from earlier years continue to build up in the ozone layer. Secondly, a growing market for illegal CFCs has developed. It is expected that only from 2035 onwards will the ozone layer begin to recover.

Climate change

Certain natural substances create a 'greenhouse effect' that makes life on Earth tenable. The current climate problems – clearly evident according to some experts in the increasing frequency, scale and power of storms and flooding – are, however, caused by the intensification of the greenhouse effect. Since the Industrial Revolution, the discharge of manmade greenhouse gases has increased to such an extent that it is now common to refer to the intensified greenhouse effect.[6] So, for example, the quantity of carbon dioxide (CO_2) in the atmosphere has risen by 31 per cent since 1750 (IPCC, 2001, p7). CO_2 is responsible for a 60 per cent share of the intensified greenhouse effect and policy is rightly focused on its reduction. The problem associated with this is that CO_2 is a substance with a strong correlation to energy consumption and therefore economic growth. End-of-pipe measures are insufficient in addressing this problem; it requires the drastic reduction of energy consumption, or an increased share of energy derived from non-fossil fuels, such as solar energy. An intensified greenhouse effect can disturb the ecological and climate system on Earth to such an extent that an unstable situation can develop. Thus, according to the Intergovernmental Panel on Climate Change (IPCC, 2001, pp13,16), the average temperature on Earth will rise by 1.4 to 5.8 degrees Celsius (the best estimate, according to the IPCC (1995, p5), is 2 degrees Celsius) and the sea level will rise by 9 to 88 centimetres (the IPCC's best estimate is 50 centimetres) between 1990 and 2100.[7] Differences between regions may be significant, with a certain rise in sea level meaning that many large areas of delta countries could be submerged; areas affected include Egypt, Bangladesh, The Netherlands, Florida in the US, and island groups such as The Maldives. Geological plates might well shift. This could result in northern Europe becoming an important viniculture area in 50 years' time, but it is equally possible that Europe could be hit by a new ice age (in the event of the warming Gulf Stream ocean current ceasing to exist or losing power). Great expanses of the US could be affected by droughts, making it impossible to produce enough grain to fill the silos. It is evident that both material and emotional damage will be extensive.

Human beings have been confronted for centuries with the consequences of their actions on ecosystems. Such problems have caused whole civilizations to die out or migrate. Examples include the salination of the irrigation system in ancient Mesopotamia, the soil erosion and degradation of the agricultural base of the Mayan civilisation, the erosion of hills in the Ancient Greek Empire, food shortages in the Roman Empire caused by the exhaustion of the soil, and the great famines and epidemics in Europe up until the 19th century. In Africa and Asia these problems continue to the present day (think, for example, of the great droughts in Ethiopia). These problems are closely related to bad agricultural practice (in addition to, for example, poor hygiene), which takes insufficient account of the capacity of natural and agricultural ecosystems.

Concern for the natural environment has also played a role in certain cultures. This is plainly reflected in the lifestyles of Native Americans, African Bushmen and Australian Aboriginals. In these cultures the environment is treated with a respect generated by these people's awareness of their total dependence upon it. The sense of community is well developed, and each individual feels part of the collective responsibility for the ecological environment. In Western society, comprehensively defined ownership rights play an important role in the care of the environment. In principle, these provide a very strong incentive for environmental care by the individual, but reduce the case for a collective responsibility for the commons. In other cultures the sense of community, individuals' awareness of their responsibility towards the community, and respect for the ecological environment all appear to be important without having to be formally laid down (such respect is most often fostered by oral traditions). Unlike property rights, however, these 'rights' cannot be enforced.

An example of the relationship between cultural patterns and environmental issues comes from Easter Island (Ponting, 1991). Pessimists often use this example to demonstrate that there are physical limitations to our actions. It is well known that the coast of the once thickly forested Easter Island is full of some 600 great stone sculptures. In order to place these sculptures, each weighing thousands of kilos, on the coast, the inhabitants cut down the islands' forests, not only to create the space to move the sculptures but also because they used tree trunks as rollers on which to move them (over a distance of 10–20 kilometres). Once the clans, in an exhausting competition to outdo one another, had each made and moved their own clan's sculptures, no tree was left standing. Erosion of the now barren ground left it infertile, and there were no trees left with which to build houses or boats to flee the island. All this resulted in the deaths of many islanders and the demise of their culture. When the explorer Johan Roggeveen became the first Westerner to land on the barren island in the Pacific Ocean in 1722, the inhabitants who were left were primitively living in caves. The lesson pessimists draw from this is that even when physical limitations are obvious, the socioeconomic system ensures that these limitations are ignored. Similar insights still inform the policy decisions of environmental and natural organizations, and provide a wise lesson for humans (compare this with the tragedy of the commons). A change in attitude and in the way in which we act is necessary if we are to escape the constraints of the present economic system and move towards one in which the external effects of economic actions are taken into account, and can direct consumption and production choice. If the economic system is to be sustainable, these effects must be internalized in market prices.

Whereas Western awareness *before* the 18th century focused on the protection and recognition of nature, after the Industrial Revolution the first signs of the new environmental awareness were perceptible. All kinds of industrial activities and the use of coal as a domestic fuel led to serious air pollution by soot, stench and sulphur oxide, particularly in the major cities. The first environmental legislation dates from 1838 and was primarily concerned with the hygiene situation in Britain.

It was principally during the 20th century that environmental problems changed in scale and extent. The population increased by a factor of 5 and energy consumption by a factor of 15 as a consequence of sharp increases in production, consumption and mobility. The century saw a significant intensification of agriculture, while the finite nature of various raw materials became apparent, as did the influence of mankind on the climate (WRR, 1992, p20). These problems first manifested themselves at local level (with such nuisances as stench and local air and water pollution), later at regional and fluvial level (eg eutrophication or the use of manure) and continental level (eg with acidification) and, finally, even at a worldwide level. It is these worldwide problems that have pushed the environment to the top of the international agenda. To date the list includes desertification, tropical deforestation, acidification, climate change resulting from the use of fossil fuels, damage to the ozone layer, problems in biodiversity, and the build-up of chemical substances in food chains and in groundwater.

An uneasiness about this encroachment on the environment started to build among larger groups within the population in the early 1960s. This was prompted in part by the conclusions drawn by Rachel's Carson's book *Silent Spring*, published in 1962.[8] The number of environmental action groups rose quickly in most Western countries, especially between 1968 and 1972. The realization that pressure on the environment could threaten human survival led to the plea for a drastic revision of the idea of growth-oriented economic progress. The basis for this was established by the Club of Rome (Meadows et al, 1972). In spite of much criticism of its conclusions and research methods, its 1972 report, *The Limits to Growth*, was an important wake-up call and played a pioneering role in countless developments in the environmental field, resulting in the introduction of structural policy measures by governments in numerous countries. The existence of 'Spaceship Earth' was under threat.[9]

Following a cyclical recession, the early and mid-1980s saw a resurgence of attention paid to the environment, this time in industry as well. This was prompted primarily by environmental catastrophes such as Seveso (a poisonous cloud that affected residents living in its vicinity in Italy in 1976), Sandoz (pollution of the River Rhine that occurred in Switzerland in 1978), Bhopal (one of the world's worst cases of toxic emissions, in India in 1984, which involved 2500 fatalities as a direct result), Chernobyl (an accident in 1986 in a nuclear reactor in the former Soviet Union, which caused many deaths and seriously affected the health of people living as far away as Western Europe) and Exxon Valdez (the huge oil disaster off the coast of Alaska in 1989). Alongside these incidents more insidious problems were raising their heads, such as the scarcity of clean drinking water, climate problems and the hole in the ozone layer. It was becoming very clear that environmental degradation, economic growth, population growth and poverty were interrelated within an international framework. This insight has led to the introduction of the sustainable development concept. This heralds a new phase in environmental consciousness; concern for the environment is no longer a discrete issue but is regarded as pivotal for human development.

SUSTAINABLE DEVELOPMENT

Sustainable development is a difficult concept that leads to great confusion.[10] It is perhaps easier to define that which is not sustainable (such as the use of exhaustible natural resources and soil erosion). In 1980 UNEP published a report in which sustainable development was officially discussed for the first time (United Nations, 1980). This report followed up on ideas proposed by nature conservation organizations such as WWF and the World Conservation Union – IUCN (formerly the International Union for Conservation of Nature and Natural Resources), in which sustainability referred to the use of ecological systems in a way which would enable their primary characteristics to remain intact.

The concept received wider attention when the World Commission on Environment and Development (WCED) published its report, entitled *Our Common Future*, in 1987. According to this report, sustainable development would enable the eradication of poverty in developing countries on the one hand and, on the other, the creation of a new balance between material wealth in developed countries and the preservation of ecological systems as the basis for life. With this conclusion, the Brundtland Commission sought a middle way between two conflicting approaches: the anthropocentric and the ecocentric. In the former system, humans are central; in the latter, the functioning of ecological systems is central. The eventual definition of the Brundtland Report, since adopted by countless countries and businesses, is however less explicit in this sense: sustainable development is a development that 'meets the needs of the present without compromising the ability of future generations to meet their own needs' (WCED, 1987, p46).

According to the Brundtland Commission sustainable development is not restrictive; its only restrictions relate to the prevailing state of technology and the social organization of natural resources, and the capacity of the ecological system to absorb the effects of human activity. It is noteworthy that no economic restrictions are mentioned. In Rio de Janeiro in 1992 at the United Nations Conference for Environment and Development (UNCED), more commonly known as the Earth Summit, restrictions were indeed a topic. In the final official documents, some 180 countries established that environment and development were interdependent, indivisible and equal in rank, which means that short-term economic interest can no longer be made a priority. The question then arises whether priority should always be attributed to the environment. The Earth Summit in Rio in 1992 stated that all human activity, and indeed social prosperity in its traditional narrow sense, should be subsumed under the principle of environmental sustainability. Development and social prosperity can only be achieved to the extent that the natural resource basis is not endangered. Principle 15 of the UNCED Declaration states therefore that:

> *in order to protect the environment, the precautionary approach shall be widely applied by States according to their capabilities; where there are threats of serious or irreversible damage, lack of full scientific certainty shall not be used as a reason for postponing cost-effective measures to prevent environmental degradation.* (UNCED, 1992, p10)

Box 2.3 focuses on a development issue affecting most of the world's population. It links the drastic need for socioeconomic development with the rapidly growing world population, and addresses how these fit into the framework of environmental sustainability proposed at Rio. In Box 2.4, the issue of the needs of future generations are addressed.

Sustainable development has, in short, a broad context, and involves the question of distribution within one generation and across generations. The heavy use of non-renewable natural resources by developed countries restricts the options of developing countries and future generations. Furthermore, current environmental pollution outstrips the ecological system's capacity to

Box 2.3 Global facts of human development

- Of the world's 6 billion people, 2.8 billion (ie almost half) live on less than US$2 per day, and 1.2 billion on less than US$1 per day (World Bank, 2000, p3).
- The share of the population living on less than US$1 per day fell from 28 per cent to 24 per cent between 1987 and 1998; however, due to a growing world population the absolute figure hardly changed (World Bank, 2000, p21).
- 70 per cent of the population living on less than US$1 per day are inhabitants of South Asia and Sub-Saharan Africa (World Bank, 2000, p23).
- The average income of the richest 20 countries in the world is 37 times that of the poorest 20; this gap has doubled in the last 40 years (World Bank, 2000, p51).
- The combined wealth of the world's 200 richest people reached US$1 trillion in 1999, while the combined incomes of the 582 million people living in the 48 least developed countries (LDCs) amounted to US$146 billion (UNDP, 2000, p82).
- The 48 LDCs attracted less than $US3 billion in foreign direct investments (FDI) in 1998; 0.4 per cent of total FDI (UNDP, 2000, p82).
- The 48 LDCs account for less than 0.4 per cent of global exports (UNDP, 2000, p82). OECD tariffs and subsidies are estimated to cause annual losses in welfare of US$20 billion in developing countries, equivalent to approximately 40 per cent of official aid to all developing countries in 1998 (World Bank, 2000, p11).
- The 48 LDCs have an adult illiteracy rate of 49 per cent (UNDP, 2000, p171).
- Civil wars have killed 5 million people worldwide in 1990–2000, and war and internal conflicts forced 50 million people to flee their homes (UNDP, 2000, p36).
- 100 million children live or work on the streets (UNDP, 2000, p4).
- More than 30,000 children a day die from mainly preventable causes, and nearly 18 million people die every year from communicable diseases (UNDP, 2000, pp8, 35).
- 30 per cent of the people in LDCs are not expected to survive beyond the age of 40; in the OECD countries this figure is 4 per cent, with 12 per cent not expected to survive beyond 60 years of age (UNDP, 2000, pp171, 172).
- Six infants out of every 100 do not see their first birthday, and eight do not survive to their fifth (World Bank, 2000, p3).
- Each year 5 million people are becoming infected with HIV/AIDS. More than 34 million people worldwide are already infected, and more than 90 per cent of infected people live in developing countries; in Botswana and Zimbabwe 25 per cent of the adults are infected. More than 18 million people have already died of AIDS (World Bank, 2000, p139).
- 27 per cent of the world population has no access to safe water; for Sub-Saharan Africa this holds true for 46 per cent of the population (UNDP, 2000, p171).
- 94 per cent of the world's 568 major natural disasters and more than 97 per cent of all natural disaster-related deaths (approximately 50,000 people) over 1990–2000 were in developing countries (World Bank, 2000, p170).
- It would cost an additional US$80 billion a year to achieve the universal provision of basic services in developing countries (UNDP, 2000, p9).

Figure 2.2 illustrates the distribution of people living on less than $US1 per day in developing regions in 1998. Figure 2.3 illustrates the absolute and relative change in this number per region between 1987 and 1998.

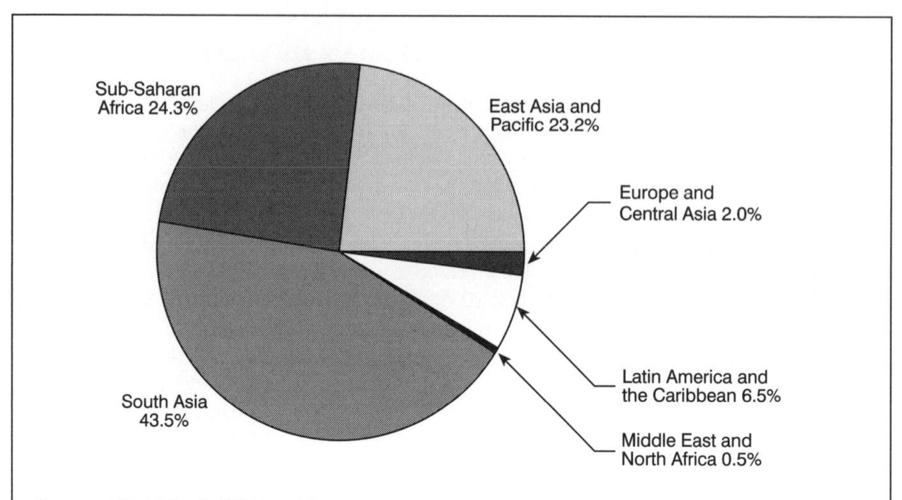

Source: World Bank (2000, p24).

Figure 2.2 *Total number of people living on less than US$1 a day per region, 1998*

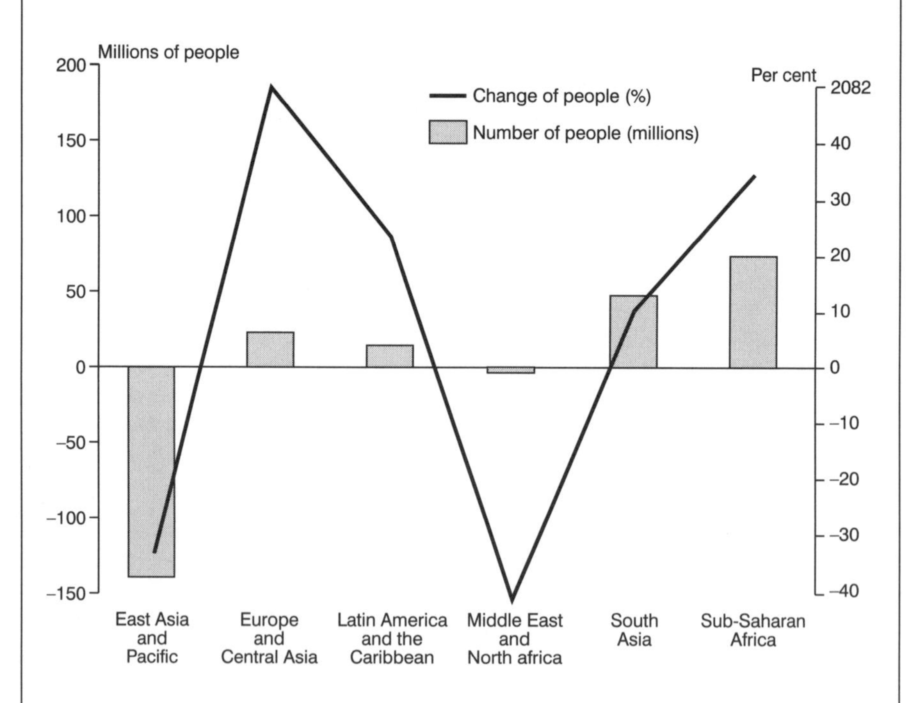

Source: World Bank (2000, p23).

Note: A positive number indicates a rise in the number of people living on less than US$1 a day; a negative number indicates a fall.

Figure 2.3 *Change in the number of people living on less than US$1 a day per region, 1987–1998*

Box 2.4 The relationship between prosperity, population size, environmental impact and technology: past and present

The problem of dire poverty and the considerable population growth anticipated in developing countries, makes the need to increase the per capita prosperity evident. The challenge will be for this to take place without drastically increasing the pressure on the environment worldwide. Insight into the influence of human activity on the environment within a country or worldwide can be gained by using the formula I = P x W x T.[11] This states that the total impact on the environment at any given time (I) is equal to the size of the population (P) times the wealth – expressed as GDP – per capita of the population (W) times the environmental impact per product (T for technology) at any given moment. This shows that population growth is just as important a factor in environmental impact as economic growth. The population growth almost entirely occurs in developing countries, where relatively few raw materials and other resources are consumed per capita. It is anticipated that by 2050 there will be 10 billion people on Earth. It is out of the question that, by then, with the continuance of the present level of damage to the environment per product, the Earth will be able to sustain the same level of consumption in the South as is enjoyed by the North. Such a scenario would entail massive environmental pollution, destruction to nature and consumption of natural resources. This is already visible in China; according to a World Bank report, China must invest over €30 billion if an environmental disaster there is to be prevented. Expectations are that the current levels of air pollution in China (as a result of the inefficient and excessive use of coal) will result in the premature death of nearly 1 million Chinese people (World Bank, 1997). The graph in Figure 2.4 clearly shows how growth in population is directly related to the consumption of raw materials.[12] The pattern of consumption in 'South II' implies roughly an eightfold increase in the consumption of raw materials and indirect environmental pollution. This does not imply that the South cannot grow or that the North must take a step back. The solution will lie in behavioural and other changes, including technological change and working together to reduce environmental impact.

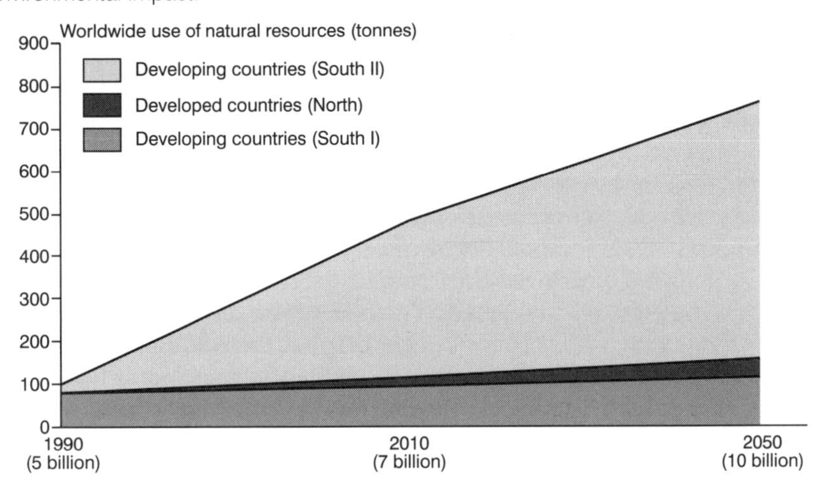

Figure 2.4 *The relationship between prosperity, population growth and environmental impact*

For religious and ethical reasons many people feel that we should consider the needs of future generations.[13] This can take place in a spirit of stewardship: we have the Earth in our charge and must care for it as best we can. One problem with the Brundtland definition is that, among other things, it does not define what is meant by 'needs'. Does this refer to purely material needs or does it include immaterial needs? The extinction of the African elephant, for example, will not decrease the material possibilities of future generations, unless they count among their needs the possibility of seeing an African elephant. This immediately raises another problem, of special relevance to biodiversity: we cannot know the needs of future generations. The concept of stewardship (whether stripped of its Christian ideological meaning or not) appears to complement the concept of sustainable development. This emphasizes the altruistic aspect and makes sustainability a cross between anthropocentric and ecoentric approaches, in which the ecocentric element concerns the stewardship (and in particular the preservation of biodiversity) in the long term. Furthermore, a similarly altruistic position can be placed in the context of self interest. One example is the growing water problem which can lead to wars and mass emigration. It is in the interest of the current generation to strive for a sustainable level of resource consumption and environmental pollution. Other issues, such as biodiversity, will however long remain largely normative and, for the most part, altruistic topics.

absorb it and future generations are bequeathed a legacy of environmental cleaning and health problems. One example of this is the cost of cleaning up nuclear waste. In 1995 the US estimated these costs at at least €180 billion, and the costs of cleaning up the environmental pollution in the former East Germany is estimated at around €190 billion.[14]

Along with questions of justice, sustainable development most definitely contains a social-ethical component. Participation in work and social cohesion are, according to many, both elements of this social component. Some people therefore talk of sustainable development as a balance of economic, social and ecological aspects. This means that there is a 'triple bottom line', and sheathed in this metaphor are the three Ps: people, profit and planet (Elkington, 1997). Society (people) are dependent upon the economy (profit), which is in turn dependent upon the worldwide ecological system (planet). The question of whether the bottom line of each of these dimensions can be reached together is only too obvious. Moreover, it prompts discussion as to what the different bottom lines are exactly, which quickly takes us into the ethical arena.[15] If they are strictly defined according to, say, scientific facts for the environmental dimension, and widely accepted norms and values for the social dimension (such as the UN human rights or the International Labour Organization (ILO) work norms) and the economic dimension (such as strong discount rates for future returns on investments), it becomes evident that there will have to be some compromise between these dimensions. The idea of a triple bottom line can be useful for business, but cannot be a guiding principle for governments or society. After all, according to the Earth Summit, the quality of the ecological system is the ultimate 'bottom line'.

With this holistic and long-term perspective as its starting point, sustainability and sustainable development in this book will focus chiefly on the ecological component, with one exception: socioeconomic issues of poverty and human rights in developing countries where such problems are a burning issue that urgently needs to be addressed. Socioeconomic issues in developed countries are entirely different and, moreover, socioeconomic development in developed countries is largely responsible for environmental pressure worldwide. Furthermore, in a broader definition of sustainability the economic and social component in developed countries can too easily be declared dominant. In fact, the reduction of environmental pollution or the preservation of nature may occur at the cost of profit, employment, competitiveness, economic growth and so on. Too often this is used as a convenient excuse for doing little or nothing to protect the environment. Given its focus on developed countries and the private sector, sustainability will therefore be defined as:

> A process of change in which the pattern of investment, the focus of technological developments, patterns of consumption and institutional change are such that the consumption of raw materials does not outstrip the regenerational capacity of the ecological system, nor does the environmental pollution outstrip its absorption capacity.[16]

Sustainable development in this book is, in short, not so much concept determining our current actions as an evoluting guiding star on the horizon, towards which we must continually look to find our way.

ENVIRONMENTAL POLICY

Introduction

The growing awareness across the world has led to varying types of environmental policy. The scale and impact of a country's policy is closely related to a number of factors: the country or region's stage of socioeconomic development; the nature and extent of the impact on the environment caused by economic activities and population density; the carrying capacity of the environment; the availability of natural resources; and the perceived necessity for change. Society's perception depends not just on socioeconomic factors, but also on cultural and religious precepts. This book primarily covers Western nations.[17] Although specific environmental questions can vary between industrialized countries, generally there exists a broad conformity in most environmental issues. Generally speaking, the aims of the resulting policies are also similar. The instruments employed, and the institutions and regulation of the environmental policy can, however, vary between countries. Some countries have, for example, a strong central environmental policy, while others follow a decentralized and market-oriented policy.

As it is not appropriate to go into this in great detail at this point, below the nature and development of one countries' environmental policy will be examined, that of The Netherlands, as many international experts regard it as having 'the most advanced environmental policy planning system worldwide' (Ministry of Housing, Spatial Planning and the Environment, 1999, p1).

Local and national environmental policy

The Nuisance Act, passed in 1875, was for almost a century the only law concerned with local environmental problems in The Netherlands; a systematic environmental policy was only established in the 1970s. Keijzers (2000) has distinguished three phases in Dutch environmental policy: shaping the ecological arena (1970–1983); encouraging pollution prevention (1984–1989); and enhancing eco-efficiency (1990–1999).

Among the highlights of the first phase were the creation of the Environmental Protection Department and the publication of the Policy Document on the Urgency of Environmental Pollution (Ministry of Public Health Care and Environmental Protection, 1972). This memorandum laid the legal basis for environmental policy in The Netherlands and arose out of public concern over the serious and growing environmental problems (water, air and soil pollution); it laid emphasis on restoring conditions conducive to public health. The policy was successful in a wide range of areas: the discharge of heavy metals was, for example, significantly reduced, and the emission of SO_2 was reduced by 25 per cent. The policy was based on the principle of 'the polluter pays', which was later – in 1975 – officially presented by the OECD.[18] The approach was top–down and command-and-control, which did not make it widely popular. Another problem was the segmented approach to environmental issues, such as measures to tackle water pollution which actually caused air pollution.

Economic crises pushed the environment to the background, but early in the 1980s there was a resurgence of interest. This was largely the result of a number of environmental disasters and the setting up of indicative multi-year programmes (Ministry of Housing, Spatial Planning and the Environment, 1985). Policy and attention shifted from being curative to being preventative, and from local to regional or international (phase two). Along with the impetus given by the Brundtland Report in 1989, the multi-year programmes led to the first National Environmental Policy Plan (NEPP; Ministry of Housing, Spatial Planning and the Environment, 1991). This, in turn, heralded the transition to the third phase of eco-efficiency. The NEPP has a two-track approach to policy. The first track focuses on the effect of environmental problems (theme areas); the second is concerned with the sources of these problems (target groups). Theme area policy shifted from being reactive to being proactive, while target group policy developed to enable greater participation and freedom for stakeholders. Eight theme areas were identified: climate change, acidification, dispersal of toxic substances, waste disposal, eutrophication (excess deposition of nutrient to soil and water leading to excess plant life), water table depletion, squandering of resources, disturbance by noise, and external safety from hazards. Target groups include agriculture, transport, industry, construction, utility companies, the retail trade and consumers.

NEPP2 (Ministry of Housing, Spatial Planning and the Environment, 1994a) and NEPP3 (Ministry of Housing, Spatial Planning and the Environment, 1998b) followed. Both these policy plans paid greater attention to the international context and the economic consequences of environmental policy. NEPP2 was intentionally focused on the implementation of policy (especially as regards the development of instrumentation to do this). At this point the first signs are evident of the move away from top–down regulation towards self-regulation by economic agents. With the arrival of the unified Environmental Management Act in 1993 the foundation of environmental policy was entirely contained within one comprehensive piece of legislation (Tonnaer, 1994).

The move towards international and long-term issues, and the increased role for stakeholders, really marked a new approach in the third phase. They created an innovation: covenants. Covenants are negotiated agreements between trade or industry, and government. They establish a sector's environmental goals and set a deadline for achieving them. While covenants are, in theory, voluntary for trade and industry, in practice participation is encouraged since companies that do not take part face stringent regulation. The advantage of covenants is that companies have the freedom to influence or determine their environmental commitments and can choose how they will tackle them. Covenants are popular and now cover 90 per cent of the country's pollution, waste, recycling and energy consumption in the sectors of industry, construction and energy. This approach is both successful and effective. One drawback is that it cannot be used where the source of pollution is diffuse, for example with small- to medium-sized enterprises (SMEs) or consumers.

Covenants have encouraged mutual trust between the government and business, as illustrated in the *Policy Document on Environment and the Economy* in 1997 (Ministry of Housing, Spatial Planning and the Environment, 1998a). This document dealt with the possibility of fusing the environment and the economy ('econology') to create a win-win situation. To this end, this document incorporated a number of groundbreaking theories and perspectives that showed that the environment and economy could indeed be merged. The further greening of the tax system is a consequence of this memorandum (Dutch Lower House, 1997). Meanwhile, around 15 per cent of the Dutch government's tax income is green. In 2001 the NEPP4 will be published. This policy document will lay greater emphasis on the relationship between economic, social and environmental issues and will address a number of intractable environmental problems (Projectgroep NMP4, 1999). Some regard this as the start of a new phase in Dutch environmental policy (Keijzers, 2000).

Dutch environmental policy is the most progressive in the world. The environmental aims and policy objectives set in the 1990s were the most ambitious. Despite this, The Netherlands achieved the highest rate of economic growth in Europe, while pollution fell dramatically.[19] The results of the Dutch policy are discussed in Box 2.5.

Regional environmental policy

Relative to other EU countries, the environment of The Netherlands is under heavy strain. Two significant factors are continual urbanization and intensive agriculture. However, the environment in Europe is not doing much better. EU research shows that there has been little improvement in the European environment over the last five years. The situation with regard to most distinct environmental problems has either deteriorated or remains unchanged. In

Box 2.5 Results of Dutch environmental policy since 1985

For the most polluting substances, The Netherlands has seen an absolute delinking of GDP growth and environmental impact over the last few decades. Figure 2.5 gives an overview of actual emissions reduction and projections per theme (RIVM, 2000a). Appendix II shows four scenarios of the effects of environmental policy for the Dutch sector structure for 2030.

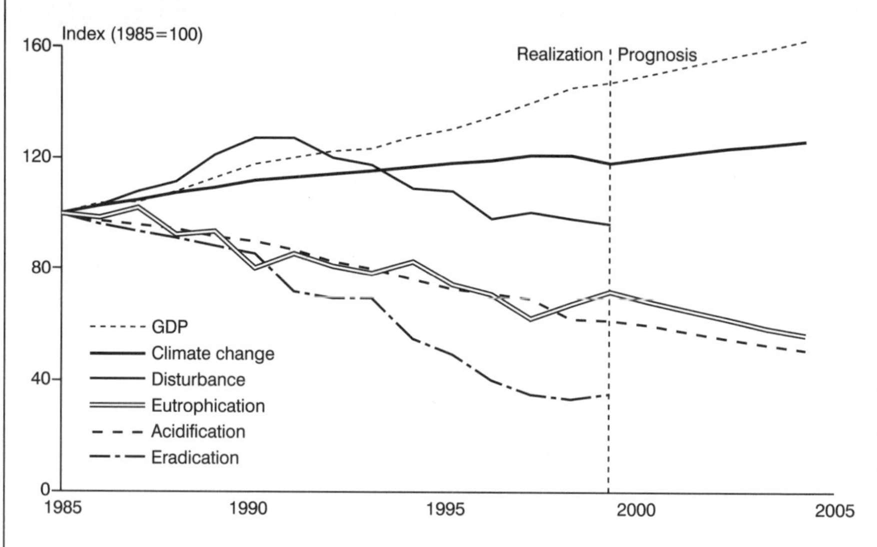

Figure 2.5 *Results, prognoses and aims of Dutch environmental policy, 1985–2005*

The delinking described above is absolute for three themes: eutrophication, acidification and eradication. Disturbance (nuisance from noise or stench) appears to be moving cautiously towards absolute decline. Climate change has achieved the essential turnaround to a relative decline but in absolute terms emissions are still on the increase. Furthermore, the biodiversity of ecosystems in The Netherlands has been under pressure for decades (RIVM, 2000b).

almost every European country, the most significant cause of environmental problems is agriculture (EEA, 2000).

Table 2.3 shows the results of policy on four environmental issues within individual EU Member States (EEA, 2000). Countries fall into three categories: front runners, average achievers and stragglers. The front runners are Germany, The Netherlands, Finland and the UK. The average achievers are Austria, Denmark, Sweden, Belgium and Luxembourg. The stragglers are the South European countries (Spain, Italy, Greece and Portugal) and France and Ireland. The distinction is rough but useful in shedding light on the effectiveness of environmental policy, and potential future tightening of environmental policy in some countries.

Table 2.3 *Performance of individual EU Member States on selected environmental issues*

	Climate Change	*Air pollution*	*Water Resources*	*Waste*
Ireland	–	–	–	=
Portugal	–	–	n/a	–
Greece	–	–	=	–
Italy	–	=	=	–
France	–	=	–	=
Spain	–	–	+	–
Luxembourg	+	=	n/a	–
Belgium	–	=	+	=
Sweden	–	=	+	=
Denmark	–	=	+	=
Austria	–	=	=	+
United Kingdom	+	+	=	–
Finland	–	+	+	=
The Netherlands	–	=	+	+
Germany		+	=	I

Note: + reflects a positive trend (moving towards target); = indicates some positive development (but either insufficient to reach target or mixed trends within the indicator); – reflects an unfavourable trend.

Until the 1980s the legal basis for environmental policy within the EU was weak. In 1973 the first European environmental programme was established. The European Act of 1985 laid the necessary legal framework for European environmental policy. The fifth Environmental Action Programme was published in 1993. *The Road to Sustainable Development* (European Union, 1993) contains a number of guiding principles for Member States:

- The precautionary principle: where knowledge is insufficient, the course of action that is chosen should be that which places the lowest level of strain on the environment.
- The principle of preventive action: to prevent irrevocable damage and to reduce the creation of waste at the source.
- The subsidiarity principle: environmental decisions and maintenance are assigned to the lowest appropriate governing body, ie from municipal, to provincial, to national, to European, to world level.
- The principles of cost-effectiveness, economic efficiency and legal efficacy.
- The principle that 'the polluter pays': those causing pollution must pay for the cost of clean-up operations and future regulations to protect the environment.[20]
- The principle of striving to reach a high level of protection: this should take regional differences into account.
- The principle that environmental requirements in other areas of European policy should be integrated.

In other parts of the developed world, similar guiding principles apply. For instance, the US is ahead in the conservation of nature in their national parks. In 1870, Yellowstone Park became the world's first national park. The US, Canada, Australia, New Zealand and even densely populated Japan are countries encompassing considerable regions of wilderness. Since they have few instances of cross-border air and river pollution, the regional environmental issues that are so prevalent in Europe play a small role in these countries. Obviously the US and Canada do experience these problems, primarily in the Great Lakes region, but they form a far less important aspect of the national environmental policy than is the case in European countries.[21] With its *Comprehensive Environmental Response, Compensation and Liability Act* (CERCLA), the US played a leading role in environmental policy for soil contamination. Chapter 6 will discuss CERCLA in more detail. For an overview of the performance and progress of various developed countries, see Appendix I.[22]

International environmental policy

International environmental policy emanates almost entirely from the UN. In 1972 in Stockholm the first UN conference on the environment was organized. Entitled 'Only One Earth',[23] one of its results was the Declaration of the United Nations Conference on Human Environment and the establishment of UNEP. The declaration gave impetus to the further development of international principles in environmental policy. It led to a number of bi- and multilateral treaties establishing the environmental policy of several countries, such as the Treaty of Lugano concerning civil liability for environmental damage caused by waste.[24] In 1992, the Earth Summit was held in Rio de Janeiro. Its aim was to develop strategies and regulations for dealing with environmental degradation within the framework of sustainable and environmentally responsible development in all countries. While some 180 government leaders and more than 10,000 representatives from NGOs gathered in a spirit of cooperation, the Earth Summit is not regarded as a success. Still, it did establish a number of treaties, for example on biodiversity, and did create Agenda 21 (and the parallel Local Agenda 21 for councils).[25] In 2002 Rio's successor, the World Summit on Sustainable Development, will be held in Johannesburg. It is widely expected that the results of Rio will not be revisited. This time the emphasis will be on the further implementation of Agenda 21, and in particular the treaties on climate change, biodiversity and desertification. The negotiations will focus on the distribution between developed and developing countries of the burden of caring for the environment. In advance of the World Summit the Earth Charter was launched at the end of 1999. It has 16 principles related to sustainability that should be adopted by the UN General Assembly in 2002.[26]

At the Earth Summit in 1992 the first political steps were taken towards an international climate policy. The first step was the creation of the United Nations Framework Convention on Climate Change (UNFCC), in which some 180 countries gave their commitment to the prevention of the emission of greenhouse gases. This framework was followed by several international government conferences which sought to turn the commitment into practice. After long negotiations among developed countries, and between them and developing countries, more than 100 countries made binding agreements on climate policy in Kyoto in 1997. It has been agreed that the discharge of greenhouse gases over the period 2008–2012 should fall by 5.2 per cent relative to 1990.[27] Individual countries received their own targets within the limitations of this worldwide aim. Country targets were based on a country's level of development and on its existing percentage contribution to the total world emissions of greenhouse gases. EU countries formulated one common aim: a reduction of 8 per cent. Within that are individual EU country targets. For example, emissions in Germany must drop by 21 per cent, while in Greece they may rise by 28 per cent. No aims were set for developing countries. The aims for the US, Japan and Russia are respectively –7 per cent, –6 per cent and 0 per cent.

Kyoto also established which methods countries could use to achieve their targets. The most obvious aim is to reduce greenhouse gases nationally. This can be achieved on the one hand by controlling CO_2 emissions by industry (by, for example, energy savings or the use of more sustainable energy sources such as wind energy), and, on the other hand, by sequestering CO_2 in, for example, forests and other forms of (agrarian) land use.[28] In addition, it was agreed that countries may use any or all of the three instruments to achieve reductions abroad that will impact on their national targets. These instruments are: joint implementation (JI), the clean development mechanism (CDM) and international emissions trading (IET). Table 2.4 represents the most likely working of these Kyoto instruments.

JI is relevant to joint investment projects between Annex I countries with the aim of reducing CO_2 emissions and sequestration of CO_2 in sinks. The resulting emissions reduction units (ERUs) can be shared between both countries or be entirely credited to the investing country, and count towards that country's Kyoto target. In fact, the CDM is a variety of JI. The CDM concerns

Table 2.4 *International climate instruments developed at Kyoto*

Name	*Clean development mechanism (CDM)*	*Joint implementation (JI)*	*International emissions trading (IET)*
Form	International investments in CO_2 emissions-reduction projects		International trade of CO_2 emissions credits and rights
Unit	certified emissions reductions	emissions reduction units	assigned amount units
Countries	Annex I ⟶ Non-Annex I	Annex I ⟷ Annex I	
Period	Internationally from 2000 onwards	Internationally from 2008 onwards (EU = 2005)	
Example	Shell invests in clean technology in Ghana	Shell invests in clean technology in Poland	Shell exports emission credits to Exxon

Source: adapted from Hugenschmidt et al (2000, p52).
Note: Annex-I countries are the OECD countries, Eastern Europe and the countries of the former Soviet Union. Non-Annex-I countries are the remaining (developing) countries, which do in fact have an obligation to keep emissions as low as possible, but for which the Kyoto Protocol 'may not be a disproportionate or excessive burden'.

projects by Annex I countries in developing countries. Under certain conditions, Annex I countries can count certified emissions reductions (CERs) resulting from these projects towards their own targets. For IET, a ceiling on the level of total emissions will be enforced. The capacity may be shared or sold to polluters, who may trade the individual assigned amount units (AAUs) among themselves.

While the framework for international climate policy was shaped in Kyoto, the treaty is as yet not ratified by enough countries to be effective. This point was on the agenda at the sixth climate conference in The Hague in November 2000, together with the methodology of the Kyoto instruments, the mechanism to enforce cooperation (sanctions) and the use of sinks such as forests to sequester CO_2. The conference, however, did not succeed in reaching agreement over the use of sinks and sanctions for countries not yet conforming to the Kyoto targets. As a result the conference was extended; government leaders will meet again in Bonn in July 2001 to try to reach an accord that enables ratification by sufficient countries. A major gap has risen between the US and the rest of the world in the beginning of 2001, when President Bush declared 'Kyoto is dead'. Major issues for participation by the US are the rigidness of the proposed framework and on the inclusion of reduction targets for developing countries in the Kyoto agreement.

Individual countries will, however, continue to develop their own climate policies, and commitment to the Kyoto Protocol is strong. This means that business should take account of a tightening of climate policy and (eventually) the possibilities offered by the Kyoto instruments. The most significant sector affected by climate policy is energy. Other important sectors include transport, agriculture and forestry. Agriculture and forestry contribute both positively and negatively to CO_2 emissions; on the one hand they form the source (eg through deforestation), on the other hand they offer possibilities for the sequestration of emissions (eg through the planting of new forests). The agricultural, forestry and recreation sectors and developing countries will suffer the consequences of climate problems. In the area of banking and finance, account will have to be taken of such climate risks. This is examined in detail in Chapter 6. Moreover, the use of the mechanisms agreed at Kyoto offers opportunities for banks. This is dealt with in Chapter 5.

CONCLUSION

Economic and environmental policy are increasingly regarded as being interdependent both nationally and internationally. This is apparent from the change in the instruments of environmental policy. In its earliest years, environmental policy focused strongly on direct regulation (the imposition of standards underpinned by sanctions) and restructuring (the use of licences, directives and prohibition). Since the end of the 1980s business and citizens have been allotted more individual freedom. There is an increasing use of financial incentives for producers and consumers, such as ecotaxes and tradable emission rights. In short, market forces are the current instrument of choice.

Can sustainable development be driven by market forces when it is essential that technology is deliberately applied to the development of environmentally friendly production processes and products? In their book *Factor Four* (1997), Ernst von Weizsäcker and Amory and Hunter Lovins discuss numerous technological solutions that could massively reduce our impact on the environment. On the basis of the formula in Box 2.4 and the above, it can now be concluded that total environmental impact can only be reduced by a sharp decrease in the environmental impact per product unit.

With the expected doubling of the world population (P) and the much-needed increase in the world wealth level (W) by a factor of 4 to 8 (necessary to equalize levels of prosperity between developed and developing countries), it will be necessary to reduce the environmental impact per product unit (T) by a factor of 16 to 32 in order to bring about a halving of the current level of world-wide environmental impact (I).

Alongside technological innovation, changing consumer preferences should engender a sharp fall in environmental impact per product unit. Seen in this light, sustainable development poses a challenge for business; growth ought to occur in those sectors with a low environmental impact per product unit. New sectors will be created and others will disappear. The factors causing this will include: changing consumer preference; environmental policy (in which environmental effects are internalized within prices), with the use of green taxes or direct regulation; and changes in supply patterns caused by competitors or business-to-business relations. Other factors that influence supply and demand changes are interest groups and banks. Anticipating trends in demand and within society is the challenge faced by business. In this way, it too will be involved in the promotion of a sustainable society.

Business plays an increasingly prominent role in the achievement of sustainability; Chapter 3 will discuss this. Governments refer increasingly to the need for the support of business and individuals in the formation of environmental policy. This implicitly includes the role of the consumer. Although consumer behaviour is the determining factor for companies when imposing environmental measures that will increase prices, the government and politicians cannot take consumer behaviour as any indication of support for environmental policy.[29] Consumer behaviour is complex and depends on many factors besides environmental friendliness, such as price, ease of access, availability and status. Rapidly increasing public membership of NGOs such as Greenpeace and WWF is perhaps a far better indicator for governments of consumer support for environmental policy, and for business perhaps a indication of forthcoming government policy and consumer demand.

The role of the government is, in essence, the setting of ecological frameworks in which companies and individuals can operate. Education and price signals figure greatly in this. Ill-advised government subsidies hinder a rapid transition to sustainability. It is estimated that, worldwide, around $US700 billion is granted in subsidies detrimental to the environment (WRI, 2000, p6; also see De Moor, 1997, and OECD, 1998, for a more general discussion of the issues). This occurs chiefly in the energy, transport and agricultural sectors. As well as using fiscal facilities and pursuing an environmental policy, the governments will have to look closely at the environmental friendliness of their investments in other policy areas and at their purchasing policies (see, for example, Canada in OECD, 1996a). The on-paper policy ought eventually to be underpinned by active care for the environment in all investment decisions and choices. The same applies to business.

3

Sustainability: The Challenges for Companies

INTRODUCTION

The environment and the economy are closely related; in fact they are mutually dependent. The government has an important role to play in this by using its instruments to internalize environmental effects into market prices so that companies and consumers make more sustainable production and consumption decisions, respectively. Obviously, more sustainable consumer purchasing behaviour will encourage companies to make more sustainable production decisions. Besides these push (government) and pull (consumer) factors, companies can also foster more sustainable development themselves, because they may have their own ideological or business reasons for striving for sustainability. Of course, it is important that the measures do not undermine the continuity of these companies in the long term, and that consumers actually buy the more sustainable products, or that governments support these companies with subsidies and tax concessions.

It often looks like a vicious circle: companies do want to market more sustainable products, yet these sustainable products usually have a higher cost price and consumers would rather not pay too much when it comes down to it.[1] The number of consumers that are actually prepared to pay more for sustainable products is still small, although the number of consumers stating a preference for more sustainable products is considerable.[2] The limited demand for these products that results is often not enough to attain the required economies of scale to be able to compete (in terms of price) with the less environmentally friendly alternatives that already exist. This vicious circle can be broken with directed government policy (such as subsidies or tax measures that reduce the price of more sustainable products), or smart innovations that are both more sustainable and cheaper. Herein lies a big challenge for companies. Low prices may exist when striving for eco-efficiency, and the many possibilities and examples that already exist are proof of this. Nevertheless, eco-efficiency does not guarantee sustainability. The challenge therefore exists to go beyond this and develop products that are actually sustainable.

It is up to business to take on this challenge. The question is: How will companies do this? Eco-efficiency and sustainability require companies to play an active and focused role. How companies take on this challenge depends on how they perceive their responsibility. This again depends on a company's norms, values, culture and structure. Seen in this light it is striking that 95 per cent of around 500 international businesses in Europe and North America regard sustainability as an important company issue.[3] The era in which large companies could operate within the legal boundaries without having to account for their actions to society at large appears to be behind us, with the appearance of the more emancipated global citizen. The discussion below on corporate governance (the accountability of organizations, pages 38–40) shows that differences exist between companies, and shows how they manifest their sense of accountability. This is particularly important with respect to the environment. The section starting on page 41 examines sustainable business and shows that the possibilities for and the limits to sustainable development are determined by the way an organization wishes to position itself in the world. In other words, this chapter does not examine the entire range of actions taken by companies with respect to the environment in detail. Many companies have already been active for quite some time, in terms of both their own processes and products. Nevertheless, there are important differences in the degree to which companies are active and have embraced sustainability. The goal of this chapter is to provide insight into these differences. Verification and transparency are very important in appraising the actual performance and progression towards sustainable business and the differences between companies, with environment or sustainability reports playing a key role. Page 45 onwards examines the general aspects of this; bank-specific aspects of reporting are presented in other chapters. After all, these reports can be an important sources of information for banks (Chapters 5 and 6) and a means for them to account for their own performance (Chapter 8).

CORPORATE GOVERNANCE AND SUSTAINABLE DEVELOPMENT

In the liberal economic climate at the turn of the 20th century, the governments of most Western countries did not get very involved in social problems. Confronted with social protests and the disadvantages of an overly liberal policy, a few companies did apply their influence and power for social objectives in those times. This can be seen as the advent of socially responsible business. It entailed that activities with respect to company strategy arose in the form of participation in and/or with the support of local social activities which contributed to solving or reducing social problems such as unemployment, criminality, decay of the urban environment, cleaning up the environment and reducing noise pollution (Boudhan et al, 1996). Nevertheless, there were major differences between individual companies and countries. The economic crisis in the 1930s and the world war that followed led to a more active role of government in most Western economies. It was largely the creation of the welfare state in continental Europe that greatly reduced the necessity and demand for business involvement in strictly social causes.

Nevertheless, the policy of governments to take a step back and the reduced influence of national governments since the early 1980s once again created the scope for social involvement by business. Still, the areas and subjects are different than they were at the turn of the 20th century. Socially responsible business practices evolved back then to prevent stricter regulations. The renewed interest in socially responsible business practices is now based on market-oriented considerations such as continuity and the ability to distinguish oneself from competitors.[4] This is related to increased awareness (due to people being, in general, better educated, the higher level of prosperity and the increased amount of communication and information technology) of people in the Western world. The following forms of socially responsible business can be distinguished: concern for employees, gifts and donations, participation of personnel in social projects, cause-related marketing (associating a product or service with a non-profit organization), idealistic advertising and sponsorship. The development of environmentally friendly products is also a component of this new involvement. Moreover, engaging in socially responsible business practices is not always a choice. The market demands it for products that are relatively homogenous, and where differentiation on social grounds is critical.

Often, socially responsible business practices will be based on a perceived social responsibility. The question is how far this responsibility extends (Van Luijk, 1997). Where are the boundaries? Which demands by various entities, such as customers, employees and environmental organizations, can the company not be reasonably expected to fulfil? These questions lead us to the matter of corporate governance, or the accountability of the company's management (eg Blair, 1995 and Renneboog, 1999). Many discussions with respect to corporate governance concentrate on the distribution (and power) of voting rights among shareholders. Nevertheless, the subject also has wider dimensions. It involves the company's operations within the social context to allow the company to occupy, through a balance of power, an enduring position in the community. Corporate governance concerns issues related to the way in which any company, listed or not, is managed and controlled and the people who can influence it in this sense. Roughly speaking, there are four categories that can be distinguished.[5]

The first category is based on shareholders. From this viewpoint, the company is defined as an exclusively economic and legal entity. This simultaneously establishes the parameters of the company's accountability. In principle, the legal and market requirements are adhered to in a strict sense and the company strives to attain the maximum return for its shareholders. It may be involved in social activity, provided that it remains voluntary and certainly if it results in a positive contribution to profit. If concern for the company's impact on the environment leads consumers to boycott it (as happened when Shell proposed letting the Brent Spar sink to the bottom of the North Sea, or with ABN AMRO Bank's involvement in hazardous mining activities in West Papua New Guinea), then shareholders will force the company to actively modify consumer perception to such a degree (that is to say to modify the policy and/or communication strategy concerning the environment) that profitability is no longer at risk.[6] This involvement arises from the inside-outwards (an internal orientation). This is in contrast to the three other categories where involvement is shaped on the outside. An example of the shareholder category is HSBC Holdings, a major bank in the UK.[7]

The second category is based on economically funded stakeholders. The thinking underlying this idea is that a company has obligations towards everyone that has vested economic interests in the company and carry their share of risk, such as suppliers, customers, shareholders, distributors and employees. This obligation is not based so much on moral or legal grounds (which is true for categories 3 and 1 respectively) as on the grounds of economic solidarity (risk sharing). An example is Fiat, which believes that concern for its suppliers is a necessary condition for its own operations (WBCSD, 1997, p11). This is no longer a purely internal orientation (category 1), but the scope for social involvement remains limited by purely internal objectives.

The third category is based on socially funded stakeholders. This viewpoint can be traced back to an idea that is gaining ever more momentum, that of 'corporate citizenship'. Society expects that companies recognize their obligations towards interests that clearly lie outside the domain of purely economic transactions. Examples of these interests are the environment, future generations, long-term unemployed and immigrants. Examples of companies with this view include The Body Shop and The Co-operative Bank.[8]

The disadvantage of this last category is its open-ended nature: anyone who claims to have a stake can demand involvement. Moreover, contradictory claims are likely to exist. In order to be able to deal with this dilemma, companies can specify their responsibilities by determining the strength of a demand based on three factors – the social legitimacy of the entity making the claim, the moral weight of the claim, and the ability of the party involved to cope with the claim on its own (Van Luijk, 1997). Specific methodologies, such as ethical accounting, have been developed for this. The decision as to how far and to what degree the circle of entities involved stretches remains a moral decision.

A fourth category is based on a customer-value approach. Actually, this approach boils down to the interests of the customer (one of the stakeholders) being placed above all others. It is not the product per se that is central, but customer needs. The Rabobank Group considers itself to belong to this category. A company in this category would like to deliver as much added value as possible with the customer being the determining factor. The exterior is prominent in this category. At this point, a comment must be made: a customer is not only interested in the added value of a product or service, but the way the company functions as a whole. The way a company functions in society can be a component of customer value. If a company only concentrates on adding value for the customer without involving this dimension, the customer-value approach has very limited substance and will not always lead to the desired result. Research shows that companies that look at the whole exterior picture (including its social qualities as perceived by the customer, such as the sustainability of its operations) and thereby make a contribution to society as a whole, are more successful than others.[9]

SUSTAINABLE BUSINESS

Chapter 2 reveals that environmental awareness has grown considerably. The awareness that technological innovation and supply and demand changes are necessary has sunk in among politicians and citizens as it has among consumers and producers. The social involvement of companies has increased, as has the perceived social responsibility of citizens, consumers and businesses. The shift has not been a gradual one, but has occurred in fits and starts. Nevertheless, the trend is clear. Companies have had to deal with changeable but always rising quality and product requirements of the market and the government, financial stimuli and conditions from banks and insurance companies, action by local residents and the environmental awareness of its own employees and management. Companies are being spurred on to change their behaviour towards the environment. It can often be split into phases: the trends can be subdivided into blocks, as it were.[10] However, the movement through the phases is not self-evident nor is it universal. Some companies have a certain immutable stance towards the environment or are already in the more progressive phases. Moreover, differences in adaptiveness or incentives exist among sectors and between businesses.

Four phases, stances or strategic concepts can be distinguished from the point of view of the drive towards sustainability.[11] These phases involve both a company's behaviour and ambition. The subdivision into phases should be seen as a description of the developments that various companies progress through, have progressed through or could progress through. The division is not static. Ultimately, every company will have to operate in a sustainable way to attain sustainability in general. The characteristics of the last phase, the sustainable phase, will be continually enriched to provide an evolving substantiation of the concept of sustainable development as held by society. This last phase or stance is therefore not necessarily the end point.

The first phase is the defensive phase. Companies initially react defensively and are surprised by the government's new environmental requirements. They focus most of their efforts on limiting the changes they need to make. Any measures taken have an end-of-pipe character and are mere additions to the existing production process. For this very reason the cost price is also often raised. Companies which hide their heads in the sand in the first phase are threatened by legal action (or are already being prosecuted), as well as uncertainty regarding their business continuity or damage to their image. They might lose orders because customers set environmental standards that they cannot meet. A good example of this is when Kimberly Clark, the producer of Kleenex, refused to extend their contract with the pulp and paper producer MacMillan Bloedel in 1994 due to its uneasiness about that company's environmental practices (*International Environmental Reporter*, 23 March 1994).

Little by little, companies alter their course and most of them take a different stance on the environment and sustainable development: they enter the preventive phase. One of the characteristics of this phase is the systematic approach to internal environmental care. Companies strive to limit, in terms of process rather than issue, emissions, waste and the consumption of material. It has been shown that considerable cost savings could be realized through these efforts. Companies accept the usefulness of legal prescriptions and no longer oppose them as a matter of course. When legislation becomes stricter, scarce resources become more expensive, polluting and waste are more costly, and social and safety aspects increase financial risks, so management will opt for the sustainable route internally, which is often shown to have long-term and even short-term economic benefits. 3M, for instance, has had a 'Pollution Prevention Pays Program' since 1975, the net cost savings of which exceed $US750 million (WBCSD, 1997, p5).

The companies which implement preventive environmental care often come to realize the potential of an offensive stance with respect to the environment. The focus shifts from the production process alone to include the product itself. During this phase, companies have a broader view of environmental care as an opportunity to generate additional revenue. They are no longer merely focused on government requirements and directives or increasing raw material prices, but also on any existing or potential wishes of customers, employees, banks, shareholders and suppliers. An example is Proctor & Gamble, which began introducing and developing concentrates to reduce the amount of packaging, among other things, as early as 1980. In 1989 the 'Ultra' product line was introduced. These products soon gained a market share of 23 per cent and by the end of 1994 it was actually 70 per cent (*Financial Times*, 7 December 1994). Confidence grows to such an extent by the third phase that companies are no longer afraid to communicate their stance systematically. They sign treaties, such as the ICC declaration on sustainable development, and they propagate their ideas among other companies through organizations like the World Business Council for Sustainable Development (WBCSD).

The considerations that go into deciding to operate preventively or offensively are therefore not only idealistic but also businesslike in nature, and can be summed up as follows (De Groene, 1995):

- environmental-hygienic: the direct impact of an incident on health;
- strategic: new products, employee motivation;
- financial: cost savings and a lower risk of environmental damage claims;
- communication: a positive and trustworthy image;
- legal: in terms of accountability, upholding environmental laws is more effective.

A growing number of companies are being swayed by these arguments. However, not all companies have reached the third phase. More than 90 per cent of private sector companies in most European countries are small- to medium-sized enterprises (SMEs) with fewer than 250 employees. In the EU, SMEs account for around 55 per cent of gross accumulative revenue and 66 per cent of the total workforce (Eurostat, 1995). Many SMEs follow the overt governmental environmental prescriptions. A study of private sector companies according to the above classification in the mid south of The Netherlands showed that at least 90 per cent of companies were still in the defensive and preventive phases. This was particularly true for SMEs; 60 per cent of companies which have advanced to the offensive phase are large companies (Jeucken and Van Tilburg, 1999). The transition from one phase to another is a complex process, however, and depends on the overall potential of technological development and the stance of the executive and management. Obviously, for banks it is important to know which phase a company is in to assess the environmental risks posed. A bank can structure its policy such that more companies take the step to the third phase, potentially lowering the risks to the bank (by improving the business continuity of the company, for instance).

In summary, the first three phases can be characterized as follows. Defensive business practices are cost-focused. Attempts are made to block legislation on the grounds of its effect on the company's competitive position (true of most companies in the 1970s and 1980s). Preventive business is also focused on costs, but this time from the perspective of achieving savings through more efficient production processes and techniques (true of most companies in the 1980s and 1990s). Offensive business is focused on profit and marketing as well (true for many companies in the 1990s). The question is whether the objectives that we, as a society, have set for sustainable development are attainable within the current economic parameters through more efficient production and the development of more environmentally friendly products. A reduction in the environmental burden does not necessarily imply sustainability. Additional steps will be needed.

Sustainable business is the fourth phase or stance a company may adopt. Whereas the sustainability of human action with respect to the environment was targeted up to the mid-1990s, the interpretation of the concept of sustainability with respect to companies has widened into what is known as the triple bottom line (see Chapter 2). This means that companies must ultimately strive for three dimensions simultaneously: people, planet and profit. Sustainable business could then be defined as:

- people (social value): value for and development of employees, customers, members, suppliers, NGOs, etc (in a socioeconomic context); examples are workplace health and safety, employee retention, labour rights and human rights;
- planet (ecological value): a sustainable level of use of natural resources and environmental burden, now and in the future, locally and globally;
- profit (financial value): a profit level necessary to guarantee the continuity of the organization's service provision.

The profit aspects are given even broader application by some (eg GRI, 2000, p1), with criteria such as: wages and benefits, labour productivity, job creation, research and development, and investments in training and other forms of human capital.

In the three Ps view, sustainable companies must attend to all of these aspects. Obviously, dilemmas will arise. A certain investment may be socially and economically beneficial but not ecologically so. That is exactly the lesson to be learned from the experience of the previous century. Obviously, a certain investment may also be ecologically beneficial but disadvantageous in socioeconomic terms. How should companies act in such situations? A stakeholder dialogue is indispensable in getting to grips with the breadth of the above definition of sustainable business and the dilemmas this might entail. Some companies in this phase have chosen this option (like The Body Shop and The Co-operative Bank). Their experiences show that dilemmas can also arise within a single aspect of sustainability. A certain stakeholder might like to obtain the maximum return on his investment while another would like to prevent massive layoffs from taking place. There is nothing new here.

The point is that the broad definition can lead to a standstill thanks to the dilemmas it presents. A narrower approach has therefore been chosen for this book, an approach in which planet (supplemented by some fundamental 'people' aspects, like human rights, freedom of speech and the reduction of poverty in developing countries) is chosen as the dominant P. The reasoning is twofold. Firstly, the environment and future generations simply do not have a vote. Secondly, socioeconomic considerations will gain the upper hand in today's social field of influence, as all stakeholders have a personal interest in this dialogue. This personal interest will easily lead to ecological interests being compromised. This is the case for everyone who is not an eco-centrist.

The ideal for the planet is the use of raw material and pollution of the environment that does not exceed the regenerative and absorption capacity of the ecological system. On the one hand this means that the supply of resources remains the same in the long run. On the other hand, it does not mean that no pollution takes place; the aim is to keep the level of pollution within the absorption capacity of the ecological system. This cannot be determined precisely at the company level, since the absorption capacity is influenced by the cumulative pollution of all companies and other economic actors with respect to a certain substance or environmental compartment. Indicators can, however, be developed to measure the sustainable level of pollution per company. Box 3.1 presents one example of such an approach being taken by a multinational.

In offensive businesses, the economic situation is the prime consideration. Opportunities and threats are weighed on economic grounds. In this sense, 'offensive' does not mean going beyond developing the market in economic terms. Sustainable and offensive business cannot mean the same thing unless prices exist for all activities and resources. Sustainable companies therefore go a step beyond what merely *appears* to be possible within the current economic framework by explicitly taking into account how semi-finished goods are produced, the extent to which resources are extracted in an environmentally friendly way, and what must happen to the end product when it is finally disposed of. Often, what is ultimately required from a business continuity perspective is a stance based on sustainability; this explicitly involves a long-term perspective and, as a result, company strategy. To illustrate: a number of companies depend entirely on finite natural resources. To ensure lasting business continuity, care for these resources is required. A producer of carbonated beverages (such as Coca-Cola) must know where it will get clean drinking water in the future. Another example is the collaboration between Unilever and WWF regarding the preservation and management of fishing grounds.

Box 3.1 Van Melle's indicator approach

Van Melle is a Dutch multinational, founded at the end of the 19th century, which concentrates on the preparation of sugar products (confectionery). Van Melle feels strongly about environmental care from the social involvement and people perspective. Van Melle's mission statement reads (1997): 'We feel we are involved in what happens to the environment and are determined to reduce our impact on it to a sustainable level.' The group wants to have attained sustainable business operations by 2005, in the sense of keeping its operating activities within the boundaries of what the environment can endure. In order to monitor the progress of the steps taken in this direction, it developed an environmental barometer.

Van Melle is concentrating on five environmental aspects it sees as relevant: the greenhouse effect, acidification, eutrophication, dehydration and waste. The factor by which emissions and extractions must fall in order not to damage the environment, the 'no-effect level', has been established through scientific research. The environmental barometer only measures the environmental burden of its own factories plus the environmental burden caused by energy extraction and transport. Environmental effects related to other aspects of the production and consumption chain are not taken into account. The environmental barometer is used both for Van Melle Group's environmental burden as a whole and for each location separately.

Between 1992 and 1995, the environmental burden of Van Melle declined by a factor of 3 or 4 according to this environmental barometer. To help achieve this, Van Melle equipped its factory in Breda, The Netherlands, with 2,500 square metres of solar collectors. To reach sustainability, Van Melle's total environmental burden must still fall by a further factor of 16. Given healthy profitability and sufficient funds, Van Melle expects to attain sustainability in 2005.

For the most part, innovative ideas and processes with a sustainable character are central, integrated chain management and closed loop production and redesign processes being examples. Each successive link in the production chain, from raw material extraction up to the disposal phase, are considered as a single whole in integrated chain management. The solution to the livestock manure problem may very well reside in the type of feed used. A company might ask itself what people need, not in the sense of a need for a certain product, but the need for *functions*. A consumer may not necessarily need a car, but needs to get from A to B as pleasantly and quickly as possible. A car could meet this need, but so could a bike or a train or a combination of the two. An even more innovative solution is to 'bring B to A' by use of modern communication technology. Videoconferencing is a good example of this.[12] Table 3.1 provides an overview of the four phases and underlying motives, substance and consequences.[13]

Accountability and reporting

If a company uses its code of conduct to communicate its stance with respect to sustainability, it should be possible to test it independently. What really happens within a company in terms of sustainability? Is progress visible? How is it doing in relation to its competitors? This kind of information is of interest

Table 3.1 *Integrative model of attitudinal stages towards sustainable business*

	Defensive business	*Preventive business*	*Offensive business*	*Sustainable business*
Strategies				
Public attitude	'Trust me'	'Tell me'	'Show me'	'Involve me'
Actors involved	Staff members	Production lines	Whole company	All (major) stakeholders
Drivers	Legislation	Efficiency	Strategic performance	Societal licence to operate
Measures	Clean-up operations	Prevention	Chain management	Sustainable measures[14]
Attitude of companies	Resistance/ indifference ('ostriches')	Passive, eco-cost minimization	Proactive, eco-revenue maximization ('smart movers')	Innovative (contribute to and stimulate sustainability)
Indicators				
Users	Environmental staff member, government	Line management, environmental stakeholders	Whole company, financial stakeholders	Top management, society, media, NGOs
Functions	Registration	Process changes, monitoring	Product redesign, communication	Integrated decision making, internalization, accountability
Expression	Emissions, costs	Efficiency of internal processes	Eco-efficiency, product characteristics	Sustainable resource use, societal costs, values
Scope	Substances, emissions	Production processes	Products, chain processes	Sustainability issues
Reference value	Regulatory targets	Other processes, previous years	Other products, competitors	Societal values, sustainability issues

to a broad group of concerned parties, such as employees and management within the company, governments, NGOs, consumers, local communities, shareholders and banks. That is the reason environmental reporting has taken flight.[15] More than 1000 companies worldwide currently publish an annual environmental report (White, 1999, p6). UNEP even expects more than 10,000 to publish an annual environmental report by 2005 (Skillius and Wennberg, 1998, p43). However, there are major differences in the way companies choose to report and major changes can occur over time even within companies.

Environmental reports first saw the light of day at the end of the 1980s. Companies reporting from the very beginning include Novo Nordisk and Monsanto (EEA, 2000, p34). A dynamic arose as a result, forcing competitors to keep up. By the early 1990s, a clear trend in voluntary environmental reporting had developed. The reports deviated considerably, in terms of content, design and target group, from modern reporting in which social and economic factors play a growing role. The first study into the quality and quantity of environmental reports took place in 1993 by Deloitte Touche Tohmatsu International, the International Institute for Sustainable Development and SustainAbility (DTTI, IISD, SustainAbility, 1993) and resulted in a five-phase model:

1. 'Green glossy': the company produces newsletters, magazines and videos with an ecological slant; a few words in financial reports are dedicated to the environment.
2. 'One-off': companies publish a one-off environmental report (or do so at long intervals), usually in conjunction with a corporate statement with respect to the environment.
3. 'Descriptive': companies publish an annual environmental report as part of an environmental care system, but primarily containing qualitative and descriptive information.
4. 'State of the art': companies publish an environmental report annually with detailed information on environmental performance and impact, both corporation-wide and at individual locations; the reports are also available online or on CD-ROM and contain references to the annual financial report.
5. 'Sustainability': the company publishes an annual sustainability report which combines and accounts for the ecological, economic and social aspects of its operations.

An international study into the state of affairs regarding environmental reporting in 1999 showed that 44 per cent of the Fortune Global Top 250 companies in the non-financial sector had produced an environmental report and that the quality of environmental reports had risen significantly. Only 18 per cent of companies in the study allowed the environmental report to be audited externally (KPMG, 1999). Moreover, many interest groups indicated that they had little faith in external auditing. A lack of auditing standards means that many interest groups do not believe that audits add to the reliability of environmental reports (GEMI, 1996). Emphatic differences exist between regions and countries of the world in this respect. In 'high trust' Scandinavian countries, external verification has little use. In Anglo-Saxon countries an unaudited report is usually no more than a PR brochure.[16] A number of studies have shown that the perceived reliability is also low because of the lack of standards for reporting. A few studies have even pointed out inconsistencies in reports from the same company over time (VBDO, 1998).

A range of initiatives therefore came about to develop standards for environmental reporting. Some of these initiatives were sector-specific (such as the VfU standard for banks or the CEFIC standard for the chemical industry) or region specific (such as KEIDANDREN in Japan and CERES in North America). Important issues included corporate-wide instead of location-specific reports and the mish-mash of eco-indicators being used. The WBCSD has therefore made an attempt to come up with region- and sector-transcending standards for eco-indicators (WBCSD, 2000). The generic management systems and standards systems ISO14000 (series) and EMAS (European Eco-Management and Audit Scheme) should also be mentioned in this respect.[17]

Most initiatives concentrated on organizational profile, environmental policy, environmental management, legislative compliance, emissions, resource efficiency, the lifecycle of product impacts, environmental liabilities and costs, and stakeholder relations (EEA, 2000, p36).

A number of governments now compel large, relatively polluting companies to publish public environmental reports. Denmark and The Netherlands are two examples. So, by implication, national standards have been created which guarantee the ability to compare the companies concerned in that country. A side benefit of that compulsory arrangement is the creation of a level playing field – a voluntary arrangement, on the other hand, would provide loopholes for free riders. However, compulsory arrangements are usually retrospective while voluntary initiatives also focus on the future.

In 1997, the Coalition for Environmentally Responsible Economies (CERES), in conjunction with UNEP, came up with an all-inclusive standard for voluntary reporting and eco-indicators, going by the name of the Global Reporting Initiative (GRI). All parties concerned are involved in this initiative and they are using the lessons of earlier initiatives and the experiences in developing a standard for financial reporting (such as generally accepted accounting principles, GAAP, and the international accounting standards, ISA).

> *The GRI is a long-term, multi-stakeholder, international undertaking whose mission is to develop and disseminate globally applicable sustainability reporting guidelines for voluntary use by organizations reporting on the economic, environmental, and social dimensions of their activities, products and services.* (GRI, 2000, p1)

The first guidelines were launched in 2000. The GRI guidelines are intended to be applicable to any size and any type of organization and improve the comparability between businesses and progress over time. In addition to the organization-wide GRI report, complementary facility, regional, or other disaggregated sub-reports may be appropriate for different stakeholders. Box 3.2 examines the GRI guidelines in greater detail.

The GRI initiative appears to be successful and is creating the foundation for a generally accepted global standard. It is supported by the many companies already following the guidelines and by the fact that the US Environmental Protection Agency (US EPA), the European Union and the Japanese Environment Agency and the Japanese Ministry of Trade and Industry (MITI) want to base their government programmes on the GRI guidelines.[18]

Conclusion

The discrepancy between short-term behaviour and what is desirable in the long term is clearly a pressure point in achieving sustainable development. Moreover, market failures exist; individuals do not take into account the effects of their actions on other individuals and society as a whole to the extent that these effects are not expressed in the prices (these effects are known as externalities).

BOX 3.2 THE GRI GUIDELINES

The goal of the GRI guidelines is to help companies report information concerning sustainability (GRI, 2000, p1):

- 'in a way that presents a clear picture of the human and ecological impact of business, to facilitate informed decisions about investments, purchases, and partnerships;
- in a way that provides stakeholders with reliable information that is relevant to their needs and interests and that invites further stakeholder dialogue and enquiry;
- in a way that provides a management tool to help the reporting organization evaluate and continuously improve its performance and progress;
- in accordance with well established, widely accepted external reporting principles, applied consistently from one reporting period to the next, to promote transparency and credibility;
- in a format that is easy to understand and that facilitates comparison with reports by other organizations;
- in a way that complements, not replaces, other reporting standards, including financial; and
- in a way that illuminates the relationship between the three linked elements of sustainability: economic (including but not limited to financial information), environmental, and social'.

The GRI guidelines provide a framework for reporting with the main goal of facilitating comparisons between companies and over time. The guidelines offer no support on actual data collection, information and reporting systems, monitoring performance and report auditing. Each GRI report must include the following topics and layout (GRI, 2000, pp23–36):

- *CEO statement*
- *Profile of the reporting organization:* This section must provide an overview of the company in a generic sense (its size, market, type of products) and explain the scope of reporting. The objective is to provide the reader with a context to interpret the report.
- *Executive summary and key indicators*
- *Vision and strategy*
- *Policies, organization and management systems:* This section must include an overview of the company's governance structure and the management systems maintained to implement the vision in the previous section. The dialogue with the parties concerned must also be accounted for.
- *Performance:* This section contains the company's performance with respect to each of the three Ps. Both quantitative and qualitative aspects are presented and the raw data must always be reported separately when using rates or graphs. The context surrounding the data and the trends must be illustrated and include commentary. The account of performance must contain a general section (the same for all companies) and can be supplemented by a separate section discussing sector or company specific aspects.

A number of companies now follow the GRI guidelines. A remarkable fact is that the guidelines are being used by companies in widely divergent sectors and in all parts of the world. Examples are: British Airways (air transport, UK), Electrolux (appliances, Sweden), Excel Industries (chemicals, India), General Motors (vehicle manufacture, US),

Henkel (chemicals and consumer products, Germany), NEC Corporation (information technology, Japan), Novo Nordisk (pharmaceuticals, Denmark), Procter & Gamble (consumer products, US), Shell (petroleum, petrochemicals, energy, UK/The Netherlands), VanCity Savings Credit Union (financial services, Canada) and South African Breweries (consumer products, South Africa).[19]

Box 3.3 Foreign direct investment (FDI) and the environment

Care for the environment is now near the top of most large companies' corporate agenda in the Western world. Through globalization and internationalization the world now has more than 40,000 multinational enterprises (MNEs). They make more than $US400 billion in FDI (the flow to FDI rose from US$193 to US$619 billion between 1990 and 1998).[20] The ten largest MNEs together generate more in annual sales than the hundred smallest countries generate in GNP (Hawken, 1993, p92). The flow of FDI now exceeds official development assistance (ODA) by a factor of five (five years ago the factor was one half; Dowell, Hart and Yeung, 2000, p1059). This makes the influence of Western MNEs on the environment and environmental policy in developing countries enormous. The way FDI is employed in developing countries therefore becomes a crucial issue. FDI and the activities of MNEs have made an important contribution to the social and economic development in many regions of the world. However, not very much is known about the environmental impact of these investment flows. This argument increasingly appears to be gaining the upper hand.

Some people refer to a race to the bottom, in which very environmentally damaging companies relocate to countries with lax environmental policies (the pollution havens hypothesis; Korten, 1995).[21] However, insufficient evidence can be produced in practice (eg Jeucken, 1998a). On the other side, the presence of MNEs – with their modern and clean technologies at their disposal – may raise the environmental standards in developing countries. However, no evidence can be found to establish this either (OECD, 2000). Some MNEs do appear to make positive contributions to sustainability in developing countries in real terms. Other MNEs nevertheless maintain obviously lower standards in developing countries than they do in their activities in the US or Europe (eg UNRISD and UNA, 1998). Growing economies will adjust their environmental requirements sooner or later. The 'Asian Tiger' economies and now even China are examples. This allows MNEs maintaining Western environmental standards for their activities in developing countries to build up a competitive edge on local companies.

In short, a race to the bottom does not appear likely. MNEs that maintain the same standards in developing countries as they do in their home countries also perform better. They are better prepared for rising environmental standards, they enjoy the cost advantages associated with environmental care and can keep their image unblemished. In a CNN and internet world it no longer matters where production takes place. Every MNE will ultimately have to account for its activities wherever they take place. Research shows that most MNEs really do maintain their environmental standards on a global basis and therefore do not differentiate according to the strictness of the environmental policies in the places they carry out their activities (Dowell, Hart and Yeung, 2000, p1072).

Economic instruments in environmental policy attempt to internalize the externalities in prices. These negative externalities such as environmental contamination could also be integrated into the economic system in a less domineering manner, namely through a revival of social involvement and responsibility. Governments can foster these elements by disseminating information, providing subsidies and other financial stimuli and ultimately formalizing these stimuli through legislation. A tendency to acknowledge and accept one's responsibility can be distinguished in individuals and organizations. Responsibility is still subject to certain economic parameters. Changes in consumer behaviour ('conscious consumption'), producer behaviour (technological, organizational and management changes in processes and/or products) and financial companies are creating a new economic environment with its own sector structure, which can provide greater scope for quality and intangible values. These changes will establish new boundaries for what is economically feasible. Box 3.3 examines this, particularly with respect to the activities of multinationals in developing countries and the positive role they can play there in terms of sustainability.

The key issue for companies on the path to sustainable business is: Are they targeting a maximum level of market breadth or profit, approaching sustainability in such a way that a maximum profit is produced within the legal strictures, or are they prepared to impose sustainability criteria on themselves of such a nature that they accept a smaller market and less profit in the short term if need be? Is this actually feasible or will profit fall to such an extent that the company ultimately fails? Examples such as Van Melle, Shell, Novo Nordisk and The Body Shop prove that that does not necessarily have to be the case. Less quantity and more quality combined with more profit is also very feasible.

4

Sustainability: A Special Role for Banks

Introduction

A clear trend emerges in the previous chapters: care for the environment and the concomitant need to build a sustainable society is not a fad but an irreversible necessity. Businesses take voluntary action or can be forced to take part in this trend, the ultimate form of which is sustainable business, as set out in Chapter 3. Banks are also being confronted with this development or are helping to shape it. Moreover, if sustainable business is to be realized at the macro level, the stance of banks will be critical. Their role in economies is an intermediary one, transforming money by space, term, scale and risk. This function thus affects the development and direction of the economy; its influence is not merely quantitative but also qualitative. In part, a bank creates opportunities for sustainable business through its financing policy and is influential through its fee activities (in giving its customers investment advice for instance). Furthermore, banks have an enormous comparative advantage with respect to the knowledge and information they have about various market sectors, legislation and market developments. Banks can deploy these instruments and knowledge in a focused way and stimulate sustainable development. It is a question of will and vision. A bank must attempt to substantiate its sustainable development policy without trying to meddle excessively in the entrepreneurs' business or becoming an extension of government (in the sense of trying to maintain legislation). Moreover, banks are also limited in their freedom by market circumstances. Not surprisingly, the real question is how far banks can go in stimulating sustainability directly.

A broad definition is used for banking in this book. 'Banks and their functions in an economy', below, examines the various kinds of financial institutions and the role of banks in the economy. This section offers readers who do not have a background in finance an introduction to the various forms financial institutions can take, their role in the economy and the developments they are undergoing. 'The drive behind sustainability at banks', page 61, then examines the points of contact between banking and sustainable development and the reason banks are getting involved in this area. This will take place by addressing the environmental impact banks can have on the one hand, and by considering the pressure banks are under from various stakeholders on the other. 'Intermediation and sustainability', page 67, reflects on the role banks can play within society in terms of directing its efforts to achieve sustainability, while 'Sustainable banking', page 71, relates this to the stance and will of banks to do so, presenting a model that attempts to shed light on this. Finally, the 'Conclusion' (page 74) provides a look ahead to Part II, in which the concrete activities of banks regarding sustainability will be considered.

BANKS AND THEIR FUNCTION IN AN ECONOMY

Types of financial institutions

Banks exist in many shapes and sizes, blurring their distinction with other financial service providers in a variety of situations still further. There are savings banks that focus solely on attracting savings. There are also very complex and broad financial service providers which are involved in consumer credit, mortgage lending, leasing, treasury, securities trading and insurance. In other instances, they are specialized financial service providers such as investment banks or insurers. In order to get a good idea of the divergent types of financial institution and banking services, a distinction can be made between institutions which create money and those that do not – in other words, depository and non-depository institutions (see Saunders, 2000, pp2–83 and Hubbard, 1994, pp272–296).

1 **Depository institutions:** institutions which lend out a significant part of the funds entrusted to them (such as savings) in the form of mortgages and loans. Among these institutions a distinction can be made between:
   - *commercial banks:* institutions which use the funds entrusted to them by their customers to extend loans to consumers and business customers, and distribute profits to the bank's shareholders;
   - *savings institutions* (savings and loans associations and mutual saving banks): an institution, usually a mutually structured organization, specialized in attracting savings and extending home financing; a mutual savings bank distributes surplus profits to its customers;
   - *cooperative banks and credit unions:* institutions which are not strictly focused on making profit, are owned by their members (who deposit their savings) and are focused on providing credit to customers; any profits are retained.
2 **Non-depository institutions:** institutions which finance their activities through customer fee contributions or by issuing securities (such as shares) on the capital market. A distinction can be made between:
   - *securities market institutions:* institutions which are involved in capital market transactions and/or offer advice and/or services to institutional investors, large companies and governments; they include investment banks (support and advise companies in mergers and acquisitions and

in raising new capital in primary markets), brokers and dealers (who bring together buyers and sellers of securities for others or for proprietary trading respectively), and traders (trading securities on a stock exchange such as the New York Stock Exchange – or via over-the-counter markets using computers and telephones);

- *investment institutions:* institutions which attract funds to invest in securities and loans with higher risks than those accepted by commercial banks; these include mutual funds (institutions which pool their customers' funds and invest them in projects or companies; bond and income funds, equity funds and money market funds are examples) and finance companies (institutions specialized in investment, participating interests and loans to businesses and consumers which are funded by issuing short- and long-term debt securities; leasing, factoring and venture capital are examples);
- *contractual savings institutions:* institutions which provide protection for individuals and businesses during less favourable times, both unforeseen and predicted; these include insurers (institutions which offer financial protection for certain unforeseen events and risks in return for periodic premium payments; life insurance and property/casualty insurance are examples) and pension funds (institutions which pool the savings of businesses and employees and invest them with the objective of providing their members with a pension when they retire);
- *multilateral and governmental financial institutions:* institutions which are financed by a single government or the governments of a number of countries whose objectives may involve stimulating certain wealth-creating activities or development; examples are government credit facilities for individuals and businesses (such as student loans and export–import banks), government guaranties for loans extended by commercial banks (such as for certain SME loans) and loans to projects in developing countries usually through multilateral development banks (such as the World Bank).

Some banks are involved in all of these areas (with the obvious exception of multilateral and governmental institutions); they are usually referred to as universal banks. There are important differences between banks in Europe and the US, where banking was broadly defined before 1933 and included most of the activities mentioned above. However, the Glass-Steagall Act put restrictions on this form of banking (see page 59). The replacement of the Glass-Steagall Act by the Gramm Leach Bliley Act in 2000 has prompted a comeback for universal banking in the US. In Europe, where universal banks are prevalent, rigid legal barriers between banking activities rarely existed.

The typical universal bank is organized according to three main clusters of activity: retail banking (commercial and private banking), investment banking (corporate or wholesale banking and market making) and asset management (management of contractual savings).[1] Activities such as leasing and insurance are also included in many cases. When people in the US talk about banking, they usually mean commercial banks (involved in the first cluster of activities). In

Europe, people usually think in terms of broad universal banks when discussing banking. Nevertheless, these concepts are constantly shifting and a certain amount of pragmatism about them is appropriate. That is why a bank is considered to be an institution or financial conglomerate that *might* be involved in any of these areas (IA-IC and IIA-IIC). Where expedient, special attention will be paid to certain specific activities (particularly in respect of mutual funds and insurance). Emphasis will be placed on the depository activities of banks.

The significance of banks in an economy

Banks extend credit and this results in money creation. By creating money, banks affect the total money supply in an economy and as a result indirectly affect the growth of the economy. Moreover, banks play an important role in allocating money across sectors of industry, thereby also indirectly influencing the nature of economic growth. Banks are also responsible for a large amount of the payment traffic in an economy. The role of banks in an economy is illustrated in Figure 4.1 using a flow circuit (based on Bronfenbrenner et al, 1990 and Klant and Van Ewijk, 1990; the dotted line shows exclusively the relationship between international and financial markets). This simplistic representation of the machinations of an economy at the macro level clearly shows the points at which the financial sector has an impact on the functioning of the economy. The arrows indicate the direction of the money flows. Government, international trade, business and household deficits and surpluses are bridged by financial transactions, thereby closing the macroeconomic circuit. This takes place in the financial markets in which banks, insurers, securities firms and other entities operate. The grey areas show these and other points at which the financial sector impacts on the economy. The importance of financial markets in an economic system is, in short, considerable. Banks play a key role in financial markets, though this can vary from country to country.

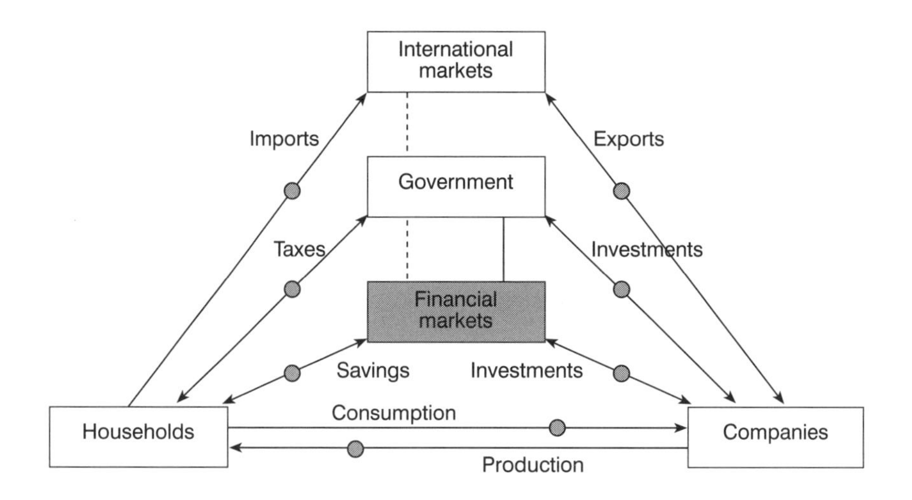

Figure 4.1 *The role of financial markets in an economic system*

The raison d'être of banks is generally explained as being related to the existence of friction in capital markets (Saunders, 2000; Hubbard, 1994; Freixas and Rochet, 1998). These frictions arise due to insufficient information and knowledge on the part of households and businesses with respect to deficits or surpluses of money. The intermediary role of banks consists of bringing together these surpluses and deficits of money in an economy through matching lenders and borrowers. Banks must screen borrowers and monitor loans. As intermediaries, banks have four key functions. Firstly, they transform money by size (denomination intermediation). After all, the amount of money one person supplies is usually not the same as the amount another person demands. Secondly, banks transform money by term (maturity intermediation). Creditors in particular usually only have short-term surpluses of money, while debtors usually have long-term capital requirements. Thirdly, banks transform money by place and/or time. A bank may allocate the money of a lender in New York to a borrower in Hong Kong. In addition, they allow money to be transferred between generations (intergenerational transfers). Finally, banks ensure that risks are transformed. Banks are generally in a better position to assess the risks of a variety of investments than are individual households which have money to invest (information services). Due to their larger size, banks are also more capable of spreading risks (price) and reducing the transaction costs. Saunders (2000, pp84–92) sums up some of the functions above in terms of general and institutional aspects:

1. **General:**
   - information services (reducing savers' monitoring costs);
   - liquidity services (increasing savers' liquidity);
   - price-risk reduction services (diversification, brokerage and asset-transformation);
   - transaction-costs-reducing services (economies of scale through pooling); and
   - maturity intermediation services (maturity mismatching of assets and liabilities).
2. **Institution-specific:**
   - money supply transmission;
   - credit allocation (between sectors);
   - intergenerational transfers;
   - payment services; and
   - denomination intermediation.

In most countries, banks are the most important financial intermediaries in the economy. The intermediary function basically brings together and links savings and investments. The intermediary role is expanding to include bringing together the lenders and borrowers of money rather than the money itself. This new activity generates fee income rather than interest income, as in the case of advice or investment funds.[2] The first form of intermediation can be gauged from the scope of the bank's assets while the total equity capitalization of an economy can be considered to be an approximation of the size of the second

Table 4.1 *Financial sector in developed countries*[3]

Country	% of total GDP	% of total bank assets	% of total equity capitalization	Bank assets, % of GDP	Equity capitalization, % of GDP
Iceland	0.0	0.0	0.1	62	219
Luxembourg	0.1	1.9	0.0	3461	70
New Zealand	0.2	0.2	0.1	125	55
Ireland	0.4	0.5	0.2	241	59
Portugal	0.5	0.7	0.2	218	60
Greece	0.5	0.2	0.7	66	163
Finland	0.6	0.4	1.5	102	367
Norway	0.7	0.4	0.2	88	41
Denmark	0.8	0.6	0.3	108	63
Austria	0.9	1.6	0.1	234	13
Sweden	1.0	0.8	1.2	109	159
Belgium	1.1	2.3	0.3	294	41
Switzerland	1.1	3.5	2.2	433	257
The Netherlands	1.7	3.1	2.5	267	196
Australia	1.7	1.3	1.3	111	106
Spain	2.5	3.0	1.2	160	61
Canada	2.7	2.0	2.5	114	121
Italy	5.2	5.0	1.9	134	51
United Kingdom	6.1	7.9	9.4	223	201
France	6.3	12.2	5.0	270	108
Germany	9.3	13.8	4.0	205	58
Japan	18.1	20.9	15.2	137	108
United States	38.5	17.5	49.8	70	166
Total	100.0	100.0	100.0		
EU-15	36.9	54.1	28.6	201	105
EMU-12	29.0	44.8	17.7	214	81

Sources: Equity capitalization and GDP figures from Datastream, Economist Intelligence Unit Database; bank assets figures from OECD (1999a).
Notes: Equity capitalization and GDP are 1999 figures; bank assets are 1997 figures. Corrections have been made for exchange rate fluctuations. EU-15 = the 15 countries within the European Union; EMU-12 = the 12 countries participating in the European Monetary Union.

form. Table 4.1 shows the significance and size of the financial sector in all developed countries (as defined in Box 1.1). The first three columns show the country's position in relation to the total of all developed countries in terms of GDP, bank assets and equity capitalization. The last two columns show the size of the banking and equity markets in each country in relation to their GDPs. The countries are ranked according to their share of total developed countries' GDP, starting with the country with the smallest GDP. This allows the three largest economic zones at the end of the table – the US, the EU and Japan – to be compared easily.

The US and the EU have approximately the same share of total developed countries' GDP. The US dominates the developed countries' equity markets with its 50 per cent share of the total. Its banking sector, by contrast, is small both in relation to its own GDP and as a share in the total banking assets of all developed countries. The opposite holds for the EU as a whole – it dominates the developed countries' banking market with a share of 54 per cent. The UK, though, is a special case within Europe. The last two columns show that the assets of the banking sector in most countries is a multiple of GDP. The US and some smaller countries are exceptions. Equity markets are less significant as a share of GDP compared to banking markets in most other countries. The US is once again an important exception in this case as are some smaller countries. Wherever banks are referred to in this book, the EU (or EMU)[4] will serve as the primary example (in the sense of 'depository institutions') and Japan as an appropriate second example. Regarding the remaining private financial institutions, the US will serve as the dominant example, with the EU (or EMU) as an appropriate second example (in the sense of non-depository institutions).

General banking trends

The development of banking reveals several discernible trends, the main ones being deregulation, internationalization, technological innovation, consolidation, concentration, conglomeration, disintermediation, off-balance-sheet activities and a blurring of the distinction between various activities. These trends were evident to a greater or lesser degree in all regions. In addition, more specific developments occurred in certain countries, such as the introduction of the euro in a number of European countries and the financial crisis in Japan (eg Danthine et al, 1999, and OECD, 1995).

An important trend is deregulation, which led to consolidation, concentration and conglomeration in the 1980s.[5] Mergers and acquisitions were first concentrated among banks themselves, but cross-sector mergers and acquisitions also became prevalent between banks, mortgage banks, insurers and/or investment banks. In addition to this blurring of clear lines between the sectors,[6] which creates universal, 'bankassurance' (combined banking and insurance) or all-finance institutions (such as Citigroup in the US or the Dutch ING Group),[7] the liberalization of capital markets and internationalization led to cross-border consolidation. One such example is the merger between the Finnish Merita Bank and Swedish Nordbanken. The phenomenon that emerged was global banking – the desire of banks to have a presence in all regions of the world in their retail activities and/or corporate banking activities (Goldman Sachs is a good example in investment banking). These shifts into universal and global banking were helped along by technological innovation, an important impetus for increasing scale and internationalization in banking.[8] All in all, several very large, wealthy banks emerged around the world. Table 4.2 shows the top ten banks in the world according to equity capital and total assets.[9] The bank's home country and its ranking in the Fortune Global 500 for all companies are indicated in brackets. It shows that banks are the largest organizations in the world when measured according to total assets. Only ten of the top one hundred businesses are not financial service providers. An entirely different picture emerges when measuring size according to equity capital – only one of the ten largest compa-

Table 4.2 *Top ten banks in the world according to total assets and equity capital, 2000*

In terms of assets	*US$ billions*	*Rank*	*In terms of equity capital*	*US$ billions*
Deutsche Bank (Germany, 1)	842	1	Citigroup (US, 9)	50
Bank of Tokyo-Mitsubishi (Japan, 2)	729	2	BankAmerica (US, 12)	44
Citigroup (US, 3)	717	3	ING Group (Netherlands, 18)	35
BNP Paribas (France, 4)	700	4	HSBC Holdings (UK, 20)	33
Bank of America (US, 5)	633	5	Bank of Tokyo-Mitsubishi (Japan, 29)	28
UBS (Switzerland, 6)	613	6	Dai-Ichi Kangyo Bank (Japan, 37)	24
Fuji Bank (Japan, 8)	568	7	Chase Manhattan (US, 38)	24
HSBC Holdings (UK, 9)	568	8	Deutsche Bank (Germany, 40)	23
Sumitomo Bank (Japan,10)	524	9	Credit Agricole (France, 41)	23
Dai-Ichi Kangyo Bank (Japan, 12)	507	10	Wells Fargo (US, 43)	22

nies is a financial service provider. The largest two banks to equity capital are American.

The degree to which trends have led to changes in various world regions diverges significantly. Trends such as technological innovation and internationalization are more universal, but the legislative framework and financial supervision has led to differences geographically.[10] For instance, the US has a far greater number of specialized financial institutions than Europe.

The rescinding of the McFadden Act in the US in the 1980s proved a strong stimulus to the cross-state merger process. Restrictions on cross-border mergers with European banks, for instance, nevertheless remained in place. The Glass Steagall Act also prohibited activities like joint ventures between banks and insurers, thus causing institutions to become powerful specialists, particularly in the field of investment banking (about seven of the top ten investment banks in the world are American, though this is also due to the enormous capital markets in the US). The Gramm Leach Bliley Act of 2000 changed this, and allowed for far-reaching collaboration between banks, insurers and brokerage firms. Since then a trend towards more universal financial institutions has emerged. This had already been happening in Europe for much longer. Universal financial institutions are still rare in Japan.

Japan's regulatory system hinders collaboration between banks, asset managers and insurers. Banks and businesses have always been closely intertwined in Japan. The Japanese financial crisis at the end of the 1980s was the impetus for a merger process and reduced, indirectly but substantially, the power of the state. Japan experienced 'bubbles' on both the equity and the property markets simultaneously while Japanese banks could legally count 45 per cent of unrealized gains as equity. These two markets crashed, causing credit portfolios and the solvency of Japanese banks to deteriorate. A number of banks went bankrupt and government intervention to support the banks and the economy was required to prevent an even worse situation. This resulted in radical restructuring and liberalization, which lasted until the end of the 1990s. Japanese banks are still trailing American and European banks in this consolidation and modernization process.[11]

In Europe, the introduction of the euro has had a huge impact on the financial sector (see Danthine et al, 1999). Individual EU Member States have been deregulating and liberalizing from the mid-1980s, prompting the emergence of bank insurers particularly in the UK and The Netherlands. The EU's Second Banking Directive of 1993 removed the most significant legal barriers to a single banking market in the EU. However, differences in language and culture as well as in fiscal and legal regulations and exchange rates still hamper cross-border mergers. On the other hand, the introduction of the euro has removed a number of barriers (directly, in terms of exchange rates, and indirectly in terms of the legal and fiscal harmonization that preceded the introduction of the euro in 1999). Consolidation will tend to continue in the asset management and investment banking areas, with cross-border mergers becoming more prevalent since 1999 (such as Dutch ING's acquisition of Belgian BBL or British HSBC's acquisition of French CCF). Mergers are also more common between the world's regions, though this is still largely limited to European banks taking over North or South American banks or American banks taking over their European counterparts.

A final trend worth mentioning is disintermediation, in which the intermediary role of banks in attracting savings money and extending loans is undermined. There are key differences in this respect between banks based on the Anglo-Saxon model and continental European and Japanese banks. The US and UK have degrees of intermediation of 20 per cent and 41 per cent respectively when bank financing is measured as a percentage of the total external finance of companies, governments and consumers, while this figure is 83 per cent for Germany and 61 per cent for The Netherlands (Albert, 1991). The process of disintermediation can be examined in two ways, from the customer's perspective and from the bank's perspective.

From the point of view of companies, an increasing number are going directly to capital markets to finance expansion by issuing new shares or commercial paper. This activity erodes the overall position of banks on the asset side of the banking balance sheet while on the liabilities side, consumers are investing more in companies directly (instead of putting their money in the bank). Of course, banks have responded to these developments by creating products and services that meet these new needs of their customers (by offering advice on capital market transactions and setting up investment funds). An increasing number of activities are taking place that no longer involve the bank's balance sheet. But banks themselves also increasingly find it necessary to have parts of their credit portfolios or their activities extend beyond the scope of their balance sheets. Important motives for this are improved capital adequacy ratios and freeing capital for other purposes. A disadvantage is that the remaining assets will have a higher risk profile. A growing number of banks transform some of their assets into off-balance sheet items through securitization. One way they do this is by pooling individual mortgages and selling the package on the capital market, through a 'special purpose company', as an investment product. Despite this, the balance sheets of banks are still growing considerably so it would be erroneous to conclude that the importance of banks in the economy is diminishing through disintermediation.

To sum up

Table 4.3 below summarizes the activities of financial institutions (based on Hubbard, 1994, p297) by classifying the activities of banks and showing the kinds of financial institution involved in each category. The borders are strictly upheld here and must be interpreted with a certain amount of circumspection; for instance, conglomeration causes the borders to become increasingly blurred.

Table 4.3 is structured in terms of products offered and fits in with the line presented in this section. ICT developments and a decline in the cost of information are prompting a strategic reorientation from a product approach to customer approach within many banks; that is, things are shifting from a supply-driven market to a demand-driven market. A division between retail, wholesale, private and institutional clients fits this trend. A customer-oriented approach does not focus on product creation but on the distribution of a wide range of services. This may even mean that banks offer the products of other financial institutions. Other types of financial institution and growth strategies also fit this approach. While increasing scale on the product side plays a far less dominant role, forms of supermarket, internet and mobile banking may become crucial.[12] Scale economies might even cause even larger financial conglomerates to emerge. The shift to a demand-driven market could stimulate the growth of powerful, specialized financial service providers. A leading British bank, Lloyds TSB, specializes in retail banking. The global investment banking market will probably be divided among a group of ten global players, the majority of whom will be American banks. In the short term, ongoing conglomeration will probably continue to hold sway, for the simple reason that increasing scale and diversification offer strategic options for future core competencies that are difficult to single out at this time. Such a trend may have played a role in Europe since the introduction of the euro (Milbourn et al, 1999).

In short, then, the banking sector is very dynamic. Developments may happen in many directions and dividing activities into certain categories between a set number of types of institutions is impossible. For this book, it is important that the diversity in banking activities, and the fact that certain trends will affect the speed and direction of sustainable banking, are understood. The growing significance of a customer-oriented approach and declining customer loyalty will emphasize the importance of improving competitiveness through differentiation – having a 'green' image, for instance.

THE DRIVE BEHIND SUSTAINABILITY AT BANKS

Why will banks integrate sustainable development in their activities? On the one hand, pressure to do so is being applied by the government, society, competitors and their customers. There are also internal drivers within banks themselves,

Table 4.3 *The activities of various financial institutions*

	Depository institutions		*Securities institutions*		*Investment institutions*		*Contractual savings institutions*		*Governmental FIs & multilateral FIs*
Services/ institutions	*Commercial banks*	*Saving banks etc*	*Investment banks*	*Securities firms*	*Mutual funds*	*Finance companies*	*Insurance companies*	*Pension funds*	
Current/ transaction accounts	✗	✗		✗	✗				
Saving	✗	✗			✗		✗	✗	
Consumer lending	✗	✗				✗			✗
Business lending	✗					✗			✗
Mortgage lending	✗	✗				✗			✗
Security insurance			✗	✗			(✗)		
Security trading			✗	✗					
Asset management	✗		✗	✗	✗		(✗)	✗	
Insurance							✗		✗

arising through a need to establish their core objectives and identity. In order to be able to understand this, it is first important to understand the environmental impact of banks.

The environmental impact of banks

To gain an impression of the environmental impact of banks, a distinction will be made between internal and external issues. Internal issues are related to the business processes within banks, while external issues are connected to the bank's products.[13]

Banks themselves are part of a relatively clean sector. The environmental burden of their energy, water and paper use is not nearly as severe as it is in many other sectors of the economy. However, the size of the banking sector in absolute terms is large enough to make the environmental impact significant. An instrument developed by Credit Suisse to measure its environmental impacts concluded that energy use has by far the most serious impact, accounting for 90 per cent of all cumulative pollution within Credit Suisse.[14] UBS came to a similar conclusion on the basis of its Environmental Performance Evaluation.[15] Other environmental reports from banks also concur that energy is the most significant aspect.[16] The potential energy savings of banks are huge as can be seen by the achievements of the most proactive ones. Between 1990 and 1993 UBS reduced its energy use by 25 per cent (UBS, 1999). Between 1991 and 1995, NatWest saved approximately $US50 million in energy costs (NatWest, 1998). The measures were taken not because of legislative pressure but because they were cost-effective. Some banks are also now using renewable energy, particularly solar energy (eg Triodos Bank). Other initiatives include the more efficient use of water and transport policies and the development of more environmentally benign credit cards. One of the leaders in such practices is The Co-operative Bank in the UK, which introduced the first biodegradable credit card in 1997, an affinity card which supports Greenpeace.[17]

If we look at the environmental burden of bank products, the problem is that, contrary to other sectors in the economy, bank products themselves do not pollute. Rather, it is the users of these products who can have a considerable impact on the environment. This makes it very hard to estimate the environmental burden of the activities the banks have funded. In addition, to date, most banks feel that external environmental care would require interference with the activities of their clients. This is one reason why banks have been reluctant to promote environmental concern on the external side of their business (even when they are likely to be exposed to risk). However, in recent years, by developing a selection of products from which a client can choose, banks have tried to cope with this dilemma.

One can take two extreme standpoints on the environmental burden of bank products. Firstly, all pollution caused by companies who are financed by banks is the responsibility of the banks. It is easy to estimate the environmental burden in this sense: it would equate to the aggregate pollution of a major part of the economy in many countries. Secondly, as the products of banks do not pollute in themselves, the users of those products – the clients – should take

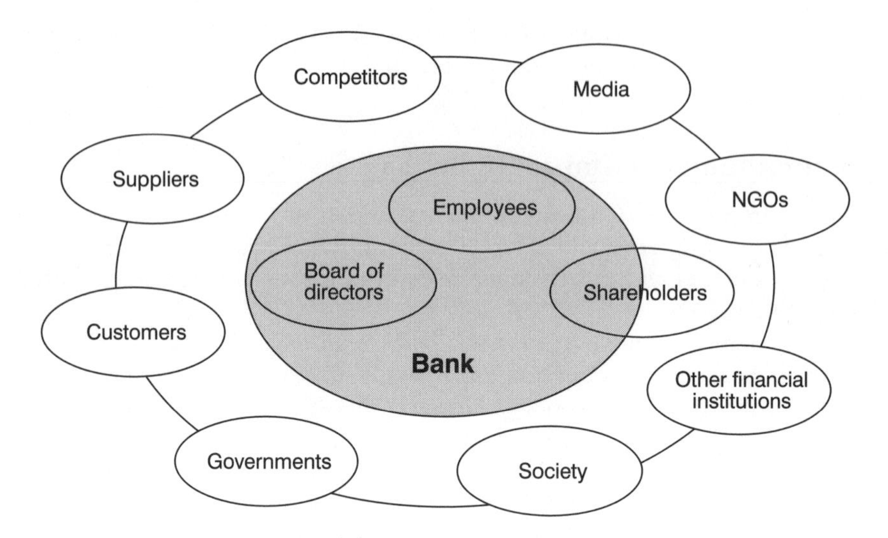

Figure 4.2 *Internal and external bank stakeholders*

sole responsibility for the pollution they create. Of course, both standpoints are absurd. The truth lies somewhere in the middle – as CERCLA (see Chapter 6) has demonstrated in the US – but still remains almost impossible to quantify.

The environmental context of banks

Moving from the more technical side of environmental issues and banks, the stakeholder model provides a clear picture of the impetus behind banks' willingness to integrate sustainability in their banking activities. Figure 4.2 depicts the internal and external environments of banks which influence the decisions regarding incorporating sustainable development into banking activities (based on Ferderick, Post and Davis, 1992).

It is important to note that on the external side, customer risks are also bank risks and can affect their own continuity. This applies particularly forcefully in the case of environmental issues. A customer's continuity may deteriorate significantly as a result of new environmental regulations. There is an important distinction to be drawn between new and existing customers, as well as between retail and business markets, in this respect. Differing groups of customers may have conflicting interests. The bank is confronted with the problem of satisfying both sides.[18] In the same way that customer risks are also bank risks, customer opportunities are also opportunities for banks. The attention being paid to sustainable development is opening entirely new markets. An example is the market for wind energy, in which large financial institutions such as ABN AMRO and UBS see increasing potential. Traditional forms of finance may be sufficient, but banks are also being challenged to develop new products which fulfil customer needs. The growing market for sustainable investment funds, such as the Storebrand Principle Global Fund (Storebrand, 2000) or the UBS Eco-Performance-Portfolios (UBS, 1999), is a good example of this trend. Most of these funds also have socioeconomic criteria. The growing importance and number of such funds illustrates that competitive pressures are driving more banks to diversify their product range towards sustainability in response to market demand. Of course, the government is also playing a key role in all of this by laying down the preconditions in which bank and borrowing activity can take place. The government is also placing increased emphasis on the potentially positive role that banks can play in achieving environmental policy objectives (eg Coopers&Lybrand, 1996 and Delphi/ Ecologic, 1997). To date, banks have reacted cautiously to government attempts to legislate on their external side and are unwilling to become the enforcers of government policies.

Banks are also affected by society's changing preferences and changing worldviews. Businesses are considered responsible for their activities and their effects on society, even in the international context. A number of studies emphasize the role of private banks in achieving sustainability in developing countries and/or economies in transition (Gentry, 1997). Businesses which act irresponsibly in the eyes of the public are also threatened by collective action undertaken by consumers in what is known as a consumer backlash. Consumers are mobilized through the media to engage in action against a certain company. Good examples of this are the boycott of Shell fuelling stations in Europe in 1995 after Shell announced that it thought it best to let its Brent Spar oil platform sink to the bottom of the North Sea, and the public attention paid to Shell's possible role in human rights abuses in Nigeria during the same period. While large-scale action has never taken place, interest groups which mobilize the public at large are likely to scrutinize the financiers of these businesses much more closely; banks too may be hit by 'consumer backlashes'.[19] This tendency will be reinforced if banks sign declarations stating that they wish to foster sustainable development and at the same time play a large role in it. Communicating social accountability and involvement is important, but it is not without danger. At the very least, the bank's own internal production processes should be able to withstand critical examination. It is therefore also important that suppliers to banks get involved in the drive for sustainability or, at any rate, can at least withstand critical examination.

NGOs can play a key role in the sustainable development framework. These can be roughly divided into action groups and business support organizations,[20] although the divisions between these categories are getting more and more blurred. Besides the confrontational role that action groups take on, NGOs can have a supportive role towards banks, sharing knowledge and experience in caring for the environment and substantiating sustainable development in the business world. An important element of sustainable banking will involve working in networks in which NGOs play an important role. The stance and role of other financial institutions also play a role. The World Bank Group (WBG), in particular through the International Finance Corporation (IFC), impose strict environmental requirements when co-financing projects in developing countries. Box 4.1 briefly examines these developmental financial institutions. The role of these institutions with respect to sustainability will be examined more specifically in Box 6.6.

Box 4.1 Global, regional and national development banks

The WBG was founded in 1944 and consists of five institutions: the International Bank for Reconstruction and Development (IBRD), International Development Association (IDA), International Finance Corporation (IFC), Multilateral Investment Guarantee Agency (MIGA) and the International Centre for Settlement of Investment Disputes (ICSID).[21] The WBG provided US$15.3 billion in loans to developing countries in 2000. At the end of 2000 there were about 100 active environmental projects amounting to approximately US$5.1 billion, and an additional US$10 billion was invested in projects with primarily environmental objectives (multi-year accumulation). The WBG is owned by more than 180 member countries. IBRD and IDA provide development assistance and loans to developing countries. The IFC promotes private sector investment in developing countries. MIGA promotes FDI by offering political risk insurance to investors and lenders. ICSID provides facilities for the settlement of investment disputes between foreign investors and their host countries. Key focal points of WBG activities include economic management and private and financial sector development, human development and environmental management.

Though the WBG acts worldwide, every economic region in the world has its own development bank, such as the Asian Development Bank (ADB), African Development Bank (AFDB), Caribbean Development Bank (CDB), Inter-American Development Bank (IADB), European Investment Bank (EIB) and the European Bank for Reconstruction and Development (EBRD). All these banks have their own objectives, which in all cases include environmental objectives, and require an environmental assessment (EA) by the borrower where adverse impacts of a project on the environment are expected or identified. Most of these EA policies use WBG guidelines as a reference or source. See Box 6.6 for a discussion of the WBG's EA policy.

Another major source of capital for projects in developing and developed countries comes from export credit agencies (ECA). The total exposure of ECAs in developing countries amounted to approximately US$463 billion in 1996 (Environmental Defense et al, 1999, p33). Almost every country has its own ECA, which facilitates and financially supports (via loans, guarantees and political risk insurance) the export of goods and services by domestic companies. A guaranteed export credit is a loan to a company in country A which is importing goods or services from country B, thus enabling it to pay for an import contract. Such a loan is usually made by the exporting company or a commercial bank in country B. Part or all of the scheduled repayments are then guaranteed by the ECA of country B against the risk of non-payment by the company or government of country A.

The US and Canada want to put WB-style environmental guidelines in place at leading ECAs to prevent a race to the bottom involving ECAs' environmental impact on financing decisions. Some European countries oppose these ideas (*Financial Times*, July 20 2000). In a report by Environmental Defense (1999) the race to the bottom is illustrated. The ECAs of, for example, Japan (Japanese Export-Import Bank, JEXIM) and Canada (Export Development Corporation, EDC) are criticized for their role in certain projects. Exceptions, however, do exist. The US Export-Import Bank already uses environmental guidelines similar to those of the WB.[22] It refused, for example, export guarantees for the China's Three Gorges Project (see page 143). The Kreditanstalt für Wiederaufbau (KfW), the German development and export bank, has developed its own guidelines as well. Moreover, it looks at the environmental impact of its internal processes and publishes an environmental report (KfW, 2000).

Bank shareholders find themselves on both sides of the border between internal and external forces.[23] The commercial added value of every environmental measure or product will be most interesting to shareholders. Even so, a growing group of shareholders is also interested in bank environmental policy from a more ideological point of view. The bank will have to find the correct balance between potential conflicting interests. The degree to which sustainability makes it onto a corporate agenda depends to a large extent on the board of directors or chief executive officer (CEO). Often, top–down legitimacy must exist for the business to really get involved in the sustainability issue internally. One way legitimacy can become firmly rooted in an organization is through a mission statement. Bottom–up ideas and processes will then arise which accelerate the actual integration of sustainability into banking activities. Employees are ultimately the best ambassadors the organization can have. They will also want to achieve certain objectives as individuals and members of an organization. A progressive environmental policy can also help in attracting and holding on to scarce highly educated personnel.

INTERMEDIATION AND SUSTAINABILITY

Introduction

Asymmetric information exists between borrowers and lenders of capital, regarding environmental aspects as well as other key information. Banks have extensive and efficient lending operations and have a comparative advantage in information as a result of the knowledge they have of economic sectors, regulations and market developments. By having an in principle solid view of environmental and financial risks, banks fulfil a key role in reducing information asymmetry between entities in the market. A bank can attach a value or price to reduce that uncertainty. Rate differentiation between sustainability aspects is justified from information and risk perspectives. The scope for rate differentiation increases if banks can raise funds more cheaply because they have a relatively well-qualified credit portfolio. Banks can go a critical step further by applying rate differentiation on the basis of the will to stimulate sustainable development. The first form of premium differentiation will be referred to as risk-related premium differentiation; the second more ideological form will be referred to as non-risk-related premium differentiation. If banks apply premium differentiation to environmental aspects, entrepreneurs who engage in activities that have a relatively small impact on the environment will be able to obtain financing at cheaper rates. Banks then stimulate a development which benefits sustainability. The question is, of course, how far this could in fact come into play, if at all.

Do banks inhibit the drive towards sustainability?

In a study carried out by the WBSCD entitled *Financing Change*, the question as to whether banks have an inhibiting or stimulating impact on achieving a sustainable society is raised. The authors (Schmidheiny and Zorraquín, 1996) posit that banks intuitively have an inhibiting effect. In order for such a society to be created, investments that do not pay off for a long time will need to be made, which goes against financial markets' preference for quick gains. Secondly, investments will have to be made in environmentally friendly products and processes which internalize important externalities and often yield lower returns; this is diametrically opposed to the high returns that financial markets seek. Considering the magnitude of international capital flows and the financial sector's function in the economic system, whether the financial sector makes sustainable development impossible or stimulates it is an important question.

> *Do the financial markets encourage a short-termist, profits-only mentality that ignores much human and environmental reality? Or are they simply tools that reflect human concerns, and so will eventually reflect disquiet over poverty and the degradation of nature by rewarding companies that treat people and the environment in a responsible manner?* (ibid, pxxi)

Based on interviews with various stakeholders within the financial sector, Schmidheiny and Zorraquín find no convincing evidence to indicate that banks impede sustainable development.

Nevertheless, this conclusion is open to debate. Whenever interviews are used, the extent to which the answers given are intended to be socially acceptable rather than sincere always arises. The environment as a factor is not incorporated into investment decisions unless it is priced in an economic system driven by profit.[24] After all, *ceteris paribus*,[25] an alternative investment exists which yields a higher return than an investment which does take this unpriced factor into account. An investment in factory A that discharges effluent into a river legally but without purification yields a higher return than an investment in factory B which has introduced costly water treatment technology (*ceteris paribus*). Factory A would also be able to raise cheaper funds than factory B. If government environmental policy is passive and unchanging, profit-maximizing banks will provide financing for environmentally unfriendly investments at lower than ideal interest rates from an ecological standpoint. Not pricing scarce resources like the environment is, then, subsidized through a lower than justified cost of capital. This viewpoint shows that short-term profit maximization impedes the attainment of a sustainable society.

This counter-argument assumes that as far as environmental risk is concerned, government environmental policy is the only thing that affects the investment risks of banks. Up to the 1990s, banks did indeed concentrate primarily on the environmental risks that relate to government policy. These risks were not as serious as they were initially perceived to be, with the significant exception of soil contamination. The power of consumers in Western countries has been growing since the mid-1990s. This has increased the scope of environmental risk considerably. It no longer matters whether or not a factory is releasing effluent *legally*, but whether or not people are still willing to buy the products of the company concerned. Risks to banks' reputations are also playing an increasingly important role.

To sum up: If banks focus solely on government environmental policy in judging environmental risk, and government policy is actually weak or passive, banks will inhibit sustainability.[26] If market reactions and perceptions are included in their decision making, and these factors have the potential to bring into question the reputation of banks or continuity of companies that are relatively heavy polluters, then banks have a neutral or even positive effect on sustainability overall. If banks consider the impact of their investment decisions on the greenness of their image, they will play a stimulating role in terms of sustainability overall. So banks can both inhibit and promote the attainment of sustainability. On the one hand, banks are subject to the policies of governments and the whims of their customers and, on the other, their own stance and willingness are critical. Obviously exceptions to these rules exist. The main point here is to explain the ways banks can inhibit or stimulate sustainability. The next section will examine the underlying dynamics of these roles.

Premium differentiation for the benefit of sustainability

How can the environment in general be incorporated into investment decisions? Firstly, through legal requirements designed by governments, such as permits and regulatory levies: the government plays a key role in enforcing the values in society at large, such as the quality of the environment. Secondly, through demands in the production and consumption chain with respect to the way companies operate: socially responsible companies will not merely strive to generate profits at any cost; an increasing number of consumers are willing to pay more for environmentally friendly products; socially responsible companies will also consider the environmental impact of the intermediate products they use and place their own requirements on the products their suppliers deliver. Thirdly, this occurs through conscious and intrinsic choices made by companies and banks themselves.

The first two approaches will spur banks on or even force them to internalize the environmental externalities into investment decisions (that is to say, factory B in our example would in principle be able to borrow more cheaply or more favourable than factory A). The government's role in this is obvious. The second way is less self-evident though. Whereas previously banks only had to consider the effect of government on profits and risks on investment decisions, a new factor has arisen which is much more difficult to gauge: the requirements imposed on them by the marketplace itself. The third factor, an extension of the second, involves companies taking the ecological ramifications of their decisions into account on their own initiative and on the basis of their own intrinsic values. This will be an important factor as well if sustainable development is ever going to be realized. Banks *could* be a natural partner to the government in implementing sustainability policy if they made a conscious decision to act as such. They, like the government itself, play a pivotal role in the macro-level necessity for sustainable development and have a unique place in substantiating this at the micro level.[27] Banks can help companies that make a conscious decision to go beyond the minimum requirements set down by governments or demanded by consumers. As suggested, banks are able to apply price differentiation because of their informational comparative advantage. When they apply price differentiation to minimize risks, it is with an eye to the first two points relating to internalizing environmental externalities (risk-related premium differentiation).[28] However, banks could also decide to apply interest-rate differentiation to situations that do not directly involve risks (non-risk-related premium differentiation). Short-term profit maximization is then transcended as the overriding goal.[29]

Den Nordske Bank and a number of Swiss banks are examining the attainability of non-risk-related premium differentiation. This does not really involve environmental risks, but a conscious decision to contribute to the attainment of a sustainable society in a direct way. After all, strong financial collateral or a healthy company may offset environmental risk. The financial risk to the bank can be quite limited, while the ecological damage is high. Moreover, even environmentally friendly companies can be confronted with environmental risks (such as when accidents occur which are extremely detrimental to the environment and for which the company is liable). The Swiss initiative goes beyond risk-related premium differentiation based on the degree of environmental risk posed by a company, because the company's environmental friendliness or the sustainability of its activities are given a higher priority (relative to the priority they are given in the industry generally).[30] Companies which are financially healthy can be confronted with higher rates or denied financing altogether.

The question is, does this jeopardize the continuity of banks? This does not necessarily need to be the case, certainly not if all banks participate in the initiative.[31] It may not be regarded as a problem that other banks do finance certain companies that are more risky in the long term. That is already happening today.

The Co-operative Bank in the UK is another example of a bank that applies premium differentiation. It wants to avoid investments in tobacco production for instance, and make more investment money available for environmentally friendly products (Delphi/Ecologic, 1997). It has developed a methodology for this which appraises companies on their contribution to sustainability (based on the 'Natural Step'[32]). These companies then qualify for favourable interest rates; higher rates when they save and lower rates when they borrow. The cumulative savings to a company can amount to 30 per cent of regular bank rates (*Green Futures*, 1998a).

Are there limits to premium differentiation?

In short, there is a choice about whether to stop financing certain relatively environmentally unfriendly projects or companies, or to supply financing to them but at higher rates. From a more positive angle, it is a question of offering certain relatively sustainable companies more favourable interest rates. The onus is not on how much profit is made but on how it is made and in what time frame. Most banks do think about which kinds of companies or business activities they wish to finance and which they do not. This is a difficult but, with respect to sustainable banking, necessary discussion.[33]

Banks are, of course, bound by minimum capital adequacy, liquidity and profitability requirements. Nevertheless, within these confines, there are a number of ways to stimulate sustainable development. However, it is sometimes difficult to see the opportunities. A blind spot for environmental aspects and sustainability does still seem to exist. The challenge to banks is to go beyond these self-defined limited boundaries to their intermediary role by developing new products and applying risk-related and/or non-risk-related premium differentiation.

SUSTAINABLE BANKING

There are a number of phases which can be distinguished with respect to sustainable development. These phases follow a fixed pattern for most banks. However, the term 'phases' is not always applicable. On the one hand, a new niche player may be able to operate in a sustainable way right from the start. On the other hand, this way of thinking may create the impression that an end point exists. As stated on page 41, sustainable business or sustainable banking should not be interpreted as a static concept. Our understanding of sustainability will undergo continual enhancement. We could also refer to stances or layers in which banks and/or departments find themselves, instead of phases. The idea of layers, stances or phases is illustrated in Figure 4.3. Each layer (from the inside out) includes all the previous layers except the defensive layer. That is to say that sustainable banking includes the elements of preventive and offensive banking.[34] Using a dynamic perspective, these stances can be divided into phases. In principle, every bank evolves through each phase. These phases apply to a bank as a whole. Incidentally, this conceals the problem of large differences in environmental performance between the various disciplines within a single bank. The point of the model below is not that all activities and disciplines within a bank meet the conditions for a certain category, but that the situation considered in its entirety fits into a certain typology. Obviously, the model might also be useful for separate divisions and activities within a bank. Moreover, financial institutions can have very different primary business focuses. Insurance products and investments will be critical to a bank-insurer with 80 per cent of activities concentrated in insurance. If insurance constitutes only 20 per cent of an institution's activities – one that has insurance and retail banking operations – bank lending would be the focus as regards sustainability.

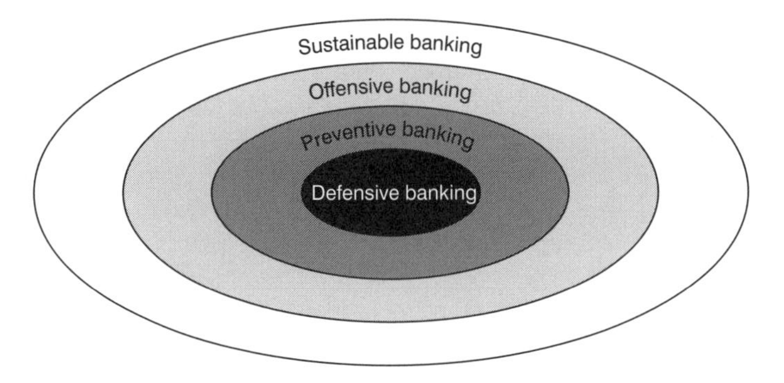

Figure 4.3 *A typology of banking and sustainable development*

The first phase is defensive banking. The bank is a follower and contests every government measure with respect to the environment and sustainable development since its direct or indirect self-interest is threatened. Cost savings in its internal environmental care are not considered and all environmental laws and regulations are thought to be threats to its business. Only curative measures are taken. In this vision, care for the environment only adds to costs and there is certainly no money to be earned from it. Although there are no banks in developed countries that actually maintain this vision any more, this stance is still present among certain individuals or disciplines within banks. However, one could also include ignorance of sustainability issues in this phase, something which is much more common among banks.

The second preventive banking phase is different from the previous phase in that potential costs savings are identified. These cost savings initially have an internal character, that is to say, they relate to the bank's own operations. Many banks are working on internal environmental care (eg paper use, energy and water use, and business travel). They also have an external character (with respect to the products involved in banking, such as loans and savings products) purely in the sense of limiting risks and investment losses related to environmental risks. So it includes looking into saving costs through fewer loss items as a result of environmental risks in credit extension.

A large group of developed countries banks have made it as far as this phase. Preventive banking is inevitable for most banks because politicians and interest groups are directly or indirectly stipulating preconditions for bank activities through environmental laws, jurisprudence and regulations. In this phase, the potential returns, risks and costs related to such environmental preconditions are integrated into the day-to-day operations of the banking business. A bank does not want to go any further than the environmental laws that exist or can be expected in the near future. This stance is no longer defensive, but it is somewhat passive, limiting external risks and liabilities and saving production costs internally. Leading banks such as NatWest Bank (a subsidiary of the Royal Bank of Scotland since 2000), ING, Bank of America and Deutsche Bank, have limited their environmental risks by using more focused investment and credit policies (such as environmental risk control lists for credit risk assessment). UBS is also integrating environmental issues in its investment banking branch (Warburg Dillon Read), a very innovative move. This Global Environmental Risk Policy for investment banking activities was implemented in February 1999 and is the first such initiative by a major bank (UBS, 1999). Although the nature of this initiative is preventive, it will only be found in banks that are also offensively oriented.

The elements of the preventive phase are also components of the offensive phase, the third phase, which goes a significant step further than preventive banking. Banks see new opportunities in the marketplace, both in the area of specific products and new markets, such as the fast-growing segment of environmental technology. Examples include the development of environmental investment funds (such as the Calvert Social Investment Fund and the Eco-Performance-Portfolios of UBS), the financing of sustainable energy (such as the so-called Solaris Project, a collaboration between Greenpeace and the

Rabobank Group; Rabobank International, 1998) and the signing of the UNEP Banking Charter by organizations such as Bank Austria, UBS, Kenya Commercial Bank Group and Salomon Inc (see Appendix VII). These banks also have the courage to make their progressive standpoint concerning sustainable development known in their communications (see the environmental reports of ABN AMRO, Bank of America, Barclays Bank and Credit Suisse for instance).[35] The bank is looking for profitable, environmentally sound opportunities in the market, which can compete with alternative investment and lending opportunities. The stance can be described as proactive, creative and innovative and is primarily focused on banks' business with customers. The extra steps are taken whenever there are win-win situations at the micro level, in which the extra investments or activities that benefit the environment have a pay-off period that lies within the required time frame and the level of risk is deemed acceptable. Incidentally, these activities can also be sustainable, though they are not always self-evident.

In sustainable banking, the bank lays down qualitative preconditions so that all its activities are sustainable. In offensive banking, the fact that a bank's activities are sustainable is coincidental, it is not targeted specifically. It is more probable that the activities are not sustainable, given that the economy and not the integration of economy and ecology predominate at this time. As stated, sustainable and offensive banking are not the same, as long as society does not price all negative environmental effects. All activities at this stage are sustainable thanks to a consciously chosen policy to that effect. These banks have, in contrast to banks with an offensive stance regarding environment, the ambition to operate sustainably in every respect. In addition to behaviour, ambition is also a key criterion. Sustainable banking can be defined as a modus operandi in which the internal activities meet the requirements of sustainable business and in which the external activities (such as lending and investments) are focused on valuing and stimulating sustainability among customers and other entities in society.

Sustainable banking will then also mean that activities are sought in entirely different directions and that certain forms of lending and participation no longer take place while others do (which largely still does not occur at this time due to commercial considerations and the techniques used for credit assessment). A bank therefore voluntarily goes a step further in sustainable banking. However, the starting point is not environment regulations or the market as such, but the vision regarding the environment, the organization's goal and the role that the organization wants to play in society. These banks are prepared to accept lower margins and/or higher risks to stimulate certain activities (they apply non-risk-related premium differentiation) that do not have a chance of succeeding in the current paradigm because the risk is considered to be too large, the profit margins too low or the pay-off period too long. The matter of how the bank wants to position itself in the world at large or is currently positioned (see pages 39–40) is essential to sustainable banking.

Banks that aspire to incorporate sustainability in all their activities, but are not yet there in all their activities, may be regarded as sustainable banks too. The most important distinctions here are ambition and conduct. Of course there should be a way to test this ambition externally and a sufficient effort in achieving sustainability must exist over time. At this time, the goal of sustainable banking appears to be feasible for only a few niche players, such as the Triodos Bank in The Netherlands or The Co-operative Bank in the UK.[36]

Micro-credit or micro-financing are examples of an innovative way in which sustainable banking can be fostered by banks. Box 4.2 examines this example further. Banks can also (once again, in developing countries) make a contribution to sustainable development by financing and facilitating sustainable energy. Box 4.3 describes the efforts of the Triodos Bank to this effect.

The point of sustainable banking is that social and commercial activities with a sustainable character are stimulated. It also involves ideas and learning processes which are not viable from a purely commercial perspective at the outset. These are generally creative initiatives and have a long-term, innovative character. Banks can operate in broad networks to stimulate these ideas or innovative processes. Thanks to their unique position in terms of the relationship based on mutual trust, their relative independence and non-partisanship, knowledge and potential application of capital with respect to governments, other market entities and interest groups, they can function as a driving force in designing these kinds of sustainability initiatives. Financial innovations that are formed when operating in broad networks that are intended to get the right parties around the table to find a solution to a particular problem are an essential element of sustainable banking. Banks can also create funds targeting innovation or simply grant money to stimulate sustainable development. An interesting example of grants being provided is debt-for-nature swaps. Box 4.4 elaborates.

Conclusion

Banks are being confronted by a variety of developments – economic, sociocultural, technological, political and legislative as well as group interests. This chapter has briefly addressed the key issues, trends and incentives in respect of sustainable banking. Part II will examine the steps that banks have been taking, from the defensive to the sustainable. Table 4.5 provides a summary (though not fully extensive) of the incentives to and elaboration of such steps.

In short, banks interact with the environment in a number of ways (Delphi/Ecologic, 1997, p1):

- as investors, supplying the investments needed for achieving sustainable development;
- as innovators, developing new financial products to stimulate sustainability;
- as valuers, pricing risks and estimating returns, from both a financial and an environmental perspective;
- as powerful stakeholders, influencing governments and the management of companies as lenders and shareholders of companies;
- as polluters, polluting the environment by their own internal processes and use of resources;
- as victims of environmental changes, eg climate change.

Box 4.2 Micro-credit or micro-financing

Of the world's 6 billion people, almost half live on less than US$2 a day (World Bank, 2000, p3). Although ODA does help to mitigate the hardship, it might create dependence on aid. An exciting alternative instrument is micro-finance. Micro-finance includes loans, deposit-taking and insurance in small amounts for the poor. Micro-finance programmes have the potential to reinforce entrepreneurial behaviour, economic self-sufficiency and smooth consumption during adverse shocks such as natural disasters.

Micro-credit is where small, short-term loans are granted to people who would like to start up or expand an economic activity. The interest rate is usually high and the loans must generally be paid back in half a year. The system of micro-credit or -financing is based on the fact that a credit is extended to a group of people who mutually guarantee each other, have similar interests, and mutually commit themselves to the realization of each others' objectives. Two models can be identified (OECD, 1997). The Grameen Bank system ties the extension of credits to personal guarantee through a group of individuals (*grameen* means 'village' in Bengali).[37] The second model is where credits are only granted to people who have invested (part of) their savings in the institution. Micro-credit involves a great deal of overhead expenses, but the percentage repaid is very high.[38] Nevertheless, micro-credits must not be settled according to banking criteria alone, although it is necessary that the operations at least break even for the continuity of such institutions. People in rural areas usually don't have safe places to store excess money. Micro-finance institutions which take deposits, even just a few dollars, are very useful in this sense. Savings and insurance are very important to enable poor people to make long-term decisions and create a buffer for difficult periods.

In 1995, there were US$7 billion of micro-credit outstanding, a large share of which was lent in Asia. Over US$19 billion was held in over 46 million savings account worldwide. 76 per cent of the sample loan volume was disbursed in Asia, compared to 21 per cent in Latin America and 23 per cent in Africa. The median loan and deposit numbers and size differ per region and institution, which is illustrated by the following figures in Table 4.4.[39]

Table 4.4 *Micro-credit activities among financial institutions and in world regions, 1995*

	Median number of loans	*Median loan size, US$*	*Median number of deposit accounts*	*Median deposit size, US$*
Banks	44.270	680	39.880	190
Savings institutions	2.870	3.010	224.180	950
Credit unions	15.320	450	38.610	410
NGOs	1.780	250	0	0
Latin America	n/a	890	n/a	420
Asia	n/a	90	n/a	40
Africa	n/a	170	n/a	70

Some big micro-finance institutions are Bank Rakyat Indonesia (with more than 16 million low-income depositors) and Grameen Bank (with more than 1 million borrowers).[40] About half of Grameen's borrowers move above the poverty line within five years. Micro-financing can involve developing countries and developed countries alike, although it is essentially more problem-ridden in developed countries (due to relatively high overheads and relatively poor returns). Micro-credits are being used by some major commercial banks in developed countries for the (relatively) poor or deteriorated neighbourhoods in Western cities. HSBC Group and Lloyds TSB are very active in the UK. Deutsche Bank has its Deutsche Bank Americas Foundation (investment portfolio of US$400 million), but is also active in developing countries (through its Deutsche Bank Micro-credit Development Fund, which is capitalized through its own monies as well as through money from rich clients of its private bank). Rabobank Group has its Rabobanks Foundation which acts as a financial donor and knowledge base for micro-finance in developed and developing countries (cumulatively $US45 million, of which three-quarters is used in developing countries).[41]

Box 4.3 Solar Home Systems

Photovoltaic (PV) solar energy is usually considered to be the only attainable and affordable way to supply schools, households, water pumps and medical buildings with energy in the developing world, especially for off-grid regions. Sustainable banks will be progressive in financing solar energy in the developing world. This may involve the use of revolving funds. These are funds that are self-sustaining once the initial amount has been deposited to finance the first PV systems. The principle works as follows. The bank or fund invests an amount of €100,000 in 100 solar home systems (autonomous PV systems – solar panels including batteries – which can supply the essential electricity needs of the residents of a house). These are distributed to interested residents who repay them over a ten-year period at an annual interest rate of 10 per cent. These repayments are invested in new systems making a gradual growth in the number of systems possible. The following ten years is used to pay off the original investment amount (most PV systems have a life span of about 20 years; NOVEM, 1996).

The Triodos Bank has set up the special Solar Investment Fund to stimulate solar energy use in the developing world, in which they are collaborating with an NGO. The fund comprised €3.6 million through 1998. In Bolivia, 10,000 households are being supplied with a complete energy system, including the solar panel, controller and battery. Deforestation can be prevented while a contribution is made to the development of rural Bolivia at the same time. At some point in the future, the systems will be acquired by a Bolivian energy company which will also be responsible for collecting user payments.

Sustainable development provides opportunities for banks in the form of offering new products to customers, like eco-credit cards and green and environmental technology funds as well as environmental insurance. Opportunities also exist in new markets such as water treatment technology or wind energy. Whether a bank can take advantage of the opportunities for new products and new financing markets depends on its level of knowledge, experience and creativity. A prerequisite for this is to accumulate this knowledge and make it accessible for those departments which are focused on new markets or product innovations as well as for the credit assessment departments in, for instance, the form of databanks. Synergy can be achieved in many activities, like sustainable savings products that are geared towards attracting money from people who specifically want to use their savings for projects that focus on

Box 4.4 Debt-for-nature swaps

A macro-level involvement can be distinguished in addition to the micro-level involvement of banks. This can be seen in the case of debt-for-nature swaps, which were based on the idea of debt-equity swaps in the financial sector. In many developing countries, the combination of a heavy debt burden with economic adjustment process has placed severe pressure on natural resources.[42] Debt-for-nature swaps were introduced in the 1980s as a tool to increase support for the environment while reducing external debt. Debt-for-nature swaps relate to high-risk loans, usually lent to developing countries' governments, that are unlikely to be recovered and are traded among Western banks at a lower price. About 40 per cent of these debts are held by private banks.

The idea behind a debt-for-nature swap is that the loan, listed far below its nominal value, is entirely written off, or can be bought back by the debtor for far less than its nominal value, with the stipulation that the debtor spends the relief in his or her own country (in its local currency) in an environmentally friendly way, like establishing a nature reserve. Such bilateral arrangements have been used by, for instance, the Swiss Debt Reduction Facility. Mostly a third party, like WWF (or another NGO), will be involved as well. The NGO purchases debt (eg US$10 million) at a discount (eg 48 per cent) from a creditor (a commercial bank, government or multilateral development bank) and negotiates separately with the debtor government for cancellation of the debt in exchange for project funding for conservation or payment in a local currency. In the case of a payment of US$10 million equivalent in local currency (ie 100 per cent payment), the multiplier or leverage for the NGO is 2.1 (10 divided by 4.8). In other words, the NGO gets 2.1 times more local currency than in an ordinary foreign-exchange transaction and in doing so can increase its conservation efforts in the debtor country in, for example, preserving parts of its rainforests, or biodiversity.

The benefit for the debtor of such a swap is a lower repayment or saving of scarce foreign currency on its foreign debt and a way of stimulating conservation efforts. The benefit for the creditor is a certain but lower price for the debt, instead of an uncertain future repayment. Obviously, reputation benefits for a commercial bank are to be expected as well. Bank of America has enabled debt-for-nature swaps in Mexico for example. Other banks which have been involved in such swaps between 1987 and 1997 are: Bank of Tokyo, JP Morgan and Fleet National Bank. The debt donated by these banks ranges from US$200,000 to US$11.5 million (as of face value debt; World Bank, 1998, p91). Deutsche Bank claims to have been the first bank involved in such swaps in 1988 (Deutsche Bank, 2000, p14), though it is not listed in the World Bank overview of actual swaps. The ultimate form of debt-for-nature swap would be a swap at a 100 per cent discount (ie a donation) in exchange for conservation. Usually, endowments or trusts are set up to guarantee a continued conservation of a natural reserve.

The first debt-for-nature swap was made in 1987 in Bolivia (debt reduction of US$650,000 for the price of US$100,000 equivalent of local currency), enabling a million-hectare buffer zone around the 135,000-hectare Beni Biosphere Reserve. Since 1987, it is estimated that over 30 countries have benefited from debt-for-nature swaps which have generated over US$1 billion in funding for the environment. Another instrument is a debt-for-development swap.[43] For example, UNICEF completed a debt-for-development swap in Senegal in 1993. UNICEF purchased US$24 million face value of debt for a purchase price of US$6 million (25 per cent of face value). The government of Senegal agreed to pay UNICEF the CFA franc equivalent of US$11 million over three years to support UNICEF projects in education, health, sanitation and water projects throughout Senegal (UNDP, 1998). Besides writing off loans that are considered to be irredeemable and the goodwill that banks can cultivate among the public through debt-for-nature or -development swaps, it is a unique expression of the drive for a sustainable world.

Table 4.5 *Sustainable banking summary*

Reasons for sustainable banking (Chapters 2–4)	*Shaping of sustainable banking (Chapters 5–8)*
Market	**Commerce**
• customers	• new products and advising
• competitors	• new markets
• shareholders	• 'corporate governance'
• suppliers	• requirements for suppliers
Image	**Involvement**
• involvement	• sponsoring, donations, participation capital
• media	• networks (NGOs, government, businesses)
Government	**Internal**
• legislation regarding customers	• databanks, control lists, etc
• legislation regarding banks	• internal environmental care
• other (eg research)	• policy/commitment from the top
NGOs	**Communication**
• UNEP, etc	• statements
• WBSCD, ICC, SVN, etc	• networks
• action groups (Amnesty International, etc)	• reports

sustainability. To facilitate this, a bank must have a system that can estimate the environmental friendliness of certain projects. This requires both financial and physical information, for which specialist knowledge is needed. A bank can develop this knowledge itself or make use of the agencies which specialize in this area. By creating standard control lists, for instance, the knowledge could then be applied in the normal kind of credit provision services or for providing customers with advice or information. Areas where this is relevant include environmental and energy subsidies, internal environmental care and potential cost savings as well as market developments. Advice of this nature may then help reduce the risks associated with credit provision and participations.

In terms of marketing (PR and image), it will be important to communicate the steps being taken to both private and corporate customers. This not only promotes knowledge among customers but may also encourage the stimulus of particular developments or ideas. Every business and its bank has to ensure, therefore, that its credibility is not threatened. If too much emphasis is placed on external activities and internal environmental care is not safeguarded, then the reactions may be the opposite of the original communication targets. There is a constant risk of adverse publicity. Communication always has to be coupled to proper, consistent and clear reporting. As regards the signing of statements like the UNEP Declaration of Banks Concerning Sustainable Development (see Appendix VI) the same applies, namely that the internal processes must first be in place. To make this happen, the will and commitment at the top of an organization along with the stance of the shareholders is crucial.

The interface between social involvement and commercial gain lies in the participation in networks. Networks are becoming increasingly important for the continuity of organizations and for the low profile communication of social involvement and ambitions in respect of sustainable development. It makes the finding of creative solutions for particular social problems or an investment issues a more straightforward matter. New forms of financing, for innovative products for instance, can be facilitated in this way. Indeed, many investments in the framework of sustainable development often need a high level of innovation. Sustainable banks will have to be prepared to take additional risks, put in extra effort and time to ensure that a project or idea succeeds and be satisfied with narrower margins if they are to give ideas and renewal processes an innovative and sustainable character. The following chapters reveal the extent to which banks are receptive to the opportunities and challenges outlined above. Their reactions to such challenges will determine the category (defensive, preventive, offensive or sustainable) to which they will eventually belong.

Part II

Banking and Sustainability

'Only after the last tree has been cut down,
Only after the last river has been poisoned,
Only after the last fish has been caught.
Only then will you discover:
money cannot be eaten.'
Cree Indian prophecy

5

Sustainability, Markets and Banking Products

Introduction

Sustainable development has an important strategic dimension for banks. The threats and opportunities for banks that arise out of the society's drive towards sustainable development can be divided into several categories by a range of criteria, from risk reduction to profit generation, and from purely business reasons to ideological reasons.

Firstly, new (product) markets develop, such as the market for various financial products on the liabilities side of banking aimed at the upcoming group of savers and investors who expressly choose to use their abundant financial resources to social or environmental ends. Banks can introduce specific products to meet this need, such as payment, savings and investment products. Activities such as these can also have interesting marketing and image-making potential. When the bank retains its traditional products as well, it is able to offer each customer a choice in the extent to which their savings or investment behaviour is sustainable.

Secondly, a bank can fulfill the role of a traditional financial intermediary. After all, society's quest for sustainable development involves the creation of new financing markets, such as markets for water purification equipment and sustainable energy. Banks can expertly step into this growing market or develop specific products such as environmental loans and leases. Innovation funds for the financing of groundbreaking technology can also be developed.

Thirdly, financial innovation can be identified in the securitization of projects that are in themselves low profit (such as solar energy), products for the financing of companies' climate policies, and public–private forms of collaboration to solve particular environmental problems. Fourthly, environmental damage or recycling insurances can be developed by the insurance company within an all-finance financial institution. Fifthly, a bank can develop specific advisory products and/or services for companies aiming to move towards sustainability. Sixthly, charitable institutions can be established within banks which donate money or knowledge to sustainable projects or developments.

This chapter will deal with these subjects. Sustainability and banking are sometimes lumped together with sustainable investing, also known as socially responsible investing (SRI), in which investments are made in a limited group of more sustainable sectors or companies. SRI is up and coming in many countries and is given a lot of attention by the media, the public and by banks. For this reason, this chapter opens with a description and analysis of the developments and popularity of SRI ('Sustainable investment funds', below). SRI is most often concerned with investment in public companies. In The Netherlands, 'green investing' in particular projects is very popular, a development which is strengthening the interface between regular banking and green financing ('Fiscal green funds', page 92). 'New, more sustainable financial products: Committed resources', page 97, covers a wide range of products to attract deposits or premiums which banks have developed with a view to sustainability and which fall outside the scope of SRI. 'New, more sustainable financial products: investments', page 102, deals with possible new financial products and markets from a lending and investment perspective and so deals with the ways in which banks choose to invest. In short, the section starting on page 97 deals with the supply of sustainable money by retail and corporate clients, and the section starting on page 102 deals with the demand of retail and corporate clients for sustainable money. 'Sustainability and advice from banks', page 109, looks at the opportunities for banks in advising customers on environmental issues and, finally, 'The financial sector and the Kyoto Protocol', page 111, deals with a new and growing market for the private sector that relates to the government's climate policy in developed countries. This is a market rich in challenges and opportunities for banks. It is clear that for all these forms of banking activity, knowledge must be built up internally or brought in. Collaboration with stakeholders can, in certain cases, contribute to this.

SUSTAINABLE INVESTMENT FUNDS

Introduction

The increasing prosperity of the last half century has increased the importance of things other than material goods and has lead to an increasing recognition that the growing prosperity also has its negative aspects. For a (small) group of environmentally aware savers and investors, financial return alone is not enough, and, moreover, this group has been rapidly growing since the 1990s. Those investors who wish to use their money to an end that will stimulate sustainable development will henceforth be referred to as the 'sustainability segment'.

The growing need for a 'broader' return on savings can be explained by the pyschologist Maslow's hierarchy of needs theory (Maslow, 1970). According to this theory five hierarchical levels of need can be distinguished. The most basic needs are physiological needs (such as food, health and shelter). The second level is safety needs. The third level consists of companionship, love and affection. The fourth level consists of esteem needs, such as self-respect and respect from others. The last level concerns self-actualization needs, such as knowledge, creativity and self-expression. The last level can only be reached if the previous levels are met. Sustainability is towards the top of this 'pyramid of needs'. A subgroup of consumers and organizations has moved on from meeting its basic needs and is striving to meet higher-order needs. When enough people strive for these needs, a market is created: the market for sustainable investment funds.[1] Various definitions of this market exist: in some countries anything to do with ethics or the environment is simply labelled 'green', and in other countries 'ethical'. In recent years SRI, the label used in the US, has increasingly been adopted elsewhere. Alongside the screening of companies for environmental and social factors in the US, shareholder influence also plays an important role. This is much less the case in Europe. In addition, the social dimension of sustainability carries greater weight in the US than the environmental component; in Europe the opposite is true. Another important difference between the US and Europe is that the rise of SRI in the US is more closely linked to institutional investors, while in Europe SRI is aimed at the retail market. These distinctions are, however, starting to fade. Some European banks, such as the SNS Bank and the Bank Pictet, have set up funds specifically aimed at institutional investors (SNS's Eco-Equity Fund and Pictet's sustainable equities). To obviate differences in terminology, the term 'sustainable investment funds' will be used in this chapter. Shareholder activism has no specific role in this but can form a part of the policy of a particular sustainable investment fund.[2]

Sustainable investment funds have progressed through three 'generations'. The complexity for assessing eligible investment increases with these generations. First generation funds use only exclusionary social and/or environmental criteria; they do not, for example, invest in the arms trade, nuclear energy or tobacco production. In the early stages of SRI (1970s), such criteria were mainly applied by Church groups and labour unions in the US. Second generation funds use positive criteria, in that they look for factors such as a progressive social or environmental policy. In third generation funds, both positive and negative criteria are applied, and attention is paid to relative performance within a sector, the 'best-of-class' method. Lufthansa may, for example, score more highly than KLM on environmental criteria, making it the preferred investment choice. Recent years have seen a strong trend towards professionalism and growth in sustainably invested capital, in the number of funds, and in the provision of information and benchmarks. In several countries the number of sustainable investment funds has doubled or quadrupled over the last five years. One area that is still virgin territory is the monitarization of environmental aspects, in which a direct link can be made between the strain placed on the environment and the annual statement of accounts.[3]

Fundamental principles

In deciding where to invest their capital, sustainable investment funds assess companies by criteria related to their environmental friendliness, social policy and, for example, their standpoint on human rights. Such criteria can in the main be positive (the extent to which the company is progressive) and/or negative (un- or less acceptable aspects of the company's operational management). The development of standardized methods has evolved rapidly and specialized bureaux exist to assess companies on such criteria . In the US, for example, a computer program has been developed to enable individual investors to make such assessments.[4] The returns achieved when such criteria are applied appear, in several cases, to be higher than the market average. Box 5.1 illustrates the working and performance of such funds, taking the ASN equity fund as an example.

A distinction should be made between the aspects used as selection criteria and the manner in which these aspects should be measured and rated. A list of environmental aspects is usually drawn up which together create a picture of how the company impacts on the environment. In making such a list the drive for completeness must be weighed against what is feasible in terms of workload. While a complete list is desirable, it requires great effort on the part of the company and the fund or the bank. Checklists usually use two types of criteria: effort and results. The former focuses on company strategy, the environmental care system, environmental programmes and planned specific environmental innovations, and so on. The latter involves assessing concrete achievements in reducing environmental impact. The assigning of values to individual subjects (eg use of energy or amount of waste) and dimensions (environmental versus social issues), and the aggregation of these values in one total score, is still subject to much controversy.

Returns and activism

A comparison with widely used benchmarks shows the excellent performance of sustainable investment funds. They earn a return at least equivalent to that of regular funds. Figure 5.1 shows the development of the US Domini 400 Social Index (DSI 400; used as the benchmark for the Domini Social Equity Fund) in relation to the Standard & Poor's 500 (S&P 500). The DSI 400 is based on the S&P 500. The approach of the DSI 400 is as follows. All businesses involved in some way or the other with tobacco, gambling, defence and nuclear energy, are filtered out of the S&P 500. Of the remaining businesses, those selected comply with a number of positive criteria, such as corporate citizenship, employee relations and environmental issues (250 in total from the S&P 500). Furthermore, 150 businesses are added which are not in the S&P 500, but score very well on positive and negative criteria. For example, Ben&Jerry's Homemade is in the DSI 400 but not in the S&P 500 and Nike is in the S&P 500 but not in the DSI 400. Other examples of businesses in the DSI 400 are Microsoft, Toys 'R' Us and Xerox. The list is periodically updated. The two indexes are in short related, but not exactly comparable. From 1990 to 1998 the median market capitalisation of the S&P 500 was about twice that of the DSI 400, and the price to earnings and the risk-return profile is higher for the DSI 400 (Robeco, 1999, p10). Moreover, the DSI 400 is overweighed in high-tech businesses. The outperformance of the Domini Index is immediately evident in Figure 5.1. However, statistically this outperformance is not significant when one makes corrections for differences between the two indexes (ibid.).

Box 5.1 The ASN equity fund

The Dutch ASN Bank was set up in 1960. A concept central to the ASN Bank is using finance to stimulate sustainable development. The bank's deposited capital doubled every five years from 1973 until 1989; in 1989 total entrusted funds amounted to around €363 million, a sum which had grown to €800 million by the end of 1999. ASN Savings invests in activities that positively affect nature and the environment, healthcare, wellbeing, developing countries and animal protection. The bank also provides ASN Environment Savings, which invests only in activities that preserve nature and the environment. Alongside these savings facilities ASN Bank has several investment funds with a total turnover of €290 million (as at the end of 1999).

The ASN Equity Fund observes a number of ethical criteria (environment, social policy and human rights) against which it assesses companies before investing in them. It is a third generation fund. This appeals to a specific market of investors. The total turnover of the ASN Equity Fund amounted at the end of 1999 to around €190 million. The fund has been placed among the best performing investment funds in The Netherlands for four years in a row; the return on the fund was no less than 52 per cent in 1999. The average annual return since its inception (1993) amounted to roughly 23 per cent. Its five-year average return is 31 per cent.

The process of choosing a company in which to invest involves two steps. Firstly, companies are selected on the basis of ethical and environmental criteria related to their intentions and present and past performance. These criteria limit the investment universe to approximately 300 companies. Secondly, SNS Asset Management (the ASN Bank is part of the SNS Reaal Group) narrows the selection to a group of companies in which the bank will invest on financial/economic criteria. This leaves about 75 companies.

The 'extraordinary' investment criteria used in the first step are roughly divisible into four categories:

1. Exclusion criteria: Companies immediately excluded are those involved in the production or distribution of nuclear energy, or the production and trade of weapons.
2. Conditional criteria: The extent to which companies have integrated their environmental policy (current situation and policy for improvement) and contribute to the sustainable, environmentally and people friendly development of a (developing) country is assessed. Companies should be able to show, among other things, that they play a positive role in the field of human rights in those countries where these rights are insufficiently observed and where they have a branch.
3. Sufficient choice criteria: Once the first two criteria have been met, and subject to the fact that sufficient companies remain to invest in, the following sectors will be excluded as well: tobacco, gambling, bioindustry and alcohol; companies using animal testing or genetic manipulation; and companies paying no active attention to social policy, like the rights of minorities and women.
4. Political criteria: In some extraordinary cases the political situation in a country can be reason enough not to invest in any company active in that country. Such a situation arises where, for example, the UN Security Council makes a ruling based on the violation of human rights.

In its research, SNS Asset Management draws on such sources of information as newspapers, specialized magazines, environmental annual reports and specialized information for sustainable investment funds. It addition it presents the companies with standardized questionnaires.[5] Sometimes a dialogue with the company or a company visit take place. Each company is given an integral score on the basis of the research.

On an overall scoreboard, company totals are compared and within a sector companies are placed in a hierarchy. The investigation is repeated periodically to verify the situation. If necessary, a company is dropped from the list of potential investments. Anything from half a day to a whole day is spent on investigating a single company. The duration partly depends on the company's cooperation in filling in the written questionnaire and with telephone or personal interviews. In practice, the company's cooperation depends on its sector and size.

Sources: ASN Bank (1998; 1999); Negenman (2001); www.asnbank.nl (March 2001).

While the DSI 400 is exclusively an US index, the introduction of the worldwide Dow Jones Sustainability Group Index (DJSGI) marked a further step forward in professionalism. The components of the DJSGI are selected on the basis of the Dow Jones Global Indices and represent the largest sustainable companies in the world. The 200 or so selected companies from 22 countries have a joint market capitalization of $US4.4 trillion. In the period 1994–1999 the DJSGI outperformed the Dow Jones Industrial Average by 163.4 per cent to 108.4 per cent (on the basis of backcasting, see Dow Jones Sustainability Group Index, 1999 and Flatz et al, 2001). Financial institutions such as HypoVereinsbank, ING, Rabobank, Credit Suisse, Gerling, Nikko, Union Investment and Westpac are licensees of the DJSGI. FTSE is launching a sustainable index in the UK in 2001 as well.[6]

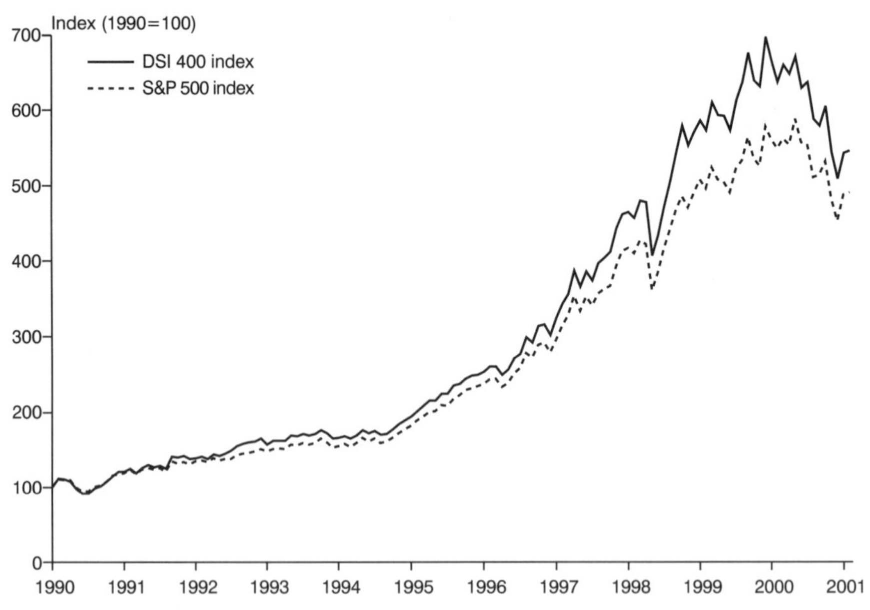

Source: Copyright © 2001 by Kinder, Lydenberg, Domini & Co, Inc. Reproduced by permission. Per $1 value changed to index numbers by permission. The Domini 400 Social Index is a service mark of Kinder, Lydenberg, Domini & Co, Inc.

Figure 5.1 *Financial performance of US sustainable companies, the Domini 400 Social Index, 1990–2001*

Various investigations have also shown that sustainable investment funds achieve an above-average return (*Environmental & Accounting Auditing*, 1997; *Tomorrow*, 1997; Salomon Smith Barney, 1998; Robeco, 1999). Although most studies and empirical data from sustainable indexes show an outperformance of funds screened on sustainability criteria, the statistical significance is problematic (see eg Louche, 2001). In short, are sustainable investment funds outperforming benchmarks like the S&P 500 or the MSCI because environmental or social advanced companies perform better because of their sustainability stance and activities or because such companies are found in certain high-growth industries, sectors (small versus large caps) or countries? One should look into the environmental–financial performance link of individual companies instead to provide a better understanding about outperforming investment portfolios. An investigation by WBCSD has done just that and concluded that ignoring the good environmental performance of companies puts banks at a competitive disadvantage: a positive correlation exists between the value of a company and the performance of the company in environmental management.[7] This is no hard and fast rule, but a proactive environmental stance is usually a strong indicator of a solid and prescient company strategy. Another investigation also reached the conclusion that strong indications exist that good environmental performance (especially in terms of innovative strategies) goes hand in hand with an above-average company return (Van der Woerd and Vellinga, 1997) and an investigation by the Hamburg Environmental Institute into the world's biggest chemical and pharmaceutical companies resulted in the conclusion that the environmentally efficient companies achieved better company results (about 28 per cent higher) than those which were not environmentally efficient (*Business and the Environment*, August 1997).

Environmental and financial criteria can, in short, amount to the same thing, given that companies with a good environmental performance usually keep their financial affairs in good order. A progressive environmental company policy is also usually found within companies which explicitly look to the long run. Such a policy leads, firstly, to direct cost savings as a result of, for example, energy saving and minimalizing waste. Secondly, it enhances image and improves profit margins, *ceteris paribus*. Thirdly, environmental liability and other environmental risks (also those resulting from a change in environmental legislation or market circumstances) are often under control.

Developments in developed countries

In the UK, Switzerland and the US sustainable investment funds have a long history. In the UK money placed in retail sustainable investment funds amounted to GB£54 million in 1984, and some GB£3.7 billion (approximately €5.9 billion) by the end of 2000. The total amount of the SRI market at the end of 1999 amounted to GB£51.7 billion (approximately €83.3 billion), ie including funds invested sustainably by churches, charities and pension funds. In the US at the end of 1999 around US$2.2 trillion were invested in sustainable investment portfolios, ie approximately 19 per cent of all professional managed assets in the US. Sustainable investments in mutual funds in the US amounted to

US$154 billion at the end of 1999. The total SRI market in Canada amounted to €35.5 billion in June 2000, about €7.1 billion of which is invested through retail funds. Figures for sustainable investments in mutual funds in other developed countries include, at year-end 2000, over €2.2 billion in Japan, over €3.6 billion in German-speaking countries (including over €1 billion in Switzerland) and approximately €1 billion in Belgium.[8] True comparisons are difficult due to differences in definitions between countries of the SRI market. Of the continental European countries, the pace of development in The Netherlands is fast, partly facilitated by the fiscal green ruling. At the beginning of 2001 about €3 billion has been sustainably invested in The Netherlands, compared with around €90 million in 1995. A number of reasons can be distinguished for the large US market:

- historical – Church groups and labour unions started widespread screening of investments in the beginning of the 1970s;
- political – a less prevalent social safety net compared to Europe;
- cultural – relatively strong consumer awareness and consumer readiness to take action; and
- financial – a relatively well-developed capital market with high retail interests.

Screening out tobacco and shareholder activism are important elements of this.[9] In particular the Anglo-Saxon countries and Switzerland and The Netherlands have seen a strong rise in the professionalism of sustainable investing, with the spin-off that a growing number of companies finds it suits their image to be chosen as an investment by a sustainable investment fund. It functions as a stamp of approval for the company's approach to the environment.

Further important developments will stem from the steps taken by institutional investors. For example, PGGM , the second-largest Dutch pension fund with over €50 billion of assets, declared that it will screen all its investments on sustainability criteria in the near future (*Financieele Dagblad*, 19 June 2000). It is already testing this by giving some sustainable investment managers considerable mandates. ABP, the largest pension fund, is now considering a similar move. In 2001, Morley Fund Management (a subsidiary of CGNU, the biggest UK insurer), which manages about €160 billion of assets including the equivalent of 2.5 per cent of the UK stock market, stated it will in future vote against top 100 companies' annual accounts unless they include an environmental report. Moreover, it expects to abstain from investing in those companies which do not publish a 'comprehensive environmental report' and are in high risk sectors (oil and gas, electricity, chemicals, automobiles, construction, healthcare and pharmaceuticals). Indications exist that this move will be followed by other UK insurers and pension funds (*Financial Times*, 4 April 2001, p1). These moves within The Netherlands and the UK are tentative for a emerging market for screened investment portfolios by professional fund managers. Banks obviously follow this trend with great interest.

An increasing number of major banks are offering best-of-class sustainable investment funds. By starting such funds in the retail segment they are gaining experience for a emerging institutional segment. Examples of such retail funds include the Föreningsbanken Environmental Fund, UBS's Environmental Capital Fund, Storebrand's Principal Global Fund, Nikko Asset Management's Eco Fund, Calvert's Social Investment Fund and HypoVereinsbank's Activest Lux EcoTech. The top three banks in Germany, The Netherlands and Switzerland all offer such best-in-class funds. French, Spanish and Italian banks lag behind most major European banks. When the bigger financial parties get involved, such products become more widely known and trust in them increases. Each fund will obviously have to find its own balance between financial and environmental/ethical criteria.

Information requirements and transparency

The financial sector has a growing need for easily accessible and concise environmental information. While banks might build up knowledge and information internally by such approaches as those outlined in Box 5.1, most mainstream banks will outsource this need. Various organizations have specialized in testing companies against environmental criteria and gathering the necessary information for sustainable investment funds. Examples are (see eg Ryall et al, 1996; Gentry, 1997) Eco-Rating International, Ethical Investment Research and Information Service (EIRIS), Eco-Risc'21, Center for the Study of Financial Innovation (CSFI), Investor Responsibility Research Center (IRRC), Kinder-Lydenberg-Domini (KLD) and Sustainable Asset Management (SAM). Some differences exist between such companies, but on the whole they fulfill the same function. Most use hard as well as soft criteria to evaluate a company. The IRRC has rated all companies in the S&P 500 index. The CSFI has copied the conventional method of credit rating, which creates a seven-point scale of environmental risk from AAA to C. So, for example, Scottish Nuclear earned an A rating, because it is '...a company with large but well-identified environmental liabilities, and sufficient financial and management strength to absorb all but exceptional risks' (Lascelles, 1994, p32). The assessment takes into account both the scale of an environmental risk and the company's ability to deal with the risk financially and in organizational terms. This method could therefore also be used for assessing credit risk and estimating market potential (see Chapter 6). In fact, the CSFI rating should run parallel to a conventional classification, given that it pertains to company risk. This is, however, not the case in practice, which implicitly shows that the financial world is (still) not fully competent at estimating environment risk: the existing credit-rating methods are deficient.

Just as sustainable investment funds require reliable, consistent, comparable information, so do their investors. The investor usually has insufficient information about the environmental return on his or her investments. Storebrand has, however, developed a method to do exactly this.[10] In addition, the investor has insufficient information about how the fund selects companies and allocates monies among the selected companies. For this reason, there has been a call for benchmarking among funds in precisely this area, against environmental and/or social criteria. Hallmarks or labels would also be an option (see Kahlenborn,

2001). Such instruments are especially needed for those investors, labelled 'cause based' investors, for whom the financial return is of no real importance and who are driven solely by social and/or ecological issues.

FISCAL GREEN FUNDS

Introduction

An especially interesting fund arose from the fiscal green regulation that the Dutch government developed in 1992–1993 in collaboration with the Dutch banking sector (in particular ASN Bank and Triodos Bank): the fiscal green fund. It was the government's objective to stimulate private funds in the direction of environmental friendly investments (Ministry of Housing, Spatial Planning and the Environment, 2000). The most important distinction of sustainable investment funds are the attractive fiscal advantages they offer the investors and the green nature of the project (whereas sustainable investment funds focus solely on companies). No similar funds exist in other countries. The Dutch initiative can provide a role model, as is apparent from the European Parliament's interest in introducing a similar ruling across the EU.

Roughly speaking, three phases can be identified. The first phase is characterized by a surplus of funds (and thus a shortage of green projects to finance). The second phase shows a surplus of projects and a shortage of deposited capital. A third phase, which took off in 2001, relates to the review of the tax system in The Netherlands. The shortage of projects in the first phase caused banks to quickly develop their knowledge of green projects; various banks set up dedicated environmental departments, which are now concerned with the development of new financial products as well. As such the fiscal green regulation has played an important role in raising awareness among the Dutch banks of sustainable banking and external environmental care.

Its role model status, and the fact that it has accelerated the development of the major banks' awareness of sustainable investment opportunities, makes it worth devoting a whole section to the Dutch fiscal green regulation.

Principles of a fiscal green fund

The Dutch government's fiscal regulation was motivated by the idea that without it environmentally friendly investments would not take off because the anticipated return of between 3 per cent and 4.5 per cent (gross average) was just too low. The regulation is innovative. Until 2001, it made the interest and dividend on green investments tax-free, as long as they are received via particular credit or investment institutions: 'gross is net' in fiscal jargon. As such, the funds are for retail investors only.

This is very attractive to savers and investors in The Netherlands given that until 2001 they were required to pay tax on any interest and dividends above €400 to €800 per year (which was tax exempt) at a marginal rate (rising to 60 per cent). Furthermore the recognized green funds are not obliged to retain the 25 per cent dividend tax applicable to other Dutch investment funds. The tax loss for the government amounts to approximately €10 million for every €450 million invested in fiscal green funds. That is, the multiplier for the government is 1:45; every €1 of public money makes €45 of private money available for green projects. The benefit for the government is that the tax ruling enables certain environmental objectives to be met more cheaply and efficiently than would be the case were subsidy regulations to be used.

The chief beneficiaries of the tax exemption have been businesses. They have seen the tax changes reflected in reduced interest on their credit for projects for which they obtained a 'green certificate'. This has been facilitated by investors' readiness to accept a lower gross return because they no longer have to pay tax. Table 5.1 gives an overview of how the fiscal green regulation works (based on Van Bellegem, 2001, p235). The table shows the difference for a business client between a standard commercial loan and a loan from a fiscal green fund, and the difference for each party involved. In this example the following assumptions have been made: the green fund in question pays a fixed return (comparable with a deposit savings account); the interest rate on a regular deposit account is 5.2 per cent; the bank's standard interest margin is 1 per cent; the tax advantage on the green investment for investor, bank and business client is split according to the ratio 1:2:10; investors pay an average of 50 per cent tax on interest or dividends earned; the tax exemption on the income tax has already been fully used. It then follows that the investors receive an extra profit of 0.2 per cent-points, the bank 0.4 per cent-points (part of which is a risk and additional transaction costs surcharge); and the business can finance its green project against an interest rate that is 2 per cent-points lower.

The split of the tax advantage over the three parties will, of course, depend on the development of the market interest, the supply and demand of projects, and the amount of 'green' money savers and investors are willing to invest. The split used in the example reflects the first phase; a surplus of green money existed combined with a legal obligation for banks to invest in 'green' projects or face a stiff penalty clause. In this phase competition between banks became so strong that even with relatively low market interest rates, the interest advan-

Table 5.1 *Principles for the Dutch fiscal green regulation*

	Standard commercial loan	*Fiscal green funds loan*	*Difference in favour of green funds*
1. Net return for saver/investor	2.6%	2.8%	+0.2%
2. Tax	2.6%	0%	–2.6%
3. Gross return for saver/investor (= 1+2)	5.2%	2.8%	–2.4%
4. Funding by bank (= 3)	5.2%	2.8%	–2.4%
5. Interest margin for bank	1%	1.4%	+0.4%
6. Interest on credit for business (= 4+5)	6.2%	4.2%	–2%

Box 5.2 The criteria for the Dutch fiscal green regulation and how it works

To be considered green financing, a business has to possess a 'green certificate'. To obtain this certification, a business must contact the bank managing the green funds. The green fund, represented by the bank, examines the economic merits of the business and the project. If this test is passed, the bank accepts the application. A detailed project plan must then be submitted to the government. This project plan must satisfy certain criteria: the scale of the project must be greater than €23,000; the project has a high environmental return; the project delivers 'some' return (ie it is economically self-supporting); the project will make a return that is, however, insufficient to attract regular financing; the project should be situated in The Netherlands or another country covered by this regulation (such as the Dutch Antilles, East European countries and a number of developing countries such as Costa Rica and China);[11] and the project falls within the one of five project categories defined by the regulations, as follows:[12]

1 forest, nature and landscape (10 per cent);
2 agriculture (31 per cent);
3 energy (24 per cent);
4 sustainable building (14 per cent);
5 other environment (21 per cent).

If the project satisfies the above criteria, the government will grant the business a green certificate within eight weeks. The certificate is valid for ten years and is independent of actual green financing. An exception is the first category; these certificates are valid for 30 years. The Ministry of the Environment or its agents, LASER and NOVEM, are responsible for the environmental monitoring. NOVEM and LASER also advise banks, individuals, companies and institutions who want to know early on if a project may later be eligible for a green certificate.

All green funds are supervised by the Dutch Central Bank and are obliged to invest at least 70 per cent of their funds in recognized green projects.[13] If a green fund invests more than 30 per cent of its capital in projects other than recognized green projects, the rest – the difference between 70 per cent of the fund capital and the invested capital – is subject to a tax penalty of 2.5 per cent. Moreover, if the funds have not succeeded in meeting the 70 per cent requirement within two years, the green status of the fund can be rescinded retroactively, with the effect that the whole fund will be subject to this penalty. The net return promised to the investor will then no longer be achievable and the bank will suffer losses (as it in most circumstances has to compensate the investor). Various green funds have already paid tax penalties.

tage for the business rose to over 2 per cent-points. The penalty clause, the sequence of events in green financing and the criteria which funds and financing have to meet in order to be certified 'green' are explained in Box 5.2.

Volume and performance of fiscal green funds

The major banks were at first hesitant about the introduction of the green regulation. The Dutch banks had not yet developed sustainable products. Yet a bank could enhance its image by having a green investment fund that would meet some of the financial problems faced by certain groups of businesses, offer savers and investors an interesting product, and make a profit. The first major Dutch bank to make use of the regulation was the Rabobank. At the end of 1995 its Green Interest Fund took about €200 million in investors' money, while the government's target had been to attract around €45 to €90 million to green investments in the course of a year (with an expected increase up to half a billion euros by the year 2000). Investor interest appeared overwhelming and exceeded all expectations: the Rabobank was forced to close its fund after nine days because it feared that it would not find sufficient projects in which to invest. The tax incentive had given idealistic investors in The Netherlands a nudge in the right direction. At the end of 2000, €2.3 billion had been invested in fiscal green funds and every major Dutch bank offered its own fiscal green fund. Table 5.2 gives an overview of the various fiscal green funds offered by the commercial banks in The Netherlands. Rabobank is market leader in the volume of funds and number of projects financed.

For equity funds, prices movements are becoming increasingly important. Over the first half of 1996 the return on several equity funds rose considerably. Six months after its introduction, enormous demand had caused the share price of ABN AMRO's 'closed-end' fund to rise by 25 per cent. Over a period of five years the share price of the ASN green project funds rose by a total of 12 per cent, while the ABN AMRO fund rose by 17 per cent. Compared to the Dutch AEX Index the green funds score badly on share price increases: between the end of 1995 and July 1997 the AEX Index rose from 100 to 200, whereas the green funds rose from 100 to an average of 105. Given the nature of green funds such a comparison is not entirely justified. A better comparison is with the Dutch CBS index for bond funds, which shows a similarly modest price increase (from 100 to 110; Scholtens, 1997). Although the low dividend for the green funds (average 3 per cent per annum) barely differs from the return offered by other

Table 5.2 *Overview of fiscal green funds in The Netherlands, year-end 2000*[14]

Name of fund	*Bank*	*Type*	*Marketability*	*Return*	*Volume*
ABN AMRO Green Fund	ABN AMRO	closed-ended equity funds	stock exchange	average 8.6%	€467m
Green interest fund/green bonds	Rabobank	fixed interest certificates	no	2.75–3.4%	€1.062m
Postbank Green Interest Certificate	ING	fixed interest certificates	no	2.1–3%	€530m
ASN Green Project Fund	ASN	open-ended equity funds	limited	average 4.3%	€95m
Triodos Green Fund	Triodos Bank	semi-open-ended equity funds	limited	average 4.4%	€124m

Source: Return figures from Noordhoek, Abbema and Ruitenbeek (1999); volume figures from personal communication with each bank.
Note: Final column gives deposited or invested amount of money from retail investors in each of the green funds as of the end of 2000.

investment institutions, the total return is somewhat lower. But this does not have to pose a problem for a green investor, given that he or she will enjoy a tax advantage *and* is probably somehow ethically motivated as well. In this respect, green funds do have a problem: they offer little or no insight into the environmental performance of the companies or projects being invested in.[15]

The future of fiscal green funds

Until 1998 green funds had to deal with a surplus of liquidity and a shortage of projects eligible for investment under the green regulation (the first phase). This lead to a veritable price war. Companies and individuals wanting to borrow green money could do so for as much as 2.25 per cent less than the market interest rate, while the government had initially aimed at 1 per cent-point under the market rate. Guaranteed returns really squeezed the margins of most green funds and many suffered losses. The situation completely turned around in 1999 when there was a shortage of green money. This was caused in principle by the fierce competition for green projects, which reduced the returns (on equity funds) enjoyed by investors. The steep rises in regular prices on the stock market at that time contributed to this trend. Another downturn came from the growing uncertainty on the survival of the green regulation under the new Dutch tax laws for the 21st century.

In the new tax system (introduced on 1 January 2001) the top tax rate has been reduced from 60 per cent to 52 per cent and the next rate from 50 per cent to 40 per cent. The first €17,000 of savings and investments are tax-exempt, above that a virtual return of 4 per cent is assumed which is then subject to tax at 30 per cent. On balance, thus, a fixed tax burden of 1.2 per cent for dividends and interest on investments and share gains. This would significantly reduce the tax advantage offered by green funds, making them barely interesting to investors and savers.

On top of that the government planned to freeze the ceiling on tax losses, which would have made further growth in green funds impossible. This would have meant the end of the regulation. This lead the Dutch banks in 1999 and 2000 to launch a tightly coordinated and successful lobby for compensatory legislation. The tax situation for the green regulation now looks like this: a taxpayer can invest a maximum of (about) €45,000 per annum under the green regulation (double for a married couple). The investment is exempt from the above-mentioned 1.2 per cent capital tax. Moreover, the investor is eligible for an additional general tax reduction of 1.3 per cent. So to an investor in a green fund which has a return of 2.5 per cent, the actual return is 5 per cent (2.5 per cent + 1.2 per cent + 1.3 per cent). Nevertheless, the banks fear that further strong growth in project capital will be restricted by the imposition of ceilings on tax losses (the government is trying to limit the maximum tax loss to around €45 million per annum). Despite this, the green regulation will continue in The Netherlands. One advantage of the new system is that it can more easily be copied by other countries.

What many projects under the regulation actually need is risk capital. The green funds are, however, not suitable for this and the Dutch government has, therefore, been thinking of introducing another fiscal green fund specially for high risk but potentially rewarding projects (Dutch Lower House, 1997). The Dutch government is considering introducing a similar ruling for social projects in developing countries.

New, more sustainable financial products: Committed resources

Introduction[16]

In previous sections attention has been given to new forms of financial products. The growing demand for them is linked to growing environmental awareness, which eventually links to other forms of awareness and values, such as an awareness of social conditions and human rights situations (Maslows's hierarchy of needs). The realization is also growing that we are just custodians of the Earth for future generations and should treat the Earth accordingly. Banks can capitalize on that growing awareness by offering products that appeal to, or even facilitate sustainability.[17] A number of such products which banks can or do use to attract the hereafter named 'sustainability segment' will be discussed below; ie this section deals with the supply of sustainable money by retail and corporate clients. The following section will deal with the demand of retail and corporate clients for sustainable money. In this section a distinction will be made between current accounts, savings products, investment products (other than SRI) and insurance products. The products discussed are segmented in that they are targeted at a particular aspect of sustainable development. The list of examples is not exhaustive; the development is far from complete.

More sustainable payment products

Although at first glance the development of products related to current accounts is not obvious, good marketing opportunities do exist, such as the World Nature Card from the Föreningssparbanken (Swedbank) in Sweden. This credit card was developed in collaboration with the WWF. Half of 1 per cent of the total payment transactions made with the card are denoted to the WWF and are used for the protection of nature in Scandinavia. The similar Royal Bank of Canada card is called the WWF Visa Affinity Card. In the US the Citigroup offers the Environmental Defense Platinum MasterCard to its clients. Bank of America has a similar affinity construction linked to its check boxes. Such arrangements cost the customer and retailer nothing: the burden is borne entirely by the bank. The commercial benefits are visible in an enhanced image and better sales of other products, particularly to young people, and it is thus a form of 'cause-related marketing'.

For current accounts there are no other financial products related to environmental care. The major banks use the resources on current accounts for their traditional investments. Exceptions to this are banks such as the Triodos Bank and The Co-operative Bank, which use the more permanently present balance on their customers' current accounts for loans and investments focused on sustainable development.

More sustainable saving products

One example of the linking of a savings account to a good cause is the VSB Panda Certificate (Fortis Group); for each such certificate sold, the VSB denotes a fixed amount to the WWF. Environmental saving has been until now principally something for niche banks such as the ASN Bank and the Triodos Bank in The Netherlands. They have specific savings products from which funds are directed to investments in and loans to companies who play a positive role in sustainable development or elements of it.[18] Several large banks, though, are also studying the possibilities for creating such specific savings accounts (Hill et al, 1997). However, in the research into 34 major international banks reported in Chapter 9 such products were encountered at just two major Dutch banks (ING and Rabobank). These are tax-driven savings products, stemming from the fiscal tax regulation discussed on pages 92–7. Rabobank is considering introducing so-called destination accounts, in which each individual customer can determine where the bank invests its money (Rabobank, 2000, p7). Technically and in terms of risk, however, such a step involves problems which might be overcome by learning from the experiences of the smaller banks mentioned.

More sustainable investment funds

More banks are active in the area of sustainable investments. A number of funds already exist that attract money for investment in companies playing a positive role in sustainable development or elements of it. This may be in terms of the company's own operations or its products. This development was covered on pages 84–92 where the emphasis was on best-of-class products. Below some special products or new developments are presented (see Schaltegger and Figge, 2001, for a typology of sustainable investment funds).

At the beginning of the 1990s some special funds emerged: environmental technology funds. These funds focused on the booming market for environmental technology in developing countries. An example is the Environmental Growth Fund at Mees Pierson (Fortis Group). This fund was cancelled after four years in 1994 due to lack of success. Still, in recent years new environmental technology funds have appeared on the market, such as the Ohra Environmental Technology Fund in 1997 and the Ohra New Energy Fund in 2001.

A special category of investment funds, sometimes marketed as insurance products, are the 'wood' funds. The first wood funds were set up in the beginning of the 1990s in various countries. Ohra linked its Teakwood Return Policy to WWF (for 5 per cent of the income), an endorsement of great marketing value. Such investments have a term of around 25 years. Rates of return of up to 25 per cent are predicted. These funds have, however, attracted a great deal of criticism. There is doubt as to whether such a high rate of return really can be achieved (given uncertain demand and the risk of natural disasters) and the environmental effect is not always positive; for example, ground can become exhausted from being subjected to a monoculture (Bulte and Van Soest, 1997).

In addition, there are funds which are totally dedicated to sustainable energy, such as the Wind Fund and the Solar Investment Fund offered by the Triodos Bank or the New Energy Technology Fund at Merrill Lynch. The smaller ethical banks, in particular, also have investment funds which attract money for directed investments in developing countries. There is a growing market for such a thematic approach, which can be seen at major banks as well. Interesting developments are sustainable index funds, such as Hypo Vereinsbank's DJSG index certificates, sustainable bond funds at Bank Sarasin, a climate fund by Dexia (see page 113), a Fuel Cell Basket (targeting professional parties), an EcoPioneer Fund from UBS, the rise of sustainable investing in the Japanese market (for example, the funds from Nikko Securities and UBS's Dr Eco Fund), sustainable investment funds with an NGO label (see eg Döbeli, 2001), developments within portfolios customized for the wealthier customer aimed at sustainable investments[19] and the development of sustainable investment by banks' own pension funds. This last development could well be an important prerequisite for the further growth of sustainable investment funds among banks. Indeed, when a bank does not wish to invest the financial resources of its pension fund sustainably (given expectations of increased risk or decreased return), it adopts a particularly poor position in the eyes of its retail investors with the message that sustainable investment does form part of a financially responsible investment strategy.

More sustainable insurance products

The insurance market too is introducing products targeted at the growing sustainability sector. Strong examples are National Provident Institution (NPI) from the UK and Uni Storebrand from Norway. Storebrand's Principle Global Fund wants to deliver a threefold return (labelled 'the triple return'). Firstly, a financial return which is higher than the benchmark index (such as the MSCI World Index). Secondly, an environmental return in which the fund's companies are more eco-efficient than the market as a whole. And thirdly, a social return in which the fund's companies operate with a better social record than the market as a whole (Storebrand, 2000, p9). As such a best-in-class approach, but with a clear and simple communication strategy. The SNS-Reaal Group in The Netherlands has brought out its Spaarbewust Polis (Aware Savings Policy) and the World Partner Policy (in collaboration with NOVIB, a member organization of Oxfam International). Investments are made on the basis of the ASN criteria or in ASN equity funds. Moreover, this type of insurance also involves donations. For the Aware Savings Policy the SNS Reaal Group donates a fixed amount per policy each year to a good cause, like an environmental education project for children in Chad. In the case of the World Partner Policy, the customer denotes 20 per cent of his or her savings to NOVIB (fully tax deductible), which uses it for development projects in developing countries.

As well as being investors, insurers are concerned with estimating risk – environmental risk, for example. In 1992 the insurance world called attention to environmental issues, having been faced with damage claims amounting to US$50 billion; all these claims stemmed from natural disasters. Hurricane Andrew in the US cost insurers more than US$17 billion and resulted in the bankruptcy of at least eight insurers. Many insurers believe the hurricane was related to the increased greenhouse effect and state that governments and industry should take action to reduce CO_2 emissions. To strengthen their cause, the (re)insurers are themselves active in prevention, making internal processes environmentally friendly, educating their customers, offering environmentally friendly companies lower premiums, and investing in such companies (Knoepfel et al, 1999). Moreover, they have published a position paper on climate change at the climate change conference in Kyoto in 1997.[20]

A distinct form of environmental insurance is recycling insurance. This involves a customer paying less for car insurance (up to 20 per cent) if, when parts are damaged and need replacing, recycled parts are used. Recycling is stimulated and both bank and customer save money. Various banks and insurers have such a product (such as Credit Suisse, Achmea and Yasuda Fire and Marine). Another insurance product for automobiles is that offered by the Japanese insurers Tokio Marine and Yasuda Fire and Marine: a 3 per cent discount is given on automobile insurance premiums for low pollution vehicles. Environmental insurances have also been set up for environmental risks in property transactions and investments, like Citigroup's Millenius.

For a long time, however, most insurers have been reluctant to insure against environmental risks. The first environmental damage insurance (EDI) from the French Garpol in 1977 was withdrawn in 1983. While this did not halt developments, it certainly contributed to the first steps being somewhat tentative. In France a new, broader insurance pool was set up, Assurpol, and in the US and UK EDIs took off. Insurers usually turned out not to know what they were getting into and quickly gave up. Environmental risk appeared complex and difficult to estimate. Major problems stemming from the past such as ground contamination and asbestos hang like a spector over the insurance market and slow down new developments. However, risk mitigation techniques and the quality of information are improving, which makes new developments possible. A striking example is the Lender Liability Insurance from Zurich Financial Services, which provides coverage for the direct environmental liability of banks.[21] Since 1995 a number of insurers have offered a reduction in their premiums to companies with a good environmental performance. This group is lead by Uni Storebrand and includes Swiss Re, Gerling, General Accident and NPI. They agreed to jointly develop methods and to share experience (*Financial Times*, 30 March 1995).

Another way around the spector lies in the initiative of the Dutch insurers in jointly creating a product. Until 1998 coverage for environment damage in The Netherlands was provided piecemeal by various insurances, such as fire, transport, and environmental liability (for certain gradually occurring damage) and businesses' liability insurance (for certain suddenly occurring damage). These insurances did not, however, form a coherent whole and moreover no insurance existed for various forms of environmental damage. In 1998 the insurers created the Integrated Environmental Damage Insurance (IEDI). This is a world first because unlike all other insurances in this area, the IEDI is not a liability insurance but a 'direct insurance'.[22] This means that only a causal relationship between the damage and the origin need be established, and not liability. The great advantage of this for both companies and insurers is that the often long and difficult path involved in asserting a liability right can be avoided and damage can be recompensed more quickly, cheaply and simply. Obviously, when liability is at stake the insurer will afterwards try to find recourse for the damage.

IEDI covers the following three risks: export risk (the insured causes damage to another); self-inflicted risk (the insured damages him or herself); and import risk (a third party damages the insured). The IEDI uses five categories of cover, each with its own level of premium. These categories range from light risk (1) to heavy risk (5): companies rated as heavy risk are charged the highest premiums and are limited in their maximum amount of cover. The classification results in the risk selection diagram shown in Figure 5.2. In principle, a system such as this can be applied to credit risks as well (see Chapter 6). Of course, the division is still rather rough, but it provides a sort of quick scan based on answers to questions that, to a company, are neutral and for an insurer or banker are easy to grasp. Other factors determining the price of the insurance are additional specific environmental information and the type and extent of cover required by the client.[23]

In practice, IEDI can mean that a business running a greater technical environmental risk, such as a chemical factory, opts for the most comprehensive type of insurance cover. The insurer is highly likely to insist that the company implements a number of preventative measures. A business with a low level of risk, such as an office-based company, will choose the most basic cover and thereby benefit from lower premiums. One great environmental advantage of IEDI is that it encourages a business to limit its environmental risk considerably.

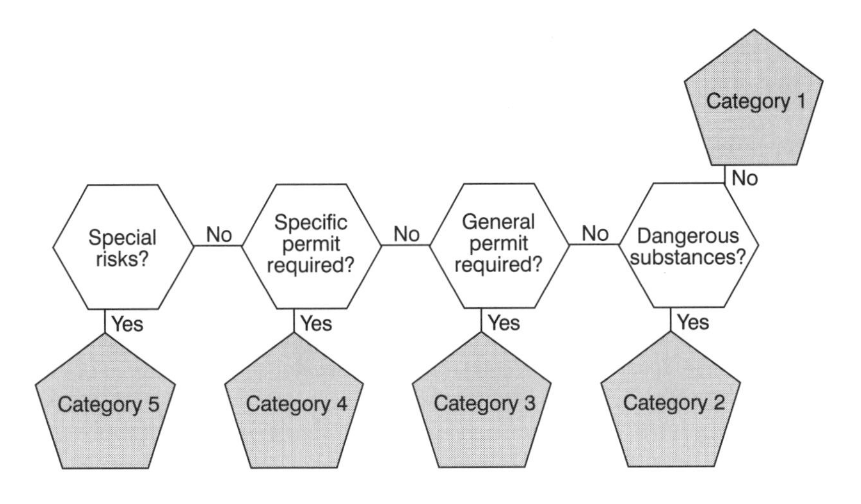

Figure 5.2 *Risk selection diagram IEDI*

NEW, MORE SUSTAINABLE FINANCIAL PRODUCTS: INVESTMENTS

Introduction

Where the section above dealt with the supply of sustainable money by retail and corporate clients, this section deals with the demand of retail and corporate clients for sustainable money. Specific products can be developed, such as environmental loans and leases. Banks might also use financial innovation or set up innovation funds for financing the more innovative segment of the sustainability market. Obviously, a bank does not need to attract specific money for these markets first by vehicles such as SRI. The regular capital is perfectly suited as long as normal financing conditions and risks apply (see Chapter 6). Sustainability is opening up whole new markets for financing, for which sometimes specific knowledge or simply attention is needed.

One such market is sustainable building. This growing new market is making a positive contribution to a sustainable society. Sustainable energy is another rapidly growing market.[24] This includes wind energy, solar energy, hydro-power, nuclear fusion and biomass. There are many innovative ideas for how to generate sustainable energy, such as the 'Archimedes water swing project' in Portugal (involving the conversion of the continuous movement of ocean waves into electricity by means of 'swings' 15 metres under water), the 'sun central project' in the Thar desert in India (heating sand with mirrors)[25] and the windmill park at sea in The Netherlands.[26] The Danes too have ambitious plans. In 2030, 50 per cent of the total electricity production should come from wind energy (a jump from 7 per cent in 1998).[27] The German investment bank Dresdner Kleinwort Wasserstein forecasts that global wind power capacity will more than triple over the next five years. All these projects require large amounts of money. Some examples may illustrate this.

According to figures by McIlvaine Company, the worldwide environmental market in 1996 was worth around €725 billion (*Milieumagazine*, April 1997). The ten Middle and Eastern European countries currently wishing to join the EU must invest around €110 billion to bring their environmental care up to the level required by the EU (*Algemeen Dagblad*, May 22 1998). The need for energy infrastructure in emerging markets worldwide is estimated at US$950 billion between 1995 and 2010 of which US$200 billion will be in renewable energy projects; ie annual investments of respectively US$63 billion and US$13 billion (IEA, 1995). For example, BBVA, as agent bank, has secured a US$855 million loan to develop wind farms in central Spain in the beginning of 2001 (*Financial Times*, 2 March 2001). To stimulate sustainable investments in developing countries, some multilateral development banks (MDB) have set up funds to pool capital and invest in sustainability projects in developing countries. Box 5.3 lists some examples. Banks participating in such funds are Deutsche Bank, Rabobank and Triodos Bank.

The environmental market is quite diverse, however, and banks need to know which market segments are likely to grow and how resilient that growth

Box 5.3 Multilateral funds for sustainability in developing countries

Internationally, the need has been recognized to support developing countries in financing the cost of addressing global environmental problems. In this spirit, international funds are established, such as GEF, REEF, SDC and PCF. GEF is a public fund, REEF and SDC are private equity funds and PCF is a mixed fund. These funds are good examples of how to implement environmental care in financing developing countries. However, these funds can play only a small, catalytic role in addressing the problems. Governments of developing countries will have to do their part in creating the institutional and regulatory environment to provide market incentives toward this objective.[28]

Global Environment Facility (GEF)

The GEF was started in 1991 and had drawn capital from 36 nations in excess of US$4.75 billion for a period until 2002. It is a financial mechanism for fostering international cooperation and action to protect the global environment. It funds the additional costs incurred when a development project also addresses environmental concerns through grants and concessional financing. It is especially targeted to climate change, depletion of the Earth's ozone layer, biological diversity, land degradation and international waters. The GEF emphasizes the removal of barriers to the implementation of climate-friendly, commercially viable energy-efficient technologies and energy conservation measures; reducing the cost of prospective low greenhouse gas-emitting technologies that are not yet commercially viable to enhance their commercial viability; and reducing implementation costs of commercial and near-commercial renewable energy technologies. The GEF uses a stakeholder model in which 166 countries participate. The UNDP, UNEP and the World Bank are the GEF's implementing agencies, and develop GEF-funded projects. A wide variety of organizations is then involved to execute the projects, including private institutions, government agencies, NGOs and other multilateral organizations.

Terra Capital Fund

The Terra Capital Fund is a private equity fund of up to US$50 million which was launched by IFC in 1996. The fund invests in commercially viable and environmentally sustainable uses of biodiversity in Latin American countries that have ratified the Convention on Biological Diversity. Target sectors include ecotourism, sustainable agriculture and forestry, and other sectors with projects that will sustainably use or protect biodiversity. The objective is to attract commercial capital into biodiversity-linked projects because such participation in the sustainable use of biodiversity is indispensable for the long-term conservation of biodiversity. The fund encourages the involvement of private enterprises by demonstrating the development of economic value from biodiversity. These stakeholders will then have an incentive to protect these assets in the long run.

Renewable Energy and Energy Efficiency Fund (REEF)

REEF was started in 2000 and is an investment fund of approximately US$175 million which addresses the need for an efficient vehicle to finance small- and medium-sized sustainable energy projects in emerging markets. It's a mixed fund with approximately US$75 million equity from private investors, an optional US$100 million IFC A/B loan and up to a US$30 million concessional financing by GEF. Moreover, additional finance may come from third parties on a project-by-project basis. This fund's structure of different capital forms shortens the project financing cycle. Investments by the REEF can be in the form of equity, partnerships and debt. Approximately 80 per cent of the fund will be invested on a project finance basis. Smaller projects will rely on consumer and equipment financing. About 80 per cent of REEF's invested capital will be in large-scale on-grid (ie connected to electric grids) power projects with renewable energy as its primary source and energy services companies that supply energy conservation and efficiency equipment and services. About 20 per cent will be invested in small-scale on-grid power projects with renewable energy as its primary source, off-grid renewable energy projects and others.

Prototype Carbon Fund (PCF)

The PCF started in 1999 and is an investment fund of US$145 million aimed at obtaining CO_2 credits by investing in CDM and JI projects (see pages 33–4). The PCF has the objective of mitigating climate change and of offering participants a learning-by-doing opportunity. The participants consist of private companies and governments who will receive a pro rata share of the emission reductions, verified and certified in accordance with carbon purchase agreements reached with the respective countries 'hosting' the projects. The emission reductions will be fully consistent with the Kyoto Protocol and the emerging framework for JI and CDM. The PCF is endeavouring to achieve a balanced portfolio both geographically and technologically. Approximately half of the investments will be made in economies-in-transition (demonstrating JI) and half will be made in developing countries (demonstrating CDM). The major emphasis will be placed on renewable energy and energy efficiency projects. PCF is scheduled to terminate in 2012.

Solar Development Group (SDG)

The objective of SDG is to increase the delivery of solar home systems (SHS) and thus bring environmentally clean electricity to rural households in developing countries (off-grid). SDG started in 2001 and has a capitalization of US$50 million, with US$29 million of investment capital devoted to an private equity investment fund (Solar Development Capital) and US$20 million of grant funds devoted to a foundation (Solar Development Foundation). SDG will invest in private sector companies involved in rural, commercially sustainable photovoltaic activities (including the distribution, sale, lease, or financing of photovoltaic SHS) and provide financing to local financial institutions who will service such companies.

will be. As a rule, the market for environmental decontamination is very dependent on government requirements and finances and is, therefore, not stable. Moreover, is it wise for a bank to be exclusively associated with companies who 'clear up rubbish'? PR considerations and a bank's own desire to foster sustainable development make the market for new consumer products and, for example, sustainable energy more interesting options. The German Ökobank, for example, has decided to exclusively finance projects that are beneficial to people and the environment in practical ways. A number of banks, such as Bank of Boston, have set up specific departments or subsidiaries to grow the burgeoning environmental market. Bank of America has gone a big step further by letting it be known throughout the whole company that special attention should be paid to businesses which play a positive environmental role and by its reluctance to do business with companies whose environmental performance is unacceptable.

The annual growth rate of 5 per cent to 25 per cent for the environmental market in its narrowest sense makes it a very attractive market to finance (IBRD, 1991). All sorts of environmental and bad debt risks will, of course, play a role. These risks are in the main the same here as in other areas of lending credit. Chapter 6 will return to this. Besides, the government could act as guarantor for funding or stimulate sustainable investing with subsidies or tax facilities. In new finance markets, new financial instruments should also be applied.

More sustainable loans

The EU is trying to encourage SMEs to invest in environmental measures and so the European Investment Fund (EIF) has established a Growth and Environmental Plan which should make banks loans more accessible to environmental investments by SMEs in the EU.[29] The plan utilizes a network of selected financial intermediaries. For particular environmental investments – determined by the EIF – the EIF acts (free of charge) as guarantor for up to 50 per cent of the total loan. The plan is not open to all financial parties since the EU budget is limited and moreover the number of banks per country that can participate is limited. It is expected that a participating bank will offer the SME lower than usual credit terms. Several banks in Europe offer their customers in the SME sector an EIF loan, such as Alpha Bank, Allied Irish Banks, Bank Austria, Barclays, BBVA, Credit Agricole, Crédit Lyonnais, Deutsche Ausgleichsbank, ING, KBC Bank, Nordea and Royal Bank of Scotland. At the end of 2000, a total of 37 European banks offered their SME clients an EIF loan.[30] Each bank determines for itself how it packages the loan, but it usually consists of the following elements: businesses do not have to pay an upfront fee; the interest is 0.5 per cent to 1 per cent lower than normal; the term varies from three to ten years; the loan is paid off on a straight-line basis or in a lump sum at the end of the term; and the loan is usually between €12,000 and €1 million.

Other banks (and sometimes those offering an EIF loan) have created their own environmental loans, usually by creating a separate fund. In 1997 NatWest set €33 million aside to stimulate business projects that improve the environment by offering loans on favorable terms (NatWest, 1998, p13). Other examples are to be found at Citigroup, Bank of America, Deutsche Bank, HypoVereinsbank, Unicredito Italiano, Rabobank, Credit Suisse and Sumitomo. Unicredito Italiano offers an interesting product that consists of an investigation into the possible cost savings of environmental policy regulations in the SME combined with an environmental loan to finance it (and possibly complemented by an environmental insurance to minimize risks; DalMaso et al, 2001).

More sustainable leasing

A lease is an agreement giving one party (such as a company) the right to use assets belonging to another party (such as a bank) for a specified period and compensation.[31] In return, regular payments are made to the owner. Leasing enables companies to avoid large capital expenditure and is a relatively low risk method of financing for banks, since they retain ownership of the goods. Thus, leasing is an alternative to regular credit financing. When an investment in, for example, wind energy is difficult to finance, leasing can be an option (where the bank will be the owner of the windmill). Also involved here, of course, are issues relating to 'moral hazard' and 'adverse selection'.[32]

The Dutch government has set up tax facilities, such as a tax relief on energy investments (NL-EIA) and the opportunity to voluntarily write off environmental investments (VAMIL). By setting up lease vehicles, major Dutch banks offer their customers the opportunity to optimize their use of these facilities. The projects financed are mainly concerned with energy saving or sustainable energy generation. These facilities offer the government an inexpensive means of achieving its environmental objectives. They involve part of the work being carried out by the banks (not gratis, of course). Such lease constructions rarely occur outside The Netherlands. The Deutsche Bank is a rare example (Deutsche Bank, 2000). Leasing has the potential to become an increasingly interesting financial product though, that will enable the environmental market to further develop.

More sustainable innovation

Various innovative developments, focusing on sustainable development in society, are still in an embryonic process.[33] These are hard to finance. This is related to the fact that as a rule new technology is still not tested in practice, that no market for it exists, that the pay-back period of investments is too long, and the short-term return is (still) too low. Moreover, most innovations are small scale in their early stages.

Banks often shy away from innovative projects: in fact, risk capital is required here. Venture capital funds are sometimes suitable for this, but usually the capital requirements are too small to interest them. Such funds usually deal with a minimum requirement of a few million euro. Moreover, proved or developed technological renovations are usually involved in such funds and not sustainable innovations. Venture capital funds are good for bigger commercially attractive projects involving so much risks that opportunities for regular bank financing are limited. For small-scale innovations that are not yet fully developed another instrument is required.

Small-scale innovations can gain support from 'informal investors'. Such investors form a group which directly finances the innovative company or idea. They bring both capital and knowledge. Such groups are usually made up of ex-business people. They become a partner to the 'inventor'. The investments are not exclusively motivated by expectations of a high return, but also by the pleasure of getting something off the ground. The risks are great: just three out of ten participations are successful.[34] This form of innovation capital usually falls into the lowest category: the invested amounts vary from €25,000 to €250,000 and most of the businesses invested in are start-ups.

In spite of their small scale, most innovations usually have high development costs. They require sufficient markets to recoup their costs, but these markets do not yet exist and their development is hampered by the efforts of their competitors to hinder the successful market launch. Capital requirements are therefore extensive and long term, which deters the informal investor. The venture capital funds are in principle not suited to meet these requirements as well, because the initiative is not yet fully developed.

If a bank wants to operate sustainably, it will be prepared to finance initiatives that foster sustainable development, and be satisfied with a low return in the short term. For a bank operating in this high-risk market, knowledge is essential. Not only must the potential for sustainable development be estimated in advance, but also foreseeable risks and problems. As a rule, a company has several rounds of financing and once this is under way companies don't tend to get off the merry-go-round. In addition, the bank is usually involved in the company's thinking process. A few major international banks have developed venture capital activities for innovative environmental technology projects. These include Deutsche Bank, Fuji Bank, UBS and Rabobank. The latter has specifically allocated funds to an innovation fund (Rabobank, 2000, p7).

More sustainable securitization

Securitization makes it possible to cover previously uninsurable risks on the capital market (alternative risk financing), or to bring forward future (uncertain) cash flows (asset backed securitization). The idea is that with a corporate bond loan an investor on the capital market assumes part of the company's uncertain yield or risk. The company pays not only interest to the investor, but also a risk premium.

Asset backed securitization (or securitization) is the discounting (bringing forward in time) of future income and on the basis of this the attraction of resources or less expensive investment finance which should generate that future income. In short, a more efficient allocation of uncertain future income is reached. A bank can act as an active intermediary in the exchange of uncertainty (for the investor) for certainty (for the company or individual).[35] When a bank engages in securitization, funds are placed within a special purpose company (SPC). This issues investors with debt instruments with which they can generate the future income of the SPC against a particular risk profile.

A new form of securitization may be the trading of environmental risks and opportunities. In general securitization concerns future income. To demonstrate that opportunities for securitization exist for future savings in the field of sustainable development, a case study based on solar panels is outlined in Box 5.4. The figures in this case are taken from the Dutch situation, though the basics will be the same in all countries.

The example clearly shows that financial innovations can encourage sustainable development. The consequence of such financing enabling individuals to install PV systems on their roofs is that PV systems are used on a greater scale. Their production costs fall as demand increases so that economies of scale become achievable. The example applies to new housing developments. By integrating such a securitization with the activities of a project developer or financier so-called zero-energy homes could be created. Securitization could facilitate the large-scale production of solar panels (by eg Shell) in cases where outside funding is required as well. In a market in which supply and demand are

BOX 5.4 PV SECURITIZATION

Energy consumption is a subject in which sustainability issues are essential. There is one sustainable energy source known to the world that, until now, has been little used in a direct way: the sun. Despite the rather miserable weather and a relatively high energy consumption, the amount of solar energy received by The Netherlands annually is 500 times the amount of electricity currently consumed there annually (NOVEM, 1996). There are two forms of solar energy: thermal solar energy and photovoltaic (PV) solar energy.[36] In what follows PV alone is considered. PV is trapped in a vicious circle: it is too expensive because the market is too small (insufficient economies of scale) and the market is too small because PV is too expensive. In this case study the use of securitization as a possible means of stimulating an economic use of PV to make it viable to a broad range of consumers is discussed.

The lengthy pay-back period and the high initial costs deter most individuals and companies from choosing solar panels. Doing so would however lead to enormous savings in electricity costs. The use of solar panels has a significant contribution to make to a sustainable society. As a rule, a family home has an average energy consumption of 3133 kWh per annum (NOVEM, 1996). This example is based on a newly built energy-efficient family home with an average energy consumption of 2000 kWh per annum. To supply this energy requirement entirely from solar energy, a PV system of around 2600 W peak is required. This implies a requirement of a solar panel covering a sun-facing roof surface of 26m^2. The relevant data and assumptions are presented in Table 5.3.

Table 5.3 *Solar energy data*

Aspect	*Price/cost/term*
Solar panel system (PV) at 2600 W peak	€18,000 (average)
Technical lifetime	20 years
Subsidy	max. 50 per cent of total project costs
Maintenance costs over economic lifetime	€0
Regular energy price (excluding VAT)	10 euro-cents per kWh
Energy price with solar energy (excluding purchase)	0 euro-cents per kWh
Remuneration for passing surplus to energy company	10 euro-cents per kWh

The individual home owner is taken to be the energy producer. If the oil price rises, the price of regular energy rises and with it the remuneration by the energy company to the individual. In the most basic form of, as it were, PV securitization, the bank finances without charge the purchase of the PV system – if there are sufficient people involved. Thus these individuals have no interest or payment obligations. The bank creates an SPC to manage the financing and the SPC in fact buys the PV system. The SPC then issues (for a commission) PV securities. The investor buys an interest in the SPC in certain denominations, for example, a notional value of €500.

In return, a contract is entered into with the home owners, in which it is agreed that per household 2000 kWh per annum will be delivered to the electrical grid:[37] the individual is a producer of 2000 kWh electricity per annum. This production capacity goes to the SPC. This means that in normal use the individual pays only for the undercapacity of the PV system at the regular electricity price.[38] This gives the individual an incentive to make (further) energy savings. The investors in the SPC derive their return, as well as diversing their risk, from three factors. Firstly, from a possible rise in the price of oil (and thereby the surplus energy remuneration). Secondly, from a possible rise in the value of the securities by trading them – possibly on the stock exchange. The interest rate, the purchase price of the PV system and the oil price are determining factors for the net discounted value of the PV system, which is the basis for trading. Thirdly, from a proportion of the production capacity of all individuals involved. In the above system variations are, of course, possible. This might include the individual paying part of the installation cost of the PV system. On the other hand, the individual can receive a lump sum remuneration for the energy produced, or a percentage of the income from the electricity production.

out of step, it can be strategic to start production while simultaneously making it possible for individuals to buy the solar panels. Setting up an integrated SPC can facilitate this.

Similar forms of securitization are also conceivable to enable individuals or companies to make energy savings, and for other forms of environmental care within companies and environmental products. Two possibilities may be the securitization of (regular) loans to environmentally friendly projects or the securitization of invested resources held by innovation funds. What is particularly important in securitization is that the transfer of environmental risks to the investor is not always good for the environment. After all, a moral hazard arises in which companies that have shifted their environmental risks to another party with the help of securization will become less careful in how they deal with those risks. It is sensible to ensure that the burden of a risk is felt by the party causing that risk, or a party as close to them as possible. If not, the moral hazard should be covered in another way.

In general, securitization has the following requirements: the existence of risks; objects or activities on a sufficiently large scale; objective data; a standardized method; and a reliable and stable stream of income (cash flow). In addition, the problems of 'moral hazard' and 'adverse selection' (the principle that the greatest risks are attracted to such an arrangement) should be solvable.

SUSTAINABILITY AND ADVICE FROM BANKS

Banks have a vested interest in accumulating knowledge about sectors and developments taking place within them. It enables them to conduct their business profitably, whether it relates to loans, insurances, and sustainable investment funds or the development of an internal environmental care system. SMEs in particular have a latent need for information about environmental aspects that banks can meet. The bank's motivation to meet this need may be commercially related, image-related, or commitment-related. The provision of advice of information can bind new and existing borrowers to the bank and reduce borrowers' credit risk. The bank may offer its knowledge about environmental subsidies, environmental law and changes to it, possible new markets and environmental management. Various banks play such a role. Some do it with the help of consulting agencies. In certain countries banks should however beware that the advice – combined with financing – is not considered as 'exercising control or a material influence' in courts; banks could be held directly liable for environmental damage caused by its clients (see Chapter 6). Below, some examples are listed of advisory bank services.

Deutsche Bank has introduced a service for SMEs. This involves a databank with data about various branches and environmental aspects related to them. It has data on around 10,000 environmental technologies, and around 2500 consulting agencies in the field of environmental management and environmental care. It appeared that one third of the SMEs were having great difficulty finding such information. The bank also offers advice (*Retail Banker International*, August 3 1992). The ING started a similar service in 1998 with their SME Point. As well as advising local clients about, for example, tax and legislation relating to environmental matters, a clear function of this service is to attract new clients and bind existing ones to the bank. In 1996 The Co-operative Bank in England invested around €1.25 million in setting up an advice centre, together with four universities, to offer sound, affordable environmental advice and consultancy services to its customers (*Business and the Environment*, June 1996).

In addition, various banks produce research reports and brochures about environmental management and internal environmental care. Thus NatWest has produced several booklets on energy-saving techniques and has, together with the WWF, produced its Better Business Pack for the SME, offering practical ways to save costs (NatWest, 1995, 1998). The ING Bank has brought out a brochure with a step-by-step environmental plan to enable businesses to make savings (ING Bank, 1992). ABN AMRO published reports with practical ways to introduce environmental care. The first report focused on opportunities for recycling. The second report included various methods of making savings and data about environmental aspects, environmental legislation and environmental subsidies (ABN AMRO, 1989, 1990). All these reports and brochures are primarily aimed at SMEs, which themselves tend to have little time or money to investigate such areas. These businesses are usually unfamiliar with issues related to environmental management and environmental care, although this is changing. Furthermore, banks use seminars and workshops to inform and advise their customers.

One way of combining advice with a product is a 'guarantee arrangement' for SMEs. A bank can go further than offering advice as a means of commitment and support. It can offer a business a tailor-made service as well. By way of illustration, Box 5.5 lays out the basic features of a guarantee arrangement. Not a single bank is known to employ this arrangement or a form of it.

Box 5.5 Guarantee arrangement for SMEs

A guarantee arrangement can be the factor that convinces SMEs to undertake preventive measures to protect the environment. In this case, the bank acts as guarantor for the costs of quick and/or in-depth studies into just how achievable preventative environmental measures would be. Three main parties are involved: the business, the research/advice agency, and the bank. The bank is the key figure since it offers a product that enables businesses to investigate the potential benefits of introducing environmental measures for little cost or risk. A guarantee arrangement has five phases: a registration period, two research phases and two decision-making phases for the business. In the first research phase a quick scan is made to assess whether profitable environmental measures would be feasible. If they are found not to be so, the business gets its registration fee back. In the subsequent two phases, many variations are possible. The quick scan is however the essence of the arrangement. For the bank, the guarantee arrangement is, in fact, a combination of risk capital and provision activities. In addition, it serves to stimulate preventative environmental care within SMEs, and this can also reduce the credit risk for the bank.

The financial sector and the Kyoto Protocol

Introduction

According to the IPCC it is now an incontrovertible fact that people are influencing the global climate through the emission of greenhouse gases. Since this can have serious consequences, world leaders gathered in Kyoto to establish global objectives and define instrumentation to tackle the problem (see pages 32–4). The consequences of the Kyoto Protocol for trade and industry can be reduced to this: CO_2 emissions will negatively influence the cash flow of businesses while excess emission rights may bring in additional revenue.[39] The mechanism is depicted in the first part of Figure 5.3. National governments will translate the Kyoto objectives for businesses. They will do this by means of concrete regulations and market instruments. The former will obligate businesses to reduce CO_2 emissions, the latter will give them an incentive to do so. Businesses can respond to the obligations and incentives by directly reducing emissions (ie by implementing technical measures) or by buying emission rights from market players, or by investing in other companies which then deliver such rights (see pages 33–4 for an explanation of such indirect measures or the Kyoto instruments). An optimal business strategy will result from the cost-effective weighing up of the costs and benefits of direct and indirect measures.

The implications for the financial sector are directly related to the question of what the above means for the cash flows and balance sheets of (industrial) businesses and the overall impact of Kyoto. The overall impact follows from figures estimating the global annual turnover of the CO_2 market at US$35 billion to US$1.2 trillion (Streiff, 2000). This is obviously a market of interest to banks as well. Company cash flows are subject to three influences:

- cost side: most businesses will sooner or later be effected by climate policy, which will bring increasing costs for tackling the problem with it. This burden will vary per country, sector and according to the business's possibilities for CO_2 reduction;
- revenue side: businesses can generate extra revenue if they produce emission credits or have surplus emission rights to sell. A similar surplus will develop if the marginal costs of tackling the problem are lower than the (anticipated) market price of emission credits (Hugenschmidt et al, 2001, p52–53);
- climate risks: a separate category is the damage to businesses of climate change itself.

In Figure 5.3 the cost and revenue aspects mentioned above are represented in a diagram and thereafter translated into the opportunities and risks they pose for banks. In the following subparagraphs each opportunity and risk to banks (and insurers) is elaborated upon further.

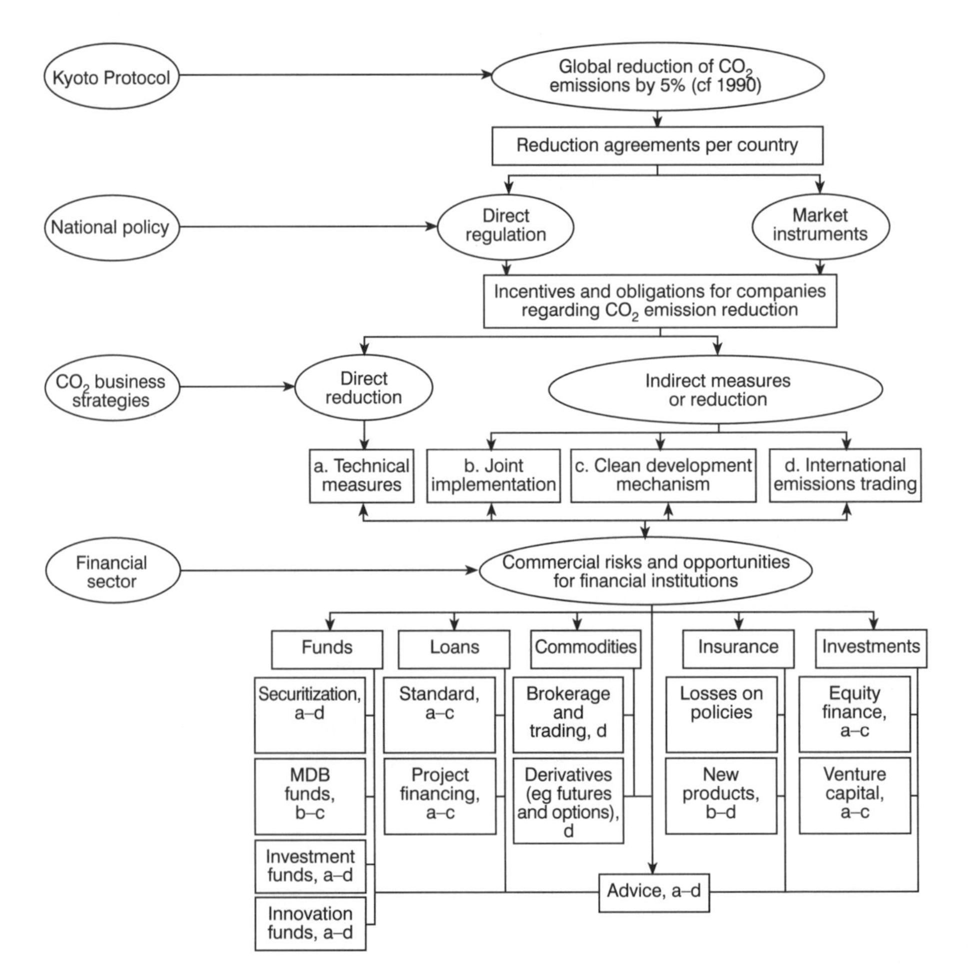

Figure 5.3 *Opportunities and risks of climate policy for banks*[40]

Formation of a fund

Investing in JI and/or CDM involves various risks, in particular country risks, but also specific economic, technological and political risks relating to developments in climate policy. These risks are greater than for regular investments in developing countries or countries in transition. For example, the institutions entailed in JI and CDM have still not been established. Diversifying these risks by forming a fund can offer a solution. The World Bank has, for example, set up the Prototype Carbon Fund with participants from the public and private sectors. Deutsche Bank and Rabobank are the only banks participating and are thereby gaining valuable early experience in JI and CDM projects.[41] Banks can themselves establish a fund and offer it as an investment product to retail or institutional investors, industry and governments as well. Various banks are close to launching similar funds, such as UBS and Credit Lyonnais. The amounts that have been given for these funds range from €65 to €400 million. Dexia, meanwhile, has established a fund in collaboration with EBRD that is exclusively directed at JI projects in Eastern Europe (volume €61 million).[42] Another possibility is for advisorial services; ING offers an advice product for JI projects.[43]

The fundamentals of such funds are simple: the funds invest in JI, CDM and/or clean technology projects (categories A–C in Figure 5.3) and pays out a financial and/or CO_2 return to investors. The certified CO_2 reductions received can be converted into money via IET (and so, for example, yield extra cash flow) or used by the industrial investors in the funds to meet their own CO_2 targets. One of the most important challenges will be the selection of efficient JI/CDM portfolios which will, for example, involve an analysis of country risks and the risk correlation between various projects.

The advantages for businesses of participating in a fund instead of themselves setting up investment projects are: a lower risk as a result of diversification, lower transaction costs and better cash flows resulting from economies of scale. Such funds enable small businesses to profit from the Kyoto mechanism as well. Banks participating early on will develop valuable knowledge and may enhance their image. In addition, the euros invested early on in JI and CDM are expected to have a higher return than euros invested subsequently.

As well as investment funds, banks can also develop innovation funds. The structure of these could be similar to that of the above-mentioned investment funds and focus on groundbreaking innovations in the fields of clean technology and sustainable energy. In its ultimate form this involves a bank raising its own capital in the form of venture capital (see 'Investments', below). Finally, banks could be active in the securitization of their own JI and CDM investments or securitizations related to IET.

Loans

Technical measures to reduce CO_2 usually require additional capital. Banks can fulfill these needs; the Kyoto Protocol will enlarge the market. Then again, banks should be able to estimate the risks involved. How far do the proposed investments go in meeting the Kyoto requirements? To what extent can the business financially bear the necessary investments? To what extent will the business be damaged by climate changes? Knowledge of the Kyoto ruling, the possibility of its being tightened up, and the consequences of climate change are essential to answering such questions. In addition, the initial division under an IET system can have consequences for business borrowers.[44]

In the light of the Kyoto Protocol both threats and opportunities emerge for project financing. For both new and existing projects costs can be reduced and cash flows improved through the optimal use of the Kyoto instruments. If the company's own marginal costs for tackling pollution are higher than those of projects abroad or the market price of reduction certificates, emission reduction can be facilitated by investment in emission reduction via JI or CDM projects or the buying of emission credits via IET. It is this strategy that will probably be most widely used to reduce the costs of existing projects, given that new projects, as a rule, are based on the most advanced available technology. Then again, new projects can be initiated or existing projects can be upgraded so that a surplus of emission credits develops, which can be sold to increase a project's cash flow. Of course, each new project should nominally take into account compliance with emission reduction objectives, their associated costs, the opportunities to generate extra cash flow and the risks of climate change for the project.

Investments

Climate policy will have a major influence on a number of sectors such as energy, chemicals and transport. The anticipated cash flows of companies active in these sectors will be influenced by the extent to which they are successful in implementing cost-efficient and value-adding adaptation strategies. Equity analysis will be strongly influenced by this. The big question over equity investments by banks is: How successful will a company be in creating emission reduction strategies and marketable emission rights or credits? Naturally, with equity investments as well a bank must be able to accurately assess the risks climate policy and climate change pose for a particular company or project.

Venture capital activities may also be influenced by climate policy. It is, after all, more interesting to invest in (energy) companies with low or zero CO_2 emissions, companies producing energy-efficient buildings or equipment and companies creating (groundbreaking) new technologies or production processes that sharply reduce or eliminate CO_2 emissions. It is also possible to use venture capital to help finance a company's changeover to production methods using fuels that produce less CO_2, like gas instead of oil. Such financing can partly be made possible by cashing in (in developing countries) or trading the emission rights or credits this yields. In this way, the financing of sustainable energy can be given a 'push' and becomes commercially interesting.

Commodities

In a system using IET, companies will trade bilaterally or via an intermediary (at their own expense through trading, or purely as an intermediary through brokerage). Companies already active in this are Cantor Fitzgerald, EcoSecurities and NatSource (all from the US).[45] Traders provide the liquidity in the market, while brokers fulfil an informative function. Banks can themselves take on such a role or act as advisor to companies. In addition to spot markets, futures markets will develop. Banks with an investment banking division may want to get involved in these areas in the functions of broker or trader or by developing and/or offering futures products. Companies will be unsure about the future prices of emissions rights, especially in the first stage of IET; this will create interest in derivatives. In the US some experience of financial innovations has already been built up through the national SO_2 policy. This policy is a form of IET, but related to SO_2 and restricted to the US. The financial sector developed 'forwards', swaps, options and more hybrid forms of options ('collars' and 'strangles') to cater for the trade that accompanied this policy. There could also be a need for a futures market for CO_2 (see for example ETEI, 1999). An already strong market is the market for weather derivatives. A market arising out of the ever greater weather extremes caused by climate change can be a substantial growth market.

The Sydney Futures Exchange (SFE) is the first stock exchange on which it is possible to trade in CO_2 reductions. SFE focuses exclusively on sinks and CO_2 reduction rights which can arise from them. The majority of investment opportunities concern forestry projects since 1990. On the Exchange, options are traded which, for example, give the right to buy or sell CO_2 reduction rights in the period 2008–2012.[46] In 1999 SBF Paris Bourse carried out a simulation for CO_2 emissions by electricity companies. As yet there are no concrete plans for the Exchange to become active in this area. The US exchanges are still extremely sceptical about whether it is technically possible to start a futures market before a number of technical and administrative requirements within the Kyoto Protocol have been fully worked out. Nevertheless, the Chicago Board of Trade, for one, is highly interested. Liquidity, the standardization of contracts and the form of the commodity (CO_2 emissions) are as yet insufficient or insufficiently developed to make the functioning of an exchange a short-term possibility. Trade in futures will remain an 'over-the-counter' market for some time to come.

Insurance

Climate problems are a double-edged sword for insurers, posing as they do both opportunities and threats. While this is also true for banks, the risks for insurers are more immediate. After all, there is a direct relationship between the risks of climate change and damage, and life insurance products. Climate change will be accompanied by increasing instability in the weather. The heavy storms from recent years are already regarded by some insurers as a portent of climate change. The financial losses are enormous. Research by Munich Re indicates that in the past decade the economic damage caused by natural disasters has increased eightfold and losses suffered by those insured have increased fifteenfold. Insurers will also be effected by an increase in death and illness (for example, as the malarial regions shift to Europe). Insurers will have to start working to new forecasting methods because these developments mean that it is no longer possible to base risk analyses of climate problems on past occurrences (Knoepfel et al, 1999). Moreover, there is an argument that climate change could lead to greater interdependencies between various risks (ie a systematization of economic risks; Figge, 2001). Systemic risks can not be eliminated by diversification. Better information instruments and higher reserve accumulation would therefore be necessary (ibid, p279).

As well as these threats, climate change offers insurers commercial possibilities. Insurers can, of course, be active in forming funds, as Gerling Re is already doing. In addition, insurers can develop products tailored to the Kyoto mechanisms, for example a liability insurance for the failure to deliver CO_2 reduction rights; through unforeseen circumstances (such as political situations) a company may not be able to fulfill the deal to sell such rights, a risk which the buyer would like to cover. Another example is an emissions credit insurance, a possibility Storebrand is investigating. This innovative insurance product would guarantee the customer that the predicted CO_2 reduction rights in a CDM or JI project would, in fact, be realized.[47]

CONCLUSION

When investors and businesses want to invest in or develop towards sustainable development, banks will strive to develop products that meet their requirements. This chapter has drawn on many examples which show that banks are having successes and, in doing so, are shaping the further development of the move towards sustainability. Both the sources from which they attract investment, and the activities and projects in which they themselves invest, have consequences for sustainable development.

To sum up, various developments are under way that offer banks opportunities and threats related to sustainable development. These can be split into four groups:

1. *Integration with traditional activities across the board:*[48]
   - invest all investment capital in a sustainable way;
   - invest all the bank's own pension portfolio sustainably;
   - finance sustainable projects using the bank's general funds;
   - finance sustainability against favourable tariffs (with or without state support); and
   - finance small-scale sustainable initiatives in developing countries (microcredit).
2. *Linking a financial project to a good cause by way of donations:*
   - traditional banking and donation of a portion of the revenue to a social and/or environmental cause (without a link to a particular product);
   - specific products for which a portion of the turnover, profit or the premium is denoted to a social and/or environmental cause;
   - debt-for-nature swaps; and

- various sorts of institutions with the aim of stimulating certain developments with money and/or knowledge.

3 *New niche products:*
  - specific products for which the money raised is invested in companies meeting certain environmental and social criteria (relatively or absolutely);
  - specific products focused on the placement of raised capital in environmental technology and other environmental markets;
  - specific insurance products, such as EDI;
  - leasing (of, for example, windmills);
  - innovation funds and stimulating innovation;
  - various forms of securitization (for example PV securitization);
  - personal advice to investors (advice about investing in more sustainable companies);
  - advice and other support to businesses (such as a guarantee arrangement); and
  - specific products to meet requirements created by climate policy and climate change.

4 *Collaboration with government parties and development banks:*
  - fiscal green funds; and
  - energy efficiency and development projects in collaboration with development banks (such as PCF and REEF).

In short, each bank must address the question of whether it will introduce sustainable products and, if so, whether a sustainable version will be made next to each existing regular product, or whether all existing products will be made sustainable. The market for sustainable financial products is becoming ever more professional. The build-up of the information, skills and capacity needed to market specific products might complement or overlap with the capacity, skills and information needed for assessing credit risks.

6

Sustainability and Financing Risks

Introduction

Efforts devoted to sustainable development and environmental management are accompanied by many opportunities and challenges for banks. Due to these efforts every lending operation, financing transaction or equity investment may also involve environmental risks for the bank. Roughly speaking, such risks can develop in three ways. Firstly, they can result from financial risks associated with the client's continuity problems. In this case, a distinction can be made between a reduction in the borrowing client's repayment capacity and a decline in the value of his collateral due to environmental issues. Secondly, such risks can develop due to a bank's direct liability for environmental damage caused by its borrowing clients. Thirdly, there are the risks to the bank's reputation and negative publicity from environmental issues. Obviously, these three risk groups can also arise simultaneously. Box 6.1 gives an overview of the risks that banks generally face. Obviously, banks are in the business of managing risks: walking away is not their objective. For each of these risks the bank determines the likelihood, extent, cost and impact should the risk actually occur. Banks will take on only those risks for which the likelihood of loss can be calculated with some degree of certainty. Risk is, in short, not the same as uncertainty.

As for environmental risks, a borrowing client can encounter continuity problems resulting from changing environmental legislation or changing market conditions. An example is that the market may be demanding or expecting an environmentally responsible production process but the borrowing client is unable to satisfy this demand or expectation and still make a profit. When a bank's client files for bankruptcy, the bank can be confronted with the problem that the client's collateral will have declined in value (eg in the case of asbestos). In the US, there have been situations in which the bank has been held directly liable for the illegal or environmentally harmful activities of its borrower. However, the negative reaction of the banks in general to this liability has forced the toning down of this legislation in the US. The EU has been struggling for decades to determine the optimal degree of lender liability and has taken the lessons learned in the US to heart. In addition to the direct liability of banks, one can also point to the indirect liability of banks.[1]

BOX 6.1 GENERAL RISKS FOR BANKS

Based on Saunders (2000, pp103–115), the following generic risks for banks can be identified:

- Interest rate risk: risks due to a mismatch between the maturity of assets and liabilities (eg refinancing risks, reinvestment risks and market value risks).
- Market risk: the risk incurred in the trading of assets and liabilities due to changes in interest rates, exchange rates and other asset prices.
- Credit risk: the risk that promised cash flows from loans and securities held by banks may not be paid in full; one can distinguish firm-specific from systematic credit risk.
- Off-balance-sheet risk: the risk incurred by a bank due to activities related to contingent assets and liabilities (eg risk of letters of credit or positions in derivatives).
- Foreign exchange risk: the risk that exchange rate changes can affect the value of bank assets and liabilities located abroad.
- Country or sovereign risk: the risk that repayments from foreign borrowers may be interrupted because of interference from foreign governments.
- Liquidity risk: the risk that a sudden surge in liability withdrawals may lead to a bank having to liquidate assets in a very short period of time and at low prices.
- Insolvency risk: the risk that a bank may not have enough capital to offset a sudden decline in the value of its assets relative to its liabilities.
- Operational risk: the risk that existing technology or support systems may malfunction or break down.
- Other risks: general macroeconomic risks, discrete risks and event risks.

Banks can react in various ways to environmental risks. Due to reasons of lender liability, the banks in the US were among the first to establish methods for evaluating such risks. The biggest reason why Europe had been lagging behind the US somewhat in this regard is that it handled liability in a different way. Although few cases are reported in the literature in which environmental risks have actually materialized, many banks have nevertheless devised checklists for the environmental risk involved in lending activities.[2] Environmental information is crucial in this regard. Broadly speaking, banks can reduce risks by rejecting the loan application outright, by adjusting the interest rate or maturity of the loan to reflect the environmental risk involved, or by inserting specific environmental compliance clauses in loan agreements.[3] A bank needs to assess environmental risk in four stages:

1. before the loan is granted;
2. during the term of the loan;
3. at the time of undertaking a financial workout with a defaulting borrower; and
4. at foreclosure and liquidation.

This chapter focuses on the environmental risks involved in lending operations, financing transactions and equity investments. Emphasis is placed on lending operations, however, since this is the core activity of banks.[4] Attention is also devoted to the environmental risks associated with project finance in developing countries. Since the environmental risks for the bank are chiefly a result of the environmental risks run by its borrower, 'Environmental risks for clients', below, provides a brief description of these. Obviously, environmental risks to clients are associated with opportunities for banks as well; companies will want to make use of such products as bank guarantees, environmental loans or environmental insurance products (see Chapter 5). Pages 129–43 address environmental risks for the banks. Indirect environmental risks are discussed in the section that begins on pages 129–35; in practice these will weigh heaviest in most countries and situations. Direct environmental risks (environmental lender liability) regarding environmental damage caused by bank clients are discussed in the section that begins on page 135. The section beginning on page 139 discusses the increasingly relevant risks to reputation involved in financing transactions due to environmental issues (risks from a communications perspective are considered on pages 178–80). Concluding the chapter, we address the instruments and information needed in determining environmental risks for credit application on page 143.

Environmental risks for clients

Introduction

In general, the client's risks are the bank's risks. For this reason, this section will explore the environmental issues that confront businesses. At one time or another, many companies have to deal with clean-up costs, conversion costs (forced on them by the government or by the market), environmental damage liabilities or additional environmental investments. These costs can mount up considerably.[5] The estimating of such costs and risks is already very difficult for entrepreneurs themselves; additionally, they are not always benefited by revealing such information to the bank. This means that it is crucial for the bank itself to acquire a clear picture of the relevant environmental risks. The bank can obtain such information by compiling general information and specific information (eg from specific government regulations) about current, past and future environmental problems in general terms as well as specifically in regard to the credit applicant's situation. The bank can outsource this or collect and assess it itself.

Considering that various sectors and branches of business can be more or less environmentally sensitive, many banks begin by gathering information related to business sectors. Being environmentally sensitive means that a sector's products, the production process itself and the emissions of the production process can be regarded as actually or potentially threatening for the environment. Such environmental risks can become financial risks for a bank. Usually, companies which operate preventively or offensively have lower or more manageable environmental risks.

This section outlines sector-related environmental sensitivity. Other economic aspects such as the market potential of certain branches are also considered (see pages 121–3). Companies are being confronted with a wide variety of legal aspects. Insight into environmental legislation, its enforcement and future changes is therefore essential for the process of extending credit (see pages 123–6). Collateral is an important factor in the lending of credit. Usually, a borrowing company's registered property (land or buildings) can function as bank collateral. But the soil can be so polluted that its value as collateral can even be negative. Due to the relative frequency and the potentially very threatening character of soil pollution for banks, this issue will be dealt with separately (see pages 126–7). But whether a company is confronted negatively by an environmental issue or whether it voluntarily embraces environmental management, it will still have to finance these kinds of activities. A bank, in its role as advisory partner or financing institution, can then point out possible fiscal concessions and subsidy schemes (see pages 127–8).

Economic aspects

Agriculture, fisheries, mineral extraction and industry are the most environmentally sensitive sectors in most countries. The environmental sensitivity of public utilities and the transport and storage sector is rated as moderate, while the environmental sensitivity of the other sectors (construction, trade, services, and governmental/other services) is considered low. But this classification is very rough. At a more specific level, when these sectors are broken down into branches more variety can be seen between the individual branches. Environmentally sensitive branches include chemicals, oil refineries, basic metal manufacturing, graphic companies, the textile industry, intensive stockbreeding, laundries, painting businesses and petrol stations.[6]

Various reasons exist for the differences between branches in regard to their environmental sensitivity:

1. The typical *elements* of a branch: eg, accounting has relatively few environmental problems, while the manufacturing of basic metals involves several different environmental problems.
2. The typical *cultures* of a branch: eg, in the construction industry, implementing environmental management is much more difficult than it is in the provision of business services.
3. The degree of *organization* typical of a branch: the chemical industry is very well organized and therefore finds it relatively easy to implement government requirements at an industry level.

Many branches such as chemicals have already taken major steps, but this is no guarantee of lower risks. Changing government requirements (such as the greening of the tax system) and changing market conditions can demand additional investments or problems. It should be kept in mind that unless the level of technology used is changed, each additional investment for energy savings, for example, will be less and less profitable. Appendix II describes a number of possible sector-related consequences (for The Netherlands) of current and future government policy.

Environmental risks vary considerably among these groups as well. Some companies within a branch are upholding a defensive position while others believe in the benefits of operating sustainably. In addition, sectors or branches that are now being confronted with relatively low environmental risks may become extremely risky in the future due to such factors as changing environmental legislation or changing market conditions. It is also necessary to look to the past. In conjunction with such matters as soil pollution, it is important to investigate which companies operated at a certain location prior to the potential borrowing company. Banks run an added risk when dealing with environmentally sensitive branches of business. These risks are then increased if such branches are sensitive to cyclical trends or are dealing with structural problems.

Yet another distinction to be made is that between SMEs and large-scale industry. Large companies can recruit or train employees especially to handle environmental issues. Often, these large companies can develop staff positions intended to amass the company's own store of knowledge on environmental issues. Their approach to environmental problems is often systematic, which is an important factor in developing a good picture of environmental risks. Such knowledge can also be employed for product innovations and finding market potentials related to society's demands to promote sustainability. In addition, a large company is better positioned to attract equity capital, whereas SMEs are usually completely assigned to bank financing. SMEs also have very little specialized knowledge themselves in regard to legislation or technological and organizational solutions for environmental problems. For this reason, their approach to environmental issues is generally unsystematic and ad hoc. These companies usually just want to comply with the requirements of government; compliance is however usually less cost-effective. Obviously, there will be certain front-runners within the SME class just as there are stragglers and members of the pack; some SMEs may outdistance large companies.[7]

In short, environmental risks will vary from company to company. The attitude of the company and its management (whether doing business defensively, preventively, offensively or sustainably) in regard to sustainability will often serve the bank as an indication of the financial risks in extending this company credit. Increasingly, a business's far-sightedness, perception of the market, knowledge of technological developments, and attitude towards such things as environmental aspects are becoming decisive factors in bank financing and in assessing financial risks.

Essentially, a preventive approach implies a systematic approach to environmental issues by implementing, for example, an environmental management system (EMS, see Chapter 8). It is precisely by applying such a preventive approach that there will be more chances for win-win situations accompanied by lower risks for future damage claims or the need for decontamination activities. With a less forward-looking approach, which is typical of many SMEs, expenditures will often be higher than the income while, *ceteris paribus*, the risks for environmental damage claims will be higher. This factor needs to be considered in the financing. Especially when dealing with SMEs, banks can offer their support by pointing out fiscal and subsidy opportunities or the possibilities and advantages offered by preventive environmental management. After all, to adequately engage in lending operations, banks should be equipped with this knowledge themselves. This information could then be systematically used for the benefit of SMEs. In principle this will lower the financial risks to banks as well. Offensive businesses will usually have a lower degree of risk and also be confronted less quickly with changing demands from the market; after all, they are the ones who are looking ahead by engaging in such activities as product innovation.

Knowledge in regard to sectors and branches labelled as environmentally sensitive and information related to the specific situation of many SMEs can function as a warning signal for extra attentiveness to environmental risks for banks. In addition to environmentally sensitive sectors, there are also sectors such as solar energy that offer potential in regard to promoting the needs of the environment. Here, too, there are risks; these are primarily linked to the innovative level of the company's activities. The business's and bank's estimation of the market is a major determining factor here. In general, the risks of offensive businesses are easier to estimate (in any case, they are easier to discuss) than the risks of defensively operating companies.

Legal aspects

Obviously, government requirements and other legal aspects are of great importance for companies. A bank should have some knowledge of such requirement and the degree to which companies have to comply with them. Such government requirements can have a national, regional or international character. Legal frameworks differ sharply between countries. The toughest environmental laws are found in the developed countries. While countries such as China also have tough environmental laws, the degree to which they are enforced differs from that of developed countries. It would of course be going too far to discuss the legal situation in all countries. For this reason (and reasons relating to The Netherlands's function as a role model, see Chapter 2) the legal aspects for entrepreneurs in The Netherlands are presented here. To finish, attention will be paid briefly to some practices in other countries and regions.

The most important piece of Dutch environmental legislation is the Environmental Management Act (Wm). Since 1993, this has been an integrated law in which an integrated environmental permit can be granted to companies. Dutch environmental policy includes these instruments:[8]

1 Instruments governed by public law:
 1.1 direct regulation (eg laws, permits and regulations);
 1.2 indirect regulation (eg authorizations for law enforcements);
 1.3 financial instruments (eg levies, subsidies and tax facilities); and
 1.4 obligations to provide information (eg Environmental Reporting Act).

2 Instruments governed by private law:
 2.1 liability law (eg strict liability);
 2.2 contract law (eg a clean soil statement); and
 2.3 covenants (eg voluntary agreements between government and industry).

3 Other:[9]
 3.1 public information and education;
 3.2 research and development;
 3.3 exemplary behaviour and exemplary projects; and
 3.4 pseudo-legislation.

From the standpoint of extending credit, the most important elements of government policy affecting companies are permits, liability, the enforcement of government requirements, financial instruments and the Soil Protection Act (Wbb). The last two elements will be discussed in two separate sections while the first three will be addressed here.

Permits are extremely important for businesses. A permit is granted by a government agency that is directed to an individual case and subject to appeal. Having been granted a permit, operations or activities that are not permitted without a permit become permissible. Permits are granted when an institution is established, expanded or changed. A permit can be granted subject to restrictions, temporarily, or can even be denied. The government has an 'updating obligation' in regard to permits. This means that at stipulated times, the proper authorities have to reconsider the permit as based on the state of technology at that time. At these times, a revision can be made or the permit can even be retracted. In cases in which it is impossible to prevent affecting the environment adversely, decisions are based on the ALARA principle.[10] According to this principle, measures should be taken that limit adverse effects as much as is reasonably possible.

The government can take administrative action against a business if it does not have the required permit or is not complying with the regulations. This means that the government can apply legal sanctions – administrative enforcement, enforced closure, imposing a default fine or retracting the permit – to force the business to comply with the rules. A default fine is the most commonly applied sanction. The businessperson can also be criminally prosecuted (punishments include custodial or prison sentences, fines, denial of certain rights and closure of the business).

Although a business might still have all the required environmental permits, it is still subject to environmental liability. Environmental liability relates to the damaging effects to the environment caused by certain acts or omissions. Tonnaer (1994) classifies two basic forms of environmental damage:

1 Damage to nature as a universal legacy ('ecological damage').
2 Environmental damage coinciding with personal damages:
  2.1 damage to environmental properties or rights derived thereof;
  2.2 damage resulting from environmental nuisances (immaterial damage); and
  2.3 damage due to dangerous situations or hazardous substances.

Due to its collective causes, ecological damage is difficult to accommodate under private liability and will usually be expressed in public liability. This form will not be handled here in any greater detail.[11] The second form of damage (hereinafter referred to simply as 'environmental damage') concerns damage that individual citizens or identifiable groups of citizens inflict on one another; in this case, the environment occupies only an intermediary position. It is just this ability to identify both the perpetrating party as well as the injured party that makes private liability possible. A company can be held liable for non-ecological environmental damage in three ways:

1 by committing an illegal act (eg by polluting its neighbour's soil);
2 by means of strict liability;[12]
3 by means of the government's right of action based on the Wbb (see pages 126–7).

The perpetrator can be attributed with an illegal act in case of his guilt or in case of 'a cause, which by virtue of the law or commonly held opinion, he is liable' (according to the Dutch Civil Code). This latter formulation makes faultless liability possible ('strict liability'). The simple fact that this risk is becoming real in those cases in which strict liability applies is enough for a business to be held liable regardless of whether it has been engaged in illegal or negligent acts. In short, in the case of strict liability, a business can be held liable more easily. This must then entail foreseeable damage in which the knowledge of the average person is assumed and not that of the specific perpetrator; culpability – not legal guilt – is the key issue in strict liability.[13] In a case of 'pure' strict liability, no illegal act or guilt has to be proven and theoretically, culpability plays no role; a causal link between damage and a hazardous substance is theoretically enough to hold a businessperson liable (Dunné, 1992).

Strict liability in regard to hazardous substances such as asbestos, dioxins and heavy metals is very unlike the strict liability related to products liability (Article 5:185 of the Dutch Civil Code). In product liability, liability also involves 'sticky liability': ie the liability lies with the party that put the substance into circulation. However, for hazardous substances strict liability is segmented: essentially, the liability rests with the link in the production chain where the substance or object was found at the time the damage occurred. Essentially, this means that although this also applies to such companies as hauliers, it may apply in a direct sense to banks as well.

The legislation relating to environmental liability is extremely complex and will only become more complicated and more stringent in the future.[14] For this reason, a business will increasingly have to call in external consultants to indicate the risks it is running and any necessity for environmental insurance.[15] A study conducted in The Netherlands in 1998 into the environmental liability experiences of companies showed that over 80 per cent of the 280 companies asked stated that their company policy took possible environmental damage into consideration. This percentage is somewhat higher (almost 90 per cent) among industry and large companies than within other sectors and for SMEs (where it is slightly less than 80 per cent). About 25 per cent of those surveyed (chiefly industrials and large companies) had been held liable for environmental damage during the period 1993–1998. The government is by far the most frequent plaintiff in these cases followed at a distance by citizens and other companies. Only 1 per cent of the cases involved an environmental organization as a plaintiff. About half of the liability cases involve costs less than €25,000; a quarter of the cases involved costs more than €250,000. Around 60 per cent of the companies have no insurance to cover environmental liability risks (VNO-NCW, 1998).

In the UK, the Environmental Protection Act 1990 introduced the concept of integrated pollution control, a concept that is also to be found in environmental policy in The Netherlands and which recognizes that the pollution in a particular environmental compartment (for example, air) can have consequences for another environmental compartment (for example, water). Permits should thus be holistic in nature to take account of such interdependencies. The UK and The Netherlands are clearly ahead in this approach. In 1999 it was laid down in EU law (integrated pollution prevention and control, IPPC).[16] IPPC will principally impact on the southern European Member States (where environmental policy was less stringent), but will also result in a larger number of companies requiring an environmental permit in order to operate legally. These companies will in principal have to take extra investment measures to meet IPPC requirements.

In North America and Australia an extraordinary complement has been added to the role of government policy as a source of information. In the US there exists a toxic release inventory (TRI) and in Canada and Australia a national pollutant release inventory (NPRI). Anyone in Canada who, for example, manufactures, processes or otherwise uses any of the NPRI-listed substances in quantities of ten tonnes or more per year, and who employs ten or more people per year, must report releases or transfers in wastes of the substances.[17] In some European countries similar databases are being developed (eg Switzerland and Denmark). Based on a right to know, these databanks provide citizens with access to pollutant release information for facilities located in their communities, so that communities have more power to hold companies accountable.

Soil pollution

Liability is very important to companies, especially when it comes to soil pollution. A distinction can be made between polluting one's own site and polluting another's site. In the latter case, legal precedents tend toward strict liability. In the former case, there has to be an indication of an illegal act with respect to the state. This involves a liability in the normal sense of the word, ie illegal activities (with respect to the state). Soil pollution may stem from dumping in the past, gradual pollution or industrial accidents. The costs for soil remediation in the US are estimated to be around US$500 billion; for The Netherlands, this amount is estimated at US$20 billion (for function-oriented soil remediation) and US$200 billion (when the soil is decontaminated at a sustainable level).[18] In The Netherlands, tens of thousands of sites are suspected of being in need of soil remediation, with costs ranging from €5000 to several million Euros per site. The Dutch system is stringent and detailed and is used by many countries around the world as a benchmark or source of information (Case, 1999, p116).

The Netherlands has had a Soil Protection Act (Wbb) in force since 1987. This means that as far as soil remediation is concerned, a distinction should be made between the pollution that was caused before 1987 and afterwards. When the Wbb came into force, it included the obligation to eliminate all sources of pollution coming into existence from that point onwards (Article 13). This liability is based on the idea that prevention is better than remediation, both for the environment and in terms of costs (soil remediation measures can be 50 to 100 times as expensive as preventive measures). The seriousness of the pollution is not a factor within the Wbb: every case of pollution that developed since 1987 has to be eliminated entirely. Trigger and target values have been established for a wide range of contaminants. If the pollution reaches the trigger value (ie is labelled as 'serious' pollution), clean-up is required up to the specific target level.

The Dutch government has selected around 200,000 industrial sites where the chance of finding soil pollution is high. Based on this list, BSB organizations are asking companies to have their soils tested voluntarily and if necessary to have them decontaminated.[19] Companies that do not participate voluntarily in the BSB activities (around 7 per cent) and which are suspected of being involved in serious cases of pollution can be confronted with a soil testing or soil remediation order imposed by the government. It is also possible that the government itself will have the soil remediation conducted (if the possibilities of issuing a soil remediation order are limited), after which the owner of the land will be charged for these services. In the case of soil pollution that was caused before 1975, the government has no right of recourse unless 'seriously culpable activities' are suspected, but it can still impose a soil remediation order. In exceptional cases, third parties such as banks can also be involved in the government's right of recourse if the government decides to have the soil remediation conducted itself. Such a case involves 'unjustified enrichment' by a bank since the cleaner soil will have a higher economic value (Article 75, Paragraph 4, Wbb).

Other countries have similar measures for soil remediation. There are important differences in approach related to whether the soil pollution is new or historic, the extent of private or collective soil remediation and the distribution of responsibilities when polluted ground is sold. In the US there exists private liability for soil pollution originating in the past, while Denmark has made the remediation of such soil the job of the community with the costs being paid by the government (and thus indirectly by tax payers). Various countries such as the US and Hungary have investment funds for past soil remediation (Superfund and Central Fund for Environmental Protection respectively). Germany has the *Gemeinlastprinzip*, a principle which means that the community pays the costs of remediation for existing and new soil contamination. The UK, by contrast, has private liability for soil remediation for 'the appropriate person who knowingly permitted polluting substance in, on or under the land' (Case, 1999, p43). In many countries, including the UK, the principle of *caveat emptor* (let the buyer beware) applies, meaning that purchasers accept full responsibility for the ground they buy. Some countries, however, have implemented protective legislative constructions for buyers. For example, a ground purchaser in Italy can rescind the purchase or insist on a discount if it appears that defects exist which reduce the value of the ground. Germany has a statutory warranty for the sale of land and in The Netherlands and France the seller of land has disclosure duties (Case, 1999, p130). Obviously, these differences impact on the position of banks and the need for due diligence.

Financial aspects of government policy

For businesses, the financial instruments of the government are as important as environment permits and environmental liabilities. Here, a distinction can be made between levies, subsidies, tax facilities and administrative compensation.[20] By implementing these instruments, the government is trying to apply financial incentives to restrain certain business activities by means of levies or to stimulate others by means of subsidies. In general, levies are more efficient and cost-effective for companies than using physical regulation to achieve the same environmental objective.[21] Emission reduction takes place at lower costs and there is also an incentive for technological development. The concept of levies is based on the 'polluter pays' principle. An example is a waste materials levy. The incentive to innovate is also high for marketable emission rights. The US acquired much experience in this regard in the 1980s and the concept is being copied in other countries as well. With this instrument, every business receives the right to emit a certain quantity of a certain substance. Instead of an obligation to pay, there is a right to pollute. Businesses then have the potential to develop technology that will sharply reduce their emissions and possibly to finance this with the sale of 'excess' emission reductions.[22] Research has shown that this instrument is extremely economically effective and efficient (high environmental gains at relatively the lowest costs, eg Tietenberg, 1985). This also applies to covenants, a non-financial government instrument with which The Netherlands has acquired much experience. Internationally, there is a great deal of interest in this instrument since government objectives are reached at relatively low costs and private industry also retains a relatively high degree of freedom in determining the way to reach the self-established objectives.

On the other side, the government has various subsidy schemes for stimulating environmentally responsible behaviour. Subsidies can involve the development and application of environmental technology, research into recycling waste materials, stimulating environmentally responsible applications, and reducing the use of environmentally harmful raw materials and additives. In addition to or instead of subsidies, the government can make use of tax facilities to stimulate environmentally responsible behaviour. A Dutch example is the Voluntary Depreciation of Environmental Investments Scheme (VAMIL). In this scheme, the costs of purchasing certain equipment (indicated on a ministerial list) can be arbitrarily (eg accelerated) depreciated. This can mean a temporary reduction in the taxable profit and thus in the corporation or income tax. Accelerated depreciation also provides a business with liquidity and interest benefits. The fiscal green regulation discussed in Chapter 5 is yet another example of a tax facility. For soil remediation carried out within SMEs, the Dutch government has devised the SMEs credit guarantee scheme. In this scheme, the government guarantees 90 per cent of the bank financing during the credit term.

These examples of levies, subsidies and facilities also exist in one form or another in other countries but it would be going too far to discuss all of them here. The purpose of this section is to provide a general idea of what borrowing companies can get involved in and what banks may want to take advantage of.

INDIRECT ENVIRONMENTAL RISKS FOR BANKS

Introduction

The environmental risks run by banks can be split into roughly three categories: direct liability (see pages 135–9), the risks to the bank's reputation and negative publicity (see pages 139–43 and 178–80) and the consequences of the environmental risks run by borrowing clients (this section). Based on the broad discussion on pages 120–8, it is possible to make a distinction between credit risks which develop due to environmental risks that endanger the continuity of the borrowing client, and credit risks that result from infringements to the bank's collateral. The bank should estimate the financial capacity of a potential client to finance its environmental risks (eg loss of markets, damage to reputation, higher production costs and environmental disasters) and the investments involved in preventive environmental management. In short, will the borrowing client's repayment capacity be hindered or possibly endangered by environmental risks? Having accurate environmental information and risk systems is essential for the bank in this regard; this is discussed on pages 143–6.

Particular care should be taken in examining the environmental sensitivity of new clients or newly founded companies. In regard to existing clients, the potential for environmental risks should be evaluated periodically along with the possible effects of these risks on the company's continuity and on the value of the collateral. After all, the client is dealing with continually changing market conditions and government requirements.

Pages 129–34 will emphasize repayment capacity. Obviously, aspects such as solvency, liquidity, the quality of the management (sound business practice), market developments and collateral are also factors in the ultimate extension of credit. Environmental risks also play a role in collateral (see page 134). The concluding section addresses the parallel that exists in regard to the information needed in the acceptance process for an environmental insurance policy and the process of extending credit. This involves 'Chinese walls', of course, but there can be an exchange of generic knowledge and experience in regard to effective environmental risk analyses and systems or new needs exhibited by client groups.

The borrowing client's continuity

Environmental aspects can threaten a company's continuity in six different ways, any of which can endanger the borrowing client's repayment capacity:

1. changing government requirements;
2. changing market environment;
3. changing external environmental conditions;
4. private liability;
5. government sanctions; and
6. criminal prosecution.

The first two aspects are more economic in nature, eg due to environmental aspects, the lifespans of production facilities or products can be reduced, thus leading to faster depreciation and replacement. If the company's financial position does not permit this expenditure, it can become a banking problem. When assessing credit risks, more is involved than looking at future developments; in regard to soil pollution, for example, the past is another important factor. The last three points apply to the legal side of the client's environmental risks which can lead to financial risks to the bank.

Government requirements

Permits are granted when an institution is established, expanded or changed. On any of these occasions (and also, for example, when the government engages in its obligatory updating of the permit), there is the possibility that the company will not receive any permit or that the permit will be retracted because the company's environmental impact does not fulfil the existing or modified government requirements for a permit. Both can seriously endanger the company's continuity. For this reason, a bank should always ascertain whether a company is obliged to have a permit, whether it actually has the necessary permit(s), whether it has had permit related problems in the past, and whether it will be able to keep them in the future. After all, the one does not automatically follow the other. In The Netherlands, 31 per cent of the companies obliged to have a permit are failing either entirely or partially to comply with the regulations related to these permits (Moret Ernst & Young, 1998). Naturally, for a bank it is important in estimating its environmental risks to determine how strict the government's enforcement is. In most developing and some developed countries this has been shown to be not that strict in the sense that it actually leads to losses to the bank (ie usually a company receives so much time to correct the problems that the bank can phase out the credit relationship if desired).

If the company has to invest extra funds to obtain its environmental permit or to satisfy the regulations involved, will these investments be sufficiently profitable or is the company's financial position such as to allow making such investments? Box 6.2 provides a real case (given anonymously) of the extension of credit to a company that was being confronted with continuity problems due to changing government requirements.

As a rule, banks only provide financing for profitable investments. Theoretically, soil remediation or other forms of decontamination are not, in and of themselves, profitable investments.[23] This also applies to environmental levies. Preventive environmental management measures or investments that result in lower levies, however, are profitable in theory (eg due to lower energy costs). Taxes on pollution will continue to rise in most countries in coming years. Even now, for example, several countries in Europe are experiencing a greening of their tax systems in which on the whole rising levies on the environment are taking place. Moreover, the Kyoto Protocol will impose costs on emitting CO_2 and create possibilities for additional returns for efficient and innovative CO_2 strategies. In short, what is questionable is whether or not the borrowing client has the financial capacity to bear these additional costs or to take preventive environmental measures.

Box 6.2 Changing governmental environmental requirements: A case study

A large company produced a certain kind of chemical substance. The company was being financed by a large European bank. At the end of the 1980s, the government clamped down on the process that the company was using to produce this substance due to the increasing environmental damage involved. Instead of a preventive attitude in regard to dealing responsibly with the environment, the company had a defensive attitude. In reaction to the new government requirements, the company then invested more than €150 million in building a new factory. Once the factory was ready, it turned out that the company could no longer produce its product profitably at the world market prices then in effect. Due to this convergence of circumstances (price pressure combined with changing environmental requirements of the government), the company instituted insolvency proceedings in the early 1990s. The company was then acquired by another company. In the end, the bank lost the majority of the outstanding loan to the tune of approximately €50 million.

In banking circles one sometimes hears that the risks associated with changing government policy will probably not be all that bad. Usually, new government requirements apply only to new situations and not to the bank's existing borrowers. If they do involve existing clients, the time frame is usually generous enough to allow the bank time to withdraw or for its client to comply. In addition, government enforcement is too lax. 'Just keep them on' is an oft-repeated saying in banking circles considering that in terms of a bank's marketing strategy, deserting one client is not always the best remedy either (as it may backfire on potential new clients). The danger, however, is that this idea will be extrapolated into the future and will be used to cover all kinds of situations. In the meantime toleration is being less and less accepted while enforcement is becoming harder and more exacting. Increasingly, companies will be forced by means of private law as well. The consequences of the burgeoning environmental risks can thus occur at a faster pace. This makes it harder for banks to gradually withdraw from a financing arrangement. Secondly, differences exist between jurisdictions. One jurisdiction will apply a more stringent environmental policy and enforcement than another. Thirdly, the bank has to deal with risks related to its reputation. Certainly after the signing of the UNEP Declaration, for example (see Chapter 8), 'just keeping them on' may not be advisable. After all, the bank can suffer a serious loss of reputation. Especially at the local level and for small-scale reputation risks, the resulting damage to its image can have serious adverse effects on the bank.

Market conditions

Besides the governmental influences on profitability and the risks related to investment decisions, another factor in recent years has been market self-control. For example, if competitors introduce more environmentally responsible products (and offer them at comparable or lower prices), added risks develop for a borrowing company that does not have such products itself. Regardless of the activities of competitors, this same concept also applies to changing consumer patterns. The loss of sales will also occur when the borrowing company is a supplier and the buyer starts demanding more environmentally friendly products or production processes (Philips increasingly selects its suppliers, for example, based on how well they meet its own environmental objectives); these could be called chain risks. In this case, repayment capacity can be seriously endangered. Items in stock, too, can decline in value due to such developments. In issues directly related to health, what can result is a domino effect. An example of this can be seen in the BSE crisis in Europe in which meat consumption among private individuals in most countries has been reduced by double-digit percentages, even when there was no evidence of a BSE problem in their own countries. The KBC Bank, for example, has had to make provision for around €44 million (KBC, 2000, p29) in regard to the dioxins crisis in Belgium (dioxins getting into the human food chain through livestock feed) and due to the bank's large market share in the financing of agriculture and its affiliated sectors. Box 6.3 provides a real case (given anonymously) of materializing environmental risk for a bank due to the insolvency of a client caused by market changes.

External environmental conditions

There are also environmental risks related to conditions outside the company which can affect the continuity of the company and which are related to environmental issues. A good example is the climate changes that will result from the intensified greenhouse effect. Agricultural areas in the US could

Box 6.3 Environmental risks and market changes: A case study

For a company in the business of industrial photocopying a balance of €100 million was financed with 40 per cent share capital and 60 per cent bank loans. The company's cash flow was sufficient to meet the interest payments and the company had no limitations in attracting new capital because investors accepted the risk profile of the company. At a certain moment, the company's management realized that the growth ratio could not be maintained, given that the demand for products had declined as a result of an increased environmental awareness among customers. Competitors had remained ahead of the company in terms of environmental measures and to protect its market position and ensure its future, the company initiated an investment programme to reduce its environmental impact (financed from reserves). As a result, the lifespan of the present assets was shortened considerably. The accompanying increased depreciation put pressure on the investors' tax-free profit and the book value of the present assets, on the basis of which the bank's financing had been extended. For both investors and banks this constituted such a significant change in the company's risk profile and its free cash flows that the company was no longer in a position to attract new capital to jack up its environmental image and so maintain its market position. Once consumers had turned their backs on the company, investors followed and the share price fell to an absolute minimum, followed eventually by the company's bankruptcy. Once the company's property was sold, the banks eventually got back 75 per cent of the credit they had extended, a loss of around €15 million.

undergo desertification and the farms located in these regions could then be driven into bankruptcy. The result would be the same for deluxe hotel chains on tropical islands should the islands or coastlines disappear below sea level. Other examples not necessarily related to climate problems are: the bankruptcy of fishermen resulting from overfishing (eg in the Mediterranean) or excessively high costs for agriculture and horticulture due to polluted soil and surface water caused by undesignated third parties. In all these situations, an external environmental problem penetrates a company without the possibility of holding anyone liable so that the company in its most extreme case goes bankrupt. Such environmental risks do not originate within the company but are still included in a company's environmental risks and are thus a bank's credit risks as well.

Private liability

The developments related to private environmental liability continue to go forward in most developed countries. With the introduction of strict liability in Europe for certain cases, the risks that a borrowing client will be held liable for environmental damage (as shown in the legal precedence related to soil pollution) are becoming even greater. A company can also be held liable if it acts in violation of an environmental permit or in the case of an accident.[24] A bank has to be informed of the risk of such liability and the financial capability or reserves (eg an environmental damage insurance policy) of the borrower to cover these risks. The bank also has to be familiar with the phenomenon of 'sticky liability' (ie products liability). A company that processes certain substances into its product or has them removed as waste remains accountable for the environmental damage that will turn up at another location in the production chain. This kind of borrowing client can even be confronted with environmental damage claims resulting from products that it no longer produces. Once more, a bank will have to look into the past.

Government sanctions

A borrowing company can be confronted with various government sanctions if it does not have the required permits or does not comply with the relevant regulations. A default fine is the most commonly applied sanction. By means of the default fine, the government is attempting to stimulate the company to take environmental measures. But will the company be able to pay this fine or to pay for the environmental measures? If the default fine instrument is ineffective, the government can retract the company's permit or confront it with a closing order. These are relatively heavy sanctions and threaten both the company and the bank.

If for various reasons the soil pollution that has been caused by a borrowing client is eliminated by the government, the government has the right of recourse against this company. A company then has not been able to plan the soil remediation itself and probably also lacks any reserves for this as well. This is accompanied by added credit risks. Seven per cent of Dutch businesspersons, for example, are refusing to participate in the BSB operation (see page 127) and can therefore be confronted with higher costs for soil remediation (resulting from such measures as a soil remediation order).

Criminal prosecution

If a businessperson has committed an environmental offence, he/she can be confronted with a criminal prosecution. This can involve fines, imprisonment or closure of the company. The bank will then be confronted with the question of the degree of financial reserves the businessperson has in order to allow his/her company to close for a period of time (eg in the case of a businessperson with a one-person business) or to pay the fines.

The borrowing client's collateral

Due to environmental aspects, the collateral pledged to the bank by a borrowing client can also be adversely affected. The production facilities listed on the borrowing client's balance sheet can actually be worth much less due to environmental aspects. A machine that can be listed on the balance sheet as worth a million euro but which is extremely environmentally polluting can be practically worthless upon sale. The value of registered property such as land or buildings can even be negative in the event of soil pollution or the presence of asbestos. The value of items in stock, too (often pledged as collateral in cases of current account credits), can drop because consumers no longer have much if any desire to buy products that are relatively damaging in environmental terms (eg aerosol cans with CFCs).[25] In short, collateral can drop significantly in value due to environmental issues. In regard to the selling out of collateral, there is a major difference between the Anglo-Saxon and the continental European (Roman) legal systems. A mortgage right is a right to sell the registered properties (land or buildings) as collateral in the event that the obligations are not met. In the event of bankruptcy, this right can be sold off. In the Roman system, a bank never becomes the legal owner of this right. In the Anglo-Saxon system, however, the bank becomes the owner of the right after foreclosure. Essentially, a bank can be held liable as owner in such cases as when polluted soil has to undergo soil remediation.[26] In this system, the collateral not only declines in value but may even acquire a negative value for the bank. Box 6.4 provides a real case (given anonymously) of materializing environmental risk for a bank due to a fall in the value of collateral.

Among sectors that exhibit great reductions in the value of their collateral as a result of environmental risks, the bank can be forced to fall back on secondary collateral. Examples of such collateral are stock, accounts receivable and surety. However, in most countries, the position of the bank in case of bankruptcy is hereby very much weakened. When a piece of collateral is land that has a negative value resulting from soil pollution, the bank will not want to collect its collateral upon bankruptcy. In doing so, it will be at the end of the line with other creditors. For such risks, banks should look for compensation in other forms: a reduction in credit limit, an increase in its rates, and requiring financial security.

In conclusion

Obviously, indirect environmental risks and financial capacity vary from company to company. The risks for banks appear to be relatively high when

Box 6.4 Environmental risks and soil contamination: A case study

A bank financed a transport company by means of a mortgage loan for the amount of one million euro with properties as collateral. For many different reasons, the company could not live up to its obligations whereupon the bank finally wanted to sell off its collateral. In the mid-1990s, the bank had the buildings assessed and was shocked to find that soil pollution was ascertained. Further soil sampling showed that only part of the site was polluted. An attempt to split the lot in order to sell at least part of the collateral met with resistance on the part of the state advocate who saw this as unjustified enrichment. In the meantime, it was shown that the transport company had inserted a passage in the sales contract with the previous owner that would hold the former owner liable for soil remediation costs. This former owner, however, did not possess sufficient funds to cover these expenses. Yet another factor was a conflict between the transport company and the bank. Ultimately, the bank agreed to the sale of its mortgage right against the discharge in full by the transport company of half a million euros. In other words, the damages for the bank amounted to half a million euros.

dealing with SMEs where financial capacity (including the possibility of attracting equity) and environmental expertise are often in shorter supply. But it is the SMEs in particular, however, that are relegated to requesting bank credit. Here, a financial security in the form of a bank guarantee or insurance policy can provide a solution. Having its own environmental damage insurance policy can offer an insurance/banking group other advantages as well. The acceptance process for the insurance policy runs largely parallel to the process of assessing credit risks. In addition, shared knowledge can be used for developing investment products with regard to sustainability. The opportunities for a close working relationship between the insurer and the bank of an All Finance institution should then become obvious.[27]

In conclusion, the bank should be alert to measures that a borrowing client is taking to bear environmental risks. Solutions are an environmental damage insurance policy, an environmental loan for taking preventive measures and subsidy schemes. In some countries, a company can make use of tax facilities in which it can accelerate the depreciation of certain environmental investments or can enter reserves on its balance sheet by reserving part of the profit. A large company such as Shell, for example, set aside provisions amounting to around two billion euro for the future clean-up of environmental damage (Shell, 1993). Setting aside reserves is a sign that a company is looking ahead and is also a sign of solid business practices.

Direct environmental risks for banks

In the main, two possible sources of direct environmental risk for banks can be identified: the enforcement of a client's security interests and the influencing of the client's entrepreneurial policy.[28] When this is accompanied by environmental damage, banks can be held directly liable for the damage by both governments and third parties ('lender liability'). By establishing a legal framework in which such liability by banks can be made an issue, banks will be prepared to reduce such risks; for the government such a framework acts as an instrument that can lead to the setting of pollution levels for companies that are optimal to welfare (Segerson, 1993). The problem for banks is the uncertainty over when exactly they are liable and the fact that the financial value of such risks bears no relationship to the financing. Thus, financing by a bank for €100,000 can lead to a liability for the bank of €1 million–€50 million. The uncertainty and the lack of a correlation between damage and financing makes it impossible to put a true price on such risks in the client's financing. The only protection for a bank lies in risk avoidance. In practice, this has lead to a retreat by banks from extremely environmentally sensitive sectors.

In 1980, a law was adopted in the US that was later interpreted by the legal authorities in such a way that banks could be held liable in certain situations for the activities that they financed. To cover the costs of the Superfund (a fund for cleaning up environmental damage), the Comprehensive Environmental Response, Compensation and Liability Act (CERCLA), based on the 'polluter pays' principle, was instituted in the US in 1980. This legislation was aimed primarily at securing financing for soil remediation. Apart from strict liability, joint and several liability for environmental damage was instituted here, even retrospectively (Beringer and Thomas, 1991).

Following a number of legal proceedings in the late 1980s, there was great consternation among banks in the US in 1991 in regard to the outcome of *The US versus Fleet Factors Corporation*. The bank involved had been held liable for the costs of decontaminating the company's soil since the bank, due to its participation in the financial management of the company, had 'the capacity to influence' the decision that resulted in the pollution, *even though* it had no actual influence on the company's activities (Bryce, 1992; the case itself is detailed in Box 6.5). Banks thought that the 'security creditor's exemption' within the CERCLA would exempt them from direct liability. Legal precedent determined otherwise.[29] Judges decided to leave some incentive for banks to take environmental risks into consideration, ie not stimulating ignorance by overstating the exemption rule.

This had far-reaching effects. A study by the American Bankers Association in the early 1990s revealed that of all commercial banks in the US, 88 per cent had by that time modified their credit policy to prevent lender liability, 63 per cent had rejected loan applications on the suspicion of possible lender liability issues, 17 per cent had abandoned property rather than taken title in fear of environmental direct risk, 14 per cent had incurred clean-up costs on property held as security and 46 per cent had discontinued the extension of credit to extremely environmentally sensitive sectors such as the chemical and agricultural sectors.[30] It was especially the smaller banks in the US that took such steps. Some small banks even went bankrupt (Schmidheiny and Zorraquín, 1996, p104). This also meant that the companies in these sectors were limited in financing the environmental measures they could take. This was obviously an undesirable side effect of this piece of legislation. Besides discontinuing loans, using differentiated rates and reducing periods to maturity, banks in the US later

Box 6.5 Direct environmental risk: *The US versus Fleet Factors*

In the late 1980s, Fleet Factors provided facilities to a company secured on its stock and equipment and by a first charge on its premises. When the company failed, Fleet foreclosed on the stock and equipment, but not on the premises. An auctioneer employed by Fleet to sell these assets at auction, moved some drums and spilled chemicals. Furthermore, a purchaser – in removing items from the premises – released asbestos into the air. The US EPA later cleaned up the chemicals and asbestos and sought recourse to Fleet for the costs involved. The court held Fleet liable, arguing that it had moved outside the protection of the secured creditor's exemption: 'Under the standard we adopt today, a secured creditor may incur section 9607(a)(2) liability, without being an operator, by participating in the financial management of a facility to a degree indicating a capacity to influence the corporation's treatment of hazardous wastes ... [A] secured creditor will be liable if its involvement with the management of the facility is sufficiently broad to support the inference that it could affect hazardous waste disposal decisions if it so chose.'

Source: Case (1999, p59).

reacted by establishing environmental auditing systems (and 'due diligence' systems for mergers and acquisitions; House, 1993). This raised the costs of extending credit so that it was especially difficult for small companies (where such costs are relatively high) to obtain financing.

Compensatory legislation in 1992 and 1995 was an attempt to reduce and clarify the direct liability of banks in the US. However, it took until September 1996 to considerably reduce the banks' insecurities in regard to direct liability. The Asset Conservation, Lender Liability and Deposit Insurance Protection Act achieved this by explicitly stating what a bank can and cannot do (Case, 1999, p60). Since 1996, a bank in the US is liable only in the event that the bank has an actual (and no longer potential) say in the environmental activities of the borrowing company or in the event of there being going-concern activities by the bank.[31] However, differences between states and in interpretation and enforcement still existed and banks remained wary of financing, for example, redevelopment projects in urban areas. Brownfield redevelopment has a positive impact on the social, economic and ecological dimensions of sustainable development. To stimulate such developments and the involvement of banks, the US EPA developed prospective purchase agreements (PPA) at the end of the 1990s for the redevelopment of Superfund sites. Under a PPA the US EPA agrees to waive future Superfund cost recovery (or not to sue for existing contamination) against the purchaser of a site once the purchaser agrees to perform or fund part or all of the current contaminated soil (US EPA, 2000c, p2).

In European countries, government proposals have been circulated to extend environmental liability to banks for situations of contaminated land from the mid-1980s to the beginning of the 1990s. Banks rigorously opposed this, partly due to the bad experiences with the CERCLA in the US. The liability regimes that were ultimately introduced in the UK and The Netherlands in

1995 excluded lenders from liability (the Environment Act in the UK and the Soil Protection Act in The Netherlands). Nevertheless, banks feel that they can still be held liable, not according to the letter of the law but by means of legal precedent. In other developed countries as well, banks feel that such direct environmental risks may in fact still exist, although legislation in almost all developed countries has exempted banks in one way or another from environmental liability where banks are holding 'indicia of ownership' purely to protect a security interest.

In the year 2000, following ten years of preparation and in spite of the fact that reactions within Europe were initially reserved in regard to how liability was handled under the CERCLA, the EU nevertheless produced a White Paper dealing with liability relating to environmental damage (EU Commission, 2000). But the EU has learned in regard to environmental liability from the experiences with CERCLA in the US. The most important lesson is that liability for environmental damage in the EU has to be regulated in such a way that there is scarce if any potential for legal precedence to hold banks liable. Unlike the CERCLA, the European White Paper does not assume a complete retroactive effect (it is left to individual Member States to assume responsibility for this). Moreover, the White Paper states that 'the person who exercises control of an activity by which the damage is caused should be the liable party under an EU environmental liability regime'; ie the proposed regime excludes banks from liability as long as they bear no operational responsibility (ibid, p18, p28). With regard to CERCLA, what is new is that in addition to soil remediation the EU wants to add 'ecological damage' to its liability regime. In both cases joint and several liability is an issue, with the exception of ecological damage resulting from 'innocuous activities'; for this, 'fault-based liability' applies (ibid, p17). Finally, environmental groups can act as plaintiffs and the EU will not introduce an obligation for financial surety.

In short, lender liability is still applied to banks that have (or have had) a say in the management policy of a company that has caused environmental damage (eg in a situation involving going-concern activities).[32] Moreover, the uncertainty for banks remains because the EU has not formulated a precise definition of what banks can and cannot do; and in the presence of joint and several liability the so-called 'deep pocket syndrome' can rear its head in the courts (this refers to the fact that judges usually target parties with the greatest financial capacity in cases of joint and several liability). The EU, however, has clearly learned from the US and listened to the arguments raised by European banks against direct liability. The risk of direct liability does, it is true, remain, but for banks the most important element of risk in the White Paper is the increased indirect risks of extending financing (see pages 129–35).[33]

When applied to banks, the meaning of environmental liability is in short still vague. It is unclear which direction European or US developments will take in regard to lender liability in the future, while financing transactions are usually long term in nature. The guidelines issued by the Bank for International Settlements (BIS) indicate that banks should quantify (in general terms) the possible liability they are running. The current vagueness regarding the character of lender liability makes this practically impossible. In the new 'Basel guidelines'

(BIS II) this even has consequences for the solvency requirement and thus for the possibilities for growth for a bank on a case-by-case basis. These kinds of developments, therefore, can imply a serious threat to the effectiveness of international financial markets. In addition, the effectiveness of liberal international financial markets is obstructed if every country has its own way of applying liability to banks. German banks operate under the Roman legal system and therefore have lower liability risks and thus lower overhead costs than UK banks (since research into the environmental sensitivity of a borrowing company needs not be so strict regarding direct risks to the bank). The exact consequences of EU policy will depend partly on its final formulation and its resulting legal precedents within Member States. The uncertainty in regard to both direct and indirect environmental liability for banks will thus continue to exist for a while in the EU and the US. Even so, it would behove banks to follow developments closely due to the immense consequences of certain details added to environmental liability at the international and national levels.

RISKS TO BANKS' REPUTATIONS

Risks to reputation are difficult to estimate in financial terms. Usually, a damage to a bank's reputation leads to it missing out on acquiring new clients and having its existing clients leave. This extends throughout the entire line of business: retail, corporate and investment banking and asset management. In contrast to other environmental risks, a reputation risk does not only involve the loan or investment itself, but rather the bank's entire lending portfolio and even its entrusted funds and all other activities. Estimating and accurately considering these risks is therefore of great importance. In case of any doubt, it would seem advisable not to get involved in a certain project or financing transaction.

Risks to reputation are more common in cases of large infrastructural investments (such as roads, railways and dams), with new technologies (such as GMOs) and when dealing with developing countries. Even a case of local reputation damage, however, can develop into something that involves the entire bank, especially when the local community is dealt visible damage and the bank is *perceived* to have acted negligently. Most risks to reputation, however, are related to project finance activities. NGOs are increasingly keeping a closer eye on banks in this regard. One NGO concentrated on water management (International Rivers Network, IRN) has even made this part of its activities and sends out its newsletter – called *BankCheck Quarterly* – to more than 2200 NGOs throughout the world. Modern information technology is making it increasingly easy for NGOs to keep track of companies and to initiate and take action.[34] But not just grassroots organizations are involved. The OECD has also announced that it wishes to actively contribute to the negative publicity of any bank that has signed certain agreements, such as the ICC and/or UNEP declaration on sustainability (see Chapter 8) and should know about the OECD guidelines for investments in developing countries, but which in practice does not adhere to these agreements and principles.[35]

Reputation risks are in no way hypothetical. A wide range of banks have already felt the pressure of NGOs and clients on the environmental impact of their financing. Some concrete cases are:

- ABN AMRO was targeted in 1998 for its financing of a company which was supposedly operating an extremely environmentally polluting copper and gold mine in West Papua New Guinea;
- the research by AIDEnvironment in 2000 which called to account all major Dutch banks (although other major international banks, such as HSBC, UBS, BNP, Citigroup, Commerzbank and Bank of Taiwan, are mentioned in the report as well) for their financial ties with palm oil plantation projects in Indonesia (the campaign was labelled Funding Forest Destruction; Wakker, 2000);
- in 2000 some NGOs called for a boycott of Morgan Stanley Dean Witter and Credit Suisse Group in connection with their substantial participation in a government bond for the Three Gorges Dam in China; and
- Sumitomo had to deal with a similar action in connection with its co-financing of the Sardar Sarovar Dam in India.[36]

These cases set precedents in that not only will NGOs call governments and polluting companies to account but increasingly they will also direct attention to the financiers and investors in certain projects and companies. If such campaigns receive widespread media attention and/or are taken seriously by the public, they can result in considerable damage for a bank. Such situations should therefore be prevented. This can be done only when the bank's environmental code is strictly observed. For this reason, simply monitoring or requiring an environmental permit – especially in developing countries – is not enough. A general rule should be that a company or project will have to be investigated in terms of its environmental aspects (especially in developing countries) before financing can be arranged.

An instrument to do just that is an environmental impact assessment (EIA). EIAs are common throughout the world and especially used for large projects and projects in developing countries. An EIA generally involves project screening, site selection, baseline data, environmental impacts and risk assessment, analysis of alternatives and environmental mitigation plans. The drafting of an EIA report is usually done by consultants on behalf of project sponsors, governments or MDBs. During the EIA process public consultation is crucial. Generally, the environmental assessment (EA) process of the World Bank, which may include an EIA, acts as a benchmark for most companies and governments. Box 6.6 illustrates the use of EA in the broad environmental risk assessment at the World Bank.

Usually an EIA will already have been conducted before banks are drawn in. Even if banks would have been involved from the start, an EIA usually does not provide an environmental risk assessment suitable for banks. Although an EIA is an important step in assessing the bank's reputational risks, an EIA doesn't fulfil requirements for analysing direct or indirect financial risks to

Box 6.6 Environmental Assessment at the World Bank Group[37]

The World Bank Group has developed the *Pollution Prevention and Abatement Handbook* for its internal EA[38] process (World Bank, 1999). It is used as an input by all members of the group, though the International Finance Corporation (IFC) and the Multilateral Investment Guarantee Agency (MIGA) complement the overall policies with somewhat different environmental analysis and review procedures.[39] Commercial banks use the handbook as reference material for financing projects in developing countries as well. In what follows the EA of the World Bank (ie the overall policy) is considered. The primary responsibility for the EA process lies with the borrower.

EA became standard procedure of the World Bank (WB) in 1989 (Operational Directive 4.00; in 1991 amended to Operational Policy 4.01). OP 4.01 made EA a flexible process in which EA became an integral part of project preparation. An EA involves five stages (WB, 1999, pp22–26).

Stage 1 involves the screening of the project once a project is identified. The WB determines the nature and magnitude of the project's potential environmental impacts. It then assigns the project to one of three categories (A, B, or C; see Appendix III). For category A projects an EA is always required but category C doesn't require an EA. Between 1989 and 1995 about 10 per cent of the screened projects were category A projects and about half were category C projects (WB, 1999, p23). In fiscal year 1999 half were category C projects (Barannik and Goodland, 2001, p342).

Stage 2 involves a scoping process to identify key issues and terms of reference for an EA. The likely environmental impacts are identified more precisely and the project's area of influence will be defined. The process also involves dissemination of information about the project and its likely environmental effects for affected local communities and NGOs, followed by consultations.

At stage 3 the EA report is prepared. This has to be the case for any category A project. Category B projects are subject to a more limited EA, the nature and scope of which are determined on a case-by-case basis. A full EA will consist of an executive summary, a policy, legal and administrative framework, a project description, baseline data, an impact assessment, an analysis of alternatives, a mitigation plan, an environmental monitoring plan and public consultation.

At stage 4 the EA report is reviewed and a project appraisal follows. The borrower has to submit its EA report to the WB, which is then reviewed by WB's environmental specialists. If the EA is found to be satisfactory, the WB's project team can go ahead with the project appraisal.

Stage 5 involves the implementation of the project by the borrower subject to the agreements which followed from the EA process. WB monitors the environmental quality of the project and the agreed mitigation measures.

Because the IFC usually works more closely with commercial banks, it uses an additional category: financial intermediaries (FIs). Where IFC finances a project through a FI which could have adverse environmental impacts, or IFC finances a project through a FI and IFC targets sub-projects which could have adverse environmental impacts, the project is classified as a category FI project. In all other circumstances IFC finances projects through a FI, the project will be classified as category C, so no EA will be required. For certain types of category FI projects, IFC requires training of FI's staff and disclosure of FI's project-related environmental performance. For all its projects IFC has an exclusion list:[40]

- Production or activities involving harmful or exploitative forms of forced labour/harmful child labour.[41]
- Production or trade in any product or activity deemed illegal under host country laws or regulations or international conventions and agreements.
- Production or trade in weapons and munitions.
- Production or trade in alcoholic beverages (excluding beer and wine).
- Production or trade in tobacco.
- Gambling, casinos and equivalent enterprises.
- Trade in wildlife or wildlife products regulated under the Convention on International Trade in Endangered Species of Wild Fauna and Flora (CITES).
- Production or trade in radioactive materials.[42]
- Production or trade in or use of unbounded asbestos fibres.
- Commercial logging operations or the purchase of logging equipment for use in primary tropical moist forest (prohibited by the forestry policy).
- Production or trade in products containing poly-chlorinated biphenyls (PCBs).
- Production or trade in pharmaceuticals subject to international phase-outs or bans.
- Production or trade in pesticides/herbicides subject to international phase-outs or bans.
- Production or trade in ozone depleting substances (like CFCs) subject to international phase-out.
- Drift net fishing in the marine environment using nets in excess of 2.5km in length.

Other MDBs, such as the European Bank for Reconstruction and Development (EBRD, 1995), have similar EAs and protocols as WB/IFC. For a description of other MDBs, see Box 4.1.

banks. More important in assessing reputational risk is understanding that (see Case, 1999, p148):

- *the degree of participation by a bank in a project is largely irrelevant*; simply being associated with the detrimental environmental effects (eg in a solely advisory function) is enough for the reputational risk to arise;
- *geographical and cultural differences exist*; a project in a developing country may have positive socioeconomic effects which do not pose a threat to the environment in the eyes of the local population, but may still backfire on a bank if clients in developed countries disagree with this environmental optimism;
- *accumulation is an issue*; while one isolated problem may be forgotten, a series of reputational issues can be very damaging to a bank;
- *the objectives and actions of NGOs are hard to predict*, while emotion plays a role alongside scientific arguments; ignoring the views of NGOs might prove devastating;
- *the greater the scale of a project*, the greater the chance of detrimental environmental effects and/or explicit attention and pressure from NGOs; and
- *project financiers will underestimate the risk when it is not quantifiable*; 'making the deal' in project finance might cut short the assessment of reputational risks.

These aspects are clearly present in the Three Gorges hydroelectric power scheme on China's River Yangtze. This project's construction runs from 1994–2009 and is expected to cost more than virtually any other single construction project in history (estimated costs run from $US28 billion up to $US75 billion (Kearins and O'Malley, 2001, p351)). Also, its size is unprecedented: the dam will be 3000 metres wide and 185 metres high and creates a 600-kilometre-long lake. The project will submerge 160 towns and cities and 770 villages, 800 ancient sites (including temples and unexcavated ruins of archaeological interest), 657 factory and mining sites, 130 power stations, 956 kilometres of roads and 23,800 hectares of crop land. Moreover, more than 1.5 million Chinese will be relocated from their homes to less fertile lands and smaller holdings; compensation per family has been set at approximately four months' earnings (Case, 1999, p147). Its sheer size and impact has led to much criticism and eventually led to the refusal of the World Bank to finance the project or parts thereof (Kearins and O'Malley, 2001, p351).

That banks in developed countries perceive reputational risk in various ways becomes evident when passages from the annual reports of the Société Generale and Bank of America about the Three Gorges project are compared. Thus the Société Generale reports with some 'pride' (2000, p41):

> *'Among the transactions arranged in 1999, SG was sole arranger for the finance of switching equipment and back-up systems materials in the third phase of the Three Gorges project in China, the world's largest dam. SG has arranged US$1 billion in financing for this project since 1996, underscoring the Group's capabilities in multi-source export finance, and its high level of customer loyalty.'*

Bank of America, by contrast, steered clear of directly financing the Three Gorges dam and similarly avoided indirect involvement in the form of loans to third parties involved in the project (Bank of America, 2000, p15). Of course, a difference in the banks' perception of the use and necessity of the project can also play a part, as can company specific norms and values. It is, however, evident that the reputational risks will also pan out in different ways.

Instruments for analysing environmental risk

To do an effective job of estimating environmental risks, having information about the environmental sensitivity of companies is essential. Banks will therefore be employing information systems and/or various resources or be making use of external expertise. Page 91 discusses several research agencies which assess companies based on their environmental aspects. Most of these agencies direct their efforts to the positive aspects of the company, ie the degree to which the company contributes to the achievement of sustainable development. These methods, therefore, are more suited to sustainable investment products. The environmental risks of such progressive companies will be essentially lower, but this is no guarantee. For the time being, the only factor in determining the extension of credit is the risk factor. A small number of research agencies do include

Box 6.7 The CSFI method for classifying environmental risk

CSFI is a research agency that combines the facts associated with a company's environmental achievements (eg environmental compliance programme, public reputation, permits, legal proceedings and known damage claims) with assessments about the soundness of its business practices, its environmental policy and its actual financial capacity to comply with environmental issues. This information can be used to answer two questions. Firstly, how big are the potential environmental risks and costs for the company? Secondly, how well is the company equipped in terms of management and financial capacity to handle this? The answers reflect the potential effect of environmental risks on the company's profit and loss account.

The CSFI has copied the conventional method of credit rating in which after these two questions are answered, a seven-point scale of environmental risks is developed that runs from AAA to C.

1. AAA: a company in an excellent financial position with a minimum of well-identified environmental risks and provided with the strength to endure any risk that arises; a company that is also capable of absorbing any more stringent government policy that may be developed within the foreseeable future.
2. AA: a company similar to AAA but with higher environmental risks.
3. A: a company similar to AA but with even higher environmental risks and less capacity for handling them.
4. BBB: a company in a strong financial position with good prospects and with environmental risks that may have a strong negative effect on the company's financial position.
5. BB: a company similar to BBB but with a real risk of a substantial environmental risk arising.
6. B: a company similar to BB but with an even higher risk of a substantial environmental risk arising.
7. C: a company in which the management quality and financial capacity for meeting environmental risks that may arise are doubtful and in which the manifestation of these risks will have a serious negative effect on the company's financial position.

this in their package of services as well. The best way of doing this as far as the extension of credit is concerned is a method in which the environmental risks are coupled with the company's financial capacity to bear those risks. The Center for the Study of Financial Innovation (CSFI) has made a move in this direction by establishing such a method for the environmental risks of companies.[43] Box 6.7 provides a brief description of this method.

The advantages of this method in comparison with others are its level of detail, the fact that the environmental risks can be estimated for each business location, and the fact that it is especially suited to financiers. The disadvantages are its laboriousness, complexity and price. In addition, certain elements in its total assessment, such as the value attached to the management quality, indicate a subjective character (Costaras, 1995).

The CSFI method is especially practical for use in large international companies. For a bank, it is nearly impossible to be informed of the environmental legislation in every country in which it does business. Being able to rely on external expertise, and particularly on a standardized method such as that of the CSFI, is a more obvious choice. Gathering environmental information itself will always have to be balanced against the costs and benefits of doing so. If a foreign bank is extremely active in, say, the US, then gathering information about American environmental legislation can offer advantages. After all, such information will be more geared to its own situation. In addition, knowledge about these matters can play a decisive role in attracting companies into a relationship with the bank. After all, through familiarity with the company's problems the bank can work alongside its client in solving its problems and by doing so gain a competitive edge.

If banks want to have their own source of information available, they can employ a wide range of sources or methods. Large companies often provide banks with a great deal of information concerning their environmental affairs. Environmental reporting is an instrument that helps to compare the level of environmental care of various companies. But it is questionable, however, as to whether environmental reporting will be done in SMEs because the costs for an environmental report will usually far surpass the benefits, including the image effects, of doing so.[44] Especially for SMEs, the use of rough checklists would seem to be the most obvious choice. Since the credit appraiser's knowledge concerning all the technical sides of various environmental aspects within companies will often be inadequate, standardized questionnaires should be simple and preferably based on questions that can be answered with 'yes', 'no' or 'not applicable/unknown'. Appendix IV includes one such example.[45] Here is a summary of possible sources of information and methods:

- standardized lists of questions;
- information from specialized agencies (like CSFI);
- computer models for environmental risk classification;
- support by environmental consultants in such forms as environmental risk appraisal, EIA (for project finance), environmental audit or due diligence;
- direct contact with the businessperson for risk assessment;
- environmental insurance;
- guarantees by MDBs or ECAs;
- environmental reports;
- standardized EMSs (like EMAS or ISO 14001; see pages 165–8);
- permits and other governmental sources (like the NPRI);
- newspapers and specialized magazines;
- past experiences or experiences of other banks in general or in regard to certain companies or groups of companies;
- policy documents or other documents issued by the government, branch organizations, etc; and
- environmental clauses in term loan facility letters (eg indemnification).[46]

What is involved, thus, is a combination of experiences and information gathered from the past as well as information about, and a concept of, the future. Will the government forbid certain substances or be introducing a severe energy levy in the near future? Will consumers be developing a marked preference for environmentally responsible products? And how will such developments affect the borrowing company (eg an energy-intensive one)? Another important question is how a bank will accumulate insight into these questions. In part, this can be done on the basis of the information sources listed above. Systematic contact with the government (possibly done through banking associations) in regard to the effects of new government policy can yield added benefits. Finally, scenario analysis can play an important role.

An interesting new method for both banks as well as insurers and investors comes from Repetto and Austin (2001). This method makes it possible to analyse the potential results of uncertain developments in environmental issues for a sector and the individual companies within that sector. Since this method deals with uncertainties, financial analysts cannot depend on financial annual reports since they essentially make no mention of them. Nevertheless, estimating possible environmental scenarios is necessary in order to make the right investment decisions. The method consists of seven steps (ibid, p281):

1. define the sector;
2. identify salient future environment issues for this sector;
3. build scenarios around each issue;
4. assign probabilities to the scenarios;
5. assess individual company exposures;
6. estimate financial impacts contingent on the scenarios; and
7. construct overall measures of expected impact and risk.

A case study of the pulp and paper industry in the US showed that companies within this sector exhibited a wide range in the degree to which they would be affected by the same possible future environmental issues and in the financial risks related to them. Some companies would hardly if at all be affected and would even be financially ahead as a result. Others, however, would be substantively damaged in their competitive position, and the costs that could result from the possible future environmental issues could climb to 15 per cent of the current share price. Even though the anticipated impact of the environmental issues for two companies was the same, differences could still develop in their financial risks. Especially for extremely environmentally sensitive sectors such as the oil and chemical industries, this scenario method can provide banks and investors with the necessary insights.

Conclusion

The previous sections have indicated the varied nature of environmental risks and the importance of having good environmental information. Standardization is important in this regard. Each client will then know exactly what he/she can expect, and the bank will have lower overhead costs. These costs can be reduced even more by cooperation between bank and insurer. This will provide both advantages of scale and scope. In Switzerland, such cooperation is not limited to a financial holding. By means of the Swiss banking association, the banks in Switzerland have developed a standardized system for assessing environmental risks. The clients of Swiss banks are reacting positively to this move and are willing to provide the banks with all the required information. This is unusual considering that private industry usually objects to providing environmental information and premium differentiation based on environmental risks. This positive attitude is probably due to the explicitness and standardization of the method at almost all the Swiss banks (Schweizerische Bankiervereinigung, 1997). Every Swiss bank naturally makes independent use of this method and the data collected. This can mean that the ultimate setting of rates and supplementary requirements (eg in regard to collateral) can vary from bank to bank.

Premium differentiation based on environmental risks is more explicit and simpler within the framework of this Swiss initiative. After all, each bank uses the same method and should theoretically arrive at a similar pricing. This limits the extent of an environmentally sensitive company's freedom to shop around for a bank. Banks will therefore be less fearful of losing a client to a competitor. It is probably unlikely, however, that such a fear is justified among various extremely environmentally sensitive clients. After all, benefits for a bank are involved when such a client is being financed by another bank. Firstly, the bank is not associated as a financier of extremely environmentally damaging companies, and this will improve its image. Secondly, in case environmental damage occurs, the competing bank will most likely have to deal with a financial blow.

The Swiss concept is also interesting because the environment and environmental risks have long been considered to be points on which banks could compete with one another. Until far into the 1990s, there was absolutely no exchange between the banks of knowledge and experience in this regard, but this attitude has almost disappeared. Banking associations in other countries did not see much point in applying the Swiss concept since it was a matter of competition among the banks. The fact that this viewpoint is disappearing can be seen by brochures in regard to environmental risks from the UK and Canadian banking associations[47] and from a book published by Phil Case (Barclays Bank) in 1999 that is entirely devoted to environmental risks for banks. He, too, has observed the competitive issue – something that has made the development of environmental risk analyses harder for Barclays (Case, 1999, px). Other examples of a more cooperative trend are the FORGE report (2000) in which various UK banks worked together and published guidelines for, among other things, assessing environmental risk, and the open dialogue within the UNEP FSI. Banks therefore no longer need to rediscover the wheel for themselves.

7

Internal Environmental Care

INTRODUCTION

Although banking is not considered to be a polluting industry, the scale of banking operations nonetheless gives rise to important environmental effects and banks are increasingly focusing on environmental care. While they impose demands on their customers' processes and products, they are responsible for similar concerns of their own. When paying lip service to sustainability in their external statements and in the media, banks run the risk of being held to account for not confronting the environmental impact inherent in their own activities. If banks are lax in their own responsibilities, they lose credibility in the eyes of their customers. Credibility is gained by a transparency in their approach, and implementing an environmental care system, whether certified to international standards or not, certainly helps. Various banks have already put such a system into effect, making a distinction between the direct environmental impact of their own production process (also known as 'operating ecology' or internal environmental care) and the indirect environmental impact of banking activities (known as 'product ecology' or external environmental care). Chapter 8 concentrates on the organization of environmental care as well as on internal and external communication, while this chapter examines the internal environmental impact and care of banks. Internal environmental care involves the environmental impact of inputs (like electricity), internal processes (like mobility) and outputs (like the emission of CO_2). Thus internal environmental care incorporates both organizational and technical measures (VfU, 1998, p18).

Internal environmental care may be implemented for voluntary, compulsory (legislation) or compelling (market) reasons. An example of a voluntary approach can be seen in the chemical industry's 'Responsible Care' programme. Initiated in 1988 by the US Chemical Manufacturers' Association, it followed the chemical-leak disaster in the Indian city of Bhopal. This programme aimed to reduce environmental pollution and risk as well as to improve safety in the production process. Since then, some 90 per cent of chemical companies in the US and Canada have subscribed to this programme. Most companies could not afford not to subscribe and something of a domino effect occurred: once a great enough number of competitors, many of whom were market leaders, had set the precedent, the rest followed in order to preserve their image. Moreover, the suppliers of the major multinationals following the Responsible Care Programme also had to comply with it. A voluntary programme has set the benchmark, and market forces compel participation.[1]

The following drivers for environmental care can be identified:

- improved cost–benefit ratio;
- greater appeal for particular investors;
- lower legal and reputational risks;
- better image for customers, NGOs and society as a whole; and
- greater attractiveness in a tight labour market.

Banks have a relatively low environmental impact per product unit, but considerable volumes make the total significant. Banks use a substantial amount of office space, devour paper, have transport needs and are consumers of energy and water. Cost savings can certainly be made, and significant ones at that. In addition, better waste management through recycling waste into resources (recycling and re-use) can generate additional revenues and improve the cost–benefit ratio.

An environmentally conscious internal policy contributes to the charisma and image of the bank. This can work positively in recruiting and attracting highly qualified personnel. For those already in employment, it means improved working conditions (Howe, 1993). What's more, internal order in environmental issues is a prerequisite to both communicating an environmentally conscious image and fulfilling a social responsibility.

Environmental care within banking also has a legal aspect. Within most developed countries banks are subject to environmental legislation, as are other sectors of the economy. Such legislation covers: energy consumption (requirements, levies and covenants); the responsibility to recycle selected wastes and goods such as packaging and electronic equipment; the discharge of harmful substances such as CO_2 and CFCs; and liability for environmental damage and contaminated land. Non-compliance with the requirements may result in significant levies, and even criminal prosecution.

Figure 7.1 sketches the various environmental categories of internal environmental care within banking. In the following sections these categories are explored.

ENERGY

Credit Suisse has developed 'Environmental Performance Indicators' to measure the environmental impact of its bank. These showed that energy consumption is by far the most important factor, accounting for 90 per cent of all cumulative pollution within Credit Suisse (Credit Suisse, 1999). UBS came to a similar conclusion on the basis of its Environmental Performance Evaluation (UBS, 1999). Other environmental reports from banks concur. Examples of high-energy consumption are heating, lighting, electronic equipment, air condi-

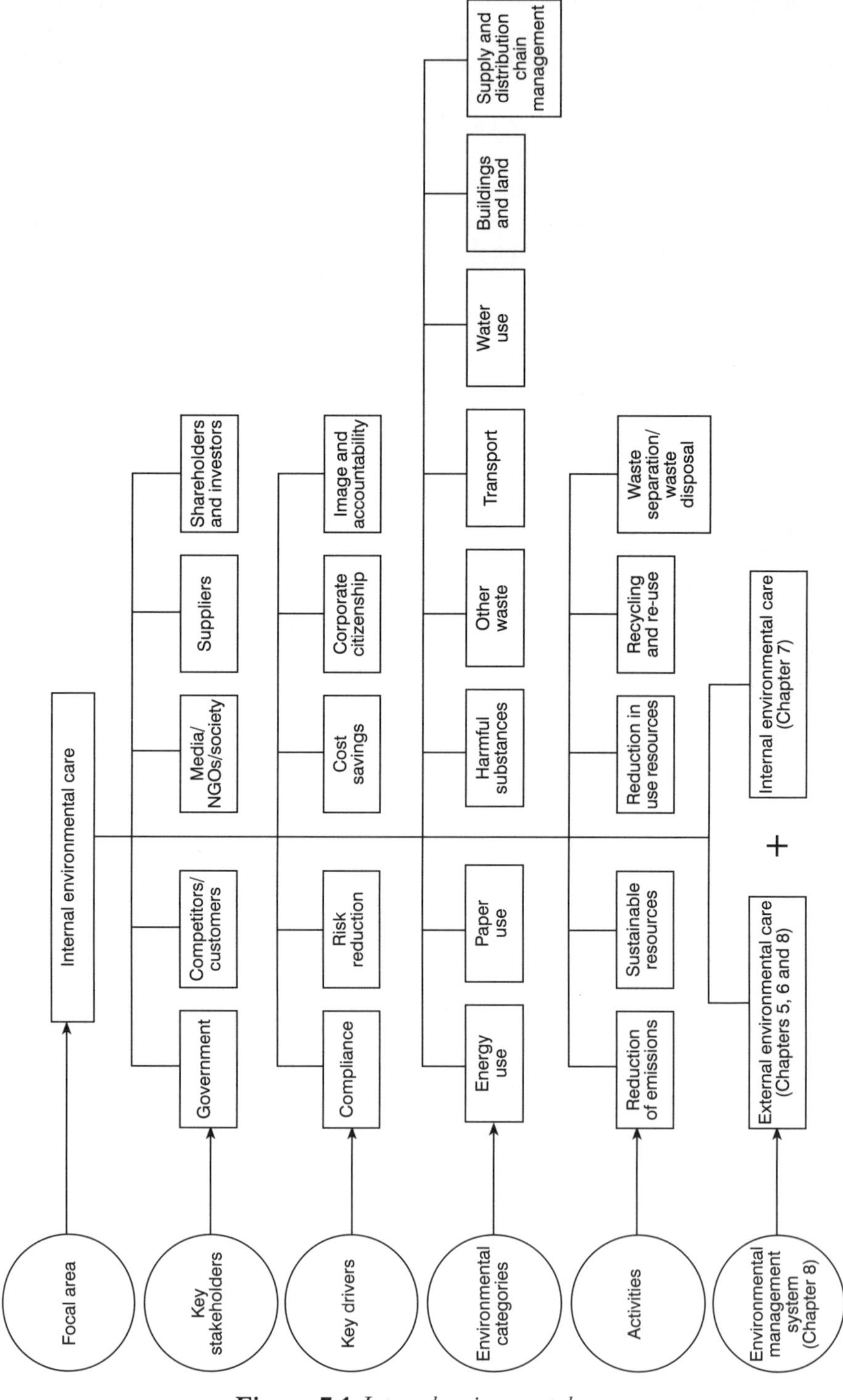

Figure 7.1 *Internal environmental care*

tioning and ventilation. In the 1980s banks were already imposing measures to reduce their office energy costs but consumption continued to rise, with expansion and increased use of office equipment being factors. Foresight and a structural approach have been important in enabling many banks to achieve an absolute fall in energy consumption – they have also been faced with government requirements (or threats). Indeed, the issue of energy consumption has the full attention of banks and governments in most countries.

Banks in Switzerland have estimated a potential energy saving of 10–40 per cent over several years (SBA, 1997, p24). In The Netherlands banks annually consume around 550 million kWh of electricity and 72 million m^3 of gas; energy savings potential is estimated at 38 per cent. Governments too are aware of this potential. By the end of 1996 the Dutch Association of Banks (NVB) agreed a long-term energy-saving plan (MJA) with the Dutch government. The NVB set itself the target of improving the situation by 25 per cent by 2006, compared to 1995 (*NVB Bulletin*, May 1996). These targets are an average for the whole sector (with the proviso that participating banks consume 80 per cent of the sector's total energy).[2] This 'deal' is, to some extent, voluntary (lack of a 'deal' would precipitate legislation) and thereby gains substantial support within the banking world. The plan provides for recovery of costs within a period of three years and gross financial savings of some €18 million (that is, exclusive of the costs of the measures and maintaining a constant ratio between gas and electricity consumption). The Dutch government has done its bit by providing special incentive rulings (like the energy investment deduction, whereby an additional fiscal deduction of 40–52 per cent of the investment costs is possible; Ministry of Economic Affairs, 1998).

Such energy savings can be achieved through a combination of organizational, technical and behavioural measures, within existing buildings as well as for new buildings and renovated premises. A major component is monitoring, which itself contributes 5–15 per cent to saving energy in practice. In this way UBS managed to reduce its energy consumption in Switzerland by 18 per cent between 1990 and 1997 (UBS, 1999, p11) and Barclays reduced its energy consumption in the UK by 33 per cent between 1979 and 1999 (Barclays, 1999, p4), despite the growth in their activities. Barclays foresees additional cost savings: a one-off investment of €0.6 million enables savings of €1.7 million per year thereafter (amounting to 12.5 per cent of annual energy costs, ibid). The NatWest Bank also achieved substantial savings despite its growth. It was the first bank in the UK to introduce an environmental care system; the cumulative net energy savings ran up to €64 million for the entire group between 1991 and 1995, representing some 24 per cent of gWh (see Figure 7.2).[3] However, these three banks are the exception rather than the rule among the large international banks. The ING Group registered an absolute rise of 15 per cent between 1996 and 1998, and only between 1998 and 1999 has there been any absolute fall, at 3.4 per cent (ING Group, 2000, p36). During the same period Credit Suisse showed an absolute rise of 12 per cent and is now targeting an absolute fall of 12.7 per cent for the period 1998 to 2004 (Credit Suisse, 1999, p4). Lloyds, after an absolute rise of 16 per cent between 1998 and 1999, see a savings potential of 0.74 per cent for 2000 (Lloyds TSB, 2000, p5).

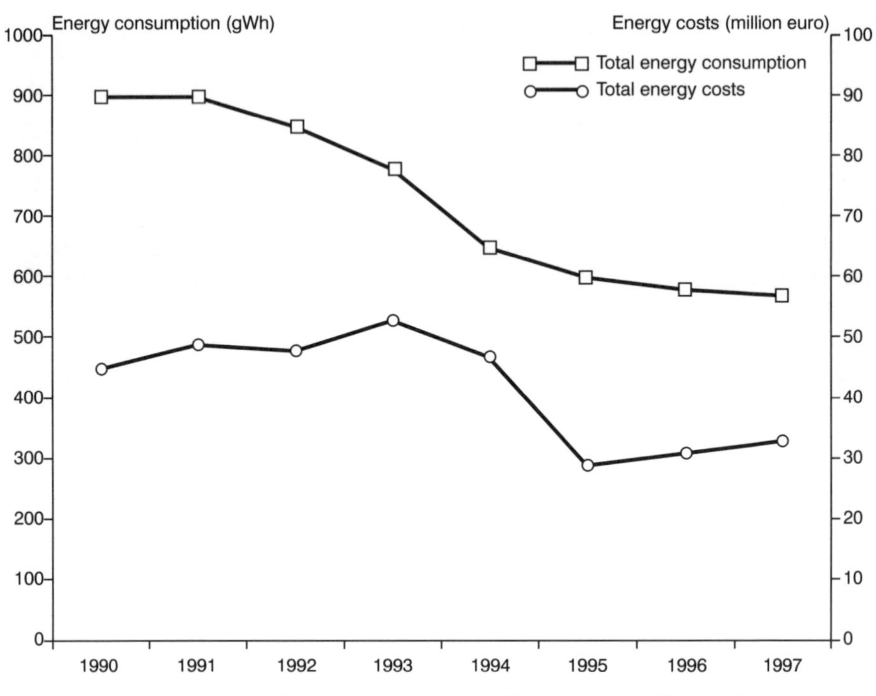

Figure 7.2 *Energy savings at NatWest Bank, 1990–1997*

Sustainable energy is also something banks are investigating, in the form of wind and solar energy, terrestrial heat and bio-energy. By using these sources alone, banks can stimulate their development. A number of banks already use solar energy (eg a German office of the Deutsche Bank, Deutsche Bank, 2000, p31). Credit Suisse aims to supply 3 per cent of its energy needs for heating through sustainable sources by 2004 (Credit Suisse, 1998, p35). In addition, progress can be stimulated by buying green electricity. The Rabobank, for instance, has 380 cash dispensers supplied by green electricity (Rabobank Group, 2000, p36) in The Netherlands, 'syphoned' from the normal electricity supply. The higher price (around 7 cents per kWh extra) is used by the electricity companies to finance sustainable energy. This is a case of the indirect use of sustainable energy: sustainable energy is distributed throughout the electricity grid together with the standard supply.

Emissions of CO_2, SO_2 and CFCs are directly related to energy consumption, so reducing energy consumption reduces emissions. Replacing equipment and technology that cause high emissions is also useful. In reporting and monitoring the discharge of harmful substances, a distinction is made between the emissions generated inside and outside the system (VfU, 1998, p52). Direct emissions are generated inside the system of a company, heat generating installations being an example, while outside the system there are indirect emissions such as those attributable to energy supplied by electricity companies, and emissions from transport. In terms of the environmental impact of the bank, indirect emissions also have to be included.

WASTE

After energy consumption, the biggest environmental issue for banks is waste, paper in particular. In most banks paper accounts for more than 50 per cent of waste – in the ING Group the figure is 66 per cent (ING, 2000, p37), at ABN AMRO it is 50 per cent (ABN AMRO, 2001, p8), Credit Suisse 57 per cent (Credit Suisse, 1999, p5) and UBS 49 per cent (UBS, 1999, p11). Hazardous waste forms 6–8 per cent of the total waste at these banks (except for ING which reports a level of only 0.5 per cent).

Waste is a logical consequence of economic trade (see Figure 2.1). Waste squanders raw materials and energy, and often results in contamination and the creation of eyesores. Minimizing waste impacts the cost–benefit ratio. In the preventive and offensive banking phases, the benefits are clear and it makes business sense to bring the quantity of waste down to a minimum. In a sustainable banking phase, additional steps will be taken to reduce the quantity of waste to a sustainable level, though this may involve a relatively long pay-back period or costs that will never be recouped.

Up to a certain degree, banks can themselves decide how much effort they wish to make to process waste. The least sustainable manner of waste processing is dumping, followed by incineration. This may recover some of the energy. It is more environmentally friendly to recycle waste, ie to re-use the waste as the raw material for a particular product. More sustainable is re-use, as is done with toner cartridges. The most sustainable option it to prevent waste in the first place. It saves money too: purchasing costs decrease. Where purchasing does occur, banks recognize its importance and tend to include an environmental clause in their purchasing contracts.

A bank employee in Switzerland uses on average 150–300kg of paper each year (SBA, 1997, p18), a figure that is likely to be similar for banks in any other developed country. Of course, a distinction can be made in terms of the environmental friendliness of paper (chlorine-bleached and new paper as opposed to recycled paper). Various banks have already switched to environmentally friendly paper for envelopes and statements, as ABN AMRO did in 1998 when it decided to send the statements for all of a customer's accounts together in one envelope (ABN AMRO, 2001, p16). For some banks marketing consumes almost a third of all paper (eg Lloyds TSB, 2000, p9). In processing waste, a distinction is made between ordinary old paper and confidential old paper which can be shredded or sent to a specialist company for processing. Minimizing the use of paper and recycling are vital if unbridled deforestation is to be prevented. In the manufacturing of a sheet of A4 paper, recycled paper uses 61 per cent less water and 54 per cent less energy consumption than new paper (based on SBA, 1997, p19). Advances like internet banking and telebanking make paperless transactions a possibility; customers receive everything in digital form. This saves substantially on paper use and the financial savings can then be passed on in part to the customer. Where paper has to be used, double-sided printing and copying are also economical options.

In reality, despite prevention measures, there will always be waste, and re-use and recycling remain important. Both entail waste separation: for example, the separate collection of paper, minor chemical waste such as ink, pens, batteries and diskettes, biodegradable waste and plastic cups. Other waste categories that are collected and processed separately include discarded computer accessories such as ink cartridges and bank passes. During the year 2000 ABN AMRO has managed to separate 68 per cent of its waste for recycling or re-use in The Netherlands (ABN AMRO, 2001, p10), though this largely reflects government regulations. An interesting development is the biodegradable bank pass. There are some 500 million bank passes in circulation worldwide, so making these biodegradable would greatly improve the environment. In the UK, The Co-operative Bank premiered the Greenpeace Visa Card in collaboration with Greenpeace and Monsanto (*The Times*, 10 May 1997). In The Netherlands almost every bank pass is recycled by a specialist company (95 per cent of all the four million passes discarded annually). Since 1998 the Rabobank has been the first large bank to separate the microchips from the plastic for re-use (Rabobank Group, 2000, p45).

Business reasons for reducing or separating waste are supplemented by legislation relating to substances that cause environmental pollution. As this is national legislation it varies from country to country. Various countries also have legislation for the use of raw materials, as in The Netherlands, where companies are compelled to minimize the volume of packaging and/or to recycle packaging for products. This legislation also applies to banks, where envelopes, for example, are an important packaging component.

SUSTAINABLE BUILDING

About half of all professionals in developed countries work in offices. Yet these offices are ultimately occupied for as little as 5 per cent of the time (in relation to a year of 365 days and 24 hours a day; Veldhoen and Piepers, 1995). This creates an enormous waste of space, space that could have been used for other purposes like nature, recreation, agriculture, business parks and housing. Economically, too, it is wasteful: inefficient use of office space (at micro level) and land (at macro level). In addition, it creates ecological wastage: at the macro level as a result of the inefficient allocation of land and natural resources (building material), and at the micro level due to the inefficient occupancy of the workplace (material, energy and water consumption). This kind of wastage features in all developed countries with strong service sectors.

The environmental impact of the office is increased by the travelling undertaken by office staff. Travel to offices sited along motorways, standard working days and car mobility all increase the impact on the environment. In addition, rush hour traffic congestion causes economic loss and lost time. Moreover, the pollution caused by emissions during the constant stop-go cycle in traffic jams is, in relative terms, enormous. Traffic management initiatives such as car pooling or the provision of bicycles for staff within a radius of, say, 10km from work and a public transport season ticket for staff who live further away can significantly contribute to reducing the environmental impact of mobility.[4] Bank of America employees, for example, are issued with a debit card. This is intended to reduce the environmental impact of commuting by giving employees an incentive to use a form of transport other than the car (BankAmerica Employee Alternative Transportation Program, Bank of America, 2000).

New office concepts and ways of working are facilitated by new management concepts and developments in the area of ICT. Innovations include teleworking and working from home, both of which help reduce the environmental impact of transportation and the need for office space. Spin-offs from these include the decrease in material use and energy consumption. In addition to office design and the organization of work in offices, sustainable building also includes the conservation of buildings themselves. The now accepted definition of sustainable building is that the environment is subjected to the lowest level of strain throughout the lifecycle of the building from design to demolition.

An organization is dynamic; an office building is not. Owing to their rigidity, buildings tend to have a short economic lifetime of around ten years, which is both economically and ecologically wasteful. By taking account of the various lifecycles of the different elements at play in buildings, they can be made more flexible and, therefore, more sustainable. This can be achieved by, for instance, devising a building in such a way that the façade can be modified, the installation used in various configurations or the interior layout easily modified by moving mobile segments. Modular building (somewhat like a Lego system) can take account of the various lifecycles of a façade (50 to 70 years), installations (15 years), layout (5 years) and facilities (IT, for instance, 2 to 3 years). In a sustainable building, the infrastructure is suited to varying use and thus has a longer lifespan.

Four other aspects come into play in the design of sustainable buildings. Firstly, the consumption of water, a increasingly scarce commodity in most developed countries. Initiatives include toilets and taps that use less water, a rainwater loop and rainwater storage. Secondly, aiming for maximum energy saving. This includes using insulation, limiting window space on north-facing façades, optimizing daylight and introducing low-energy lighting and solar panels. Thirdly, the optimal use of low environmental impact, re-usable and upgradable materials together with a cleaner form of waste processing. A fourth, more micro-level aspect of sustainable building is the fostering of the quality of the surroundings and the interior environment of offices, an area concerned with the health and welfare of the end-user. Various banks, like ING in its new headquarters opened in 1999 in Amsterdam (ING, 2000, pp23–24), are already familiar with sustainable building practice in their own office organization.

SUPPLY AND DISTRIBUTION CHAIN MANAGEMENT

Increasing numbers of banks are focusing on their core activities and outsourcing auxiliary services. Banks wanting to adopt an aggressive approach to environmental care will take account of a company's environmental impact when selecting a supplier. They may also advise an existing supplier to implement a strategy of continuous reduction in the environmental impact of the goods and services delivered. For instance, cleaning companies that service bank buildings will have to comply with environmental requirements. Banks can make significant savings by implementing a more environmentally friendly purchasing policy. Energy saving can be more cost-effective through the purchase of more energy-efficient electronic equipment (like PCs with an energy saving function) than by measures to reduce the energy consumption of existing equipment. Legislation, like that which governs CFCs in refrigeration equipment, and the lifetime of, for instance, furniture and promotional material, also plays a role. Most banks already explicitly address environmental effects by incorporating an environmental clause into their purchasing policy. And sustainable banks adopt initiatives like Fair Trade – that is, purchasing coffee and food, for instance, at prices that are fair to the farmers, and that are produced according to decent social standards in developing countries.

Although most banks pay attention to the environmental impact of the products and services of their suppliers, few banks address the environmental impact of the distribution policy of banking products. The Co-operative Bank in the UK is an exception.[5] Box 7.1 shows its environmental policy on distribution and the indicators it has developed. Indicators are both environmental and social.

CONCLUSION

This chapter has cited various examples of the environmental impact of banks. However, these examples have been taken from a limited group of banks. This has resulted from, on the one hand, the choice to focus attention on the top three banks per developed country, and on the other hand, the fact that the data provided by the banks in their environmental reports is largely not yet comparable between banks and are sometimes purely qualitative. An initiative was started in Germany in 1996 to develop indicators and guidelines for the gathering and reporting of data for the internal environmental impact of the financial sector. The result was a set of eight key indicators that are used by various banks with headquarters in Germany and Switzerland (VfU, 1998). By way of concluding this chapter and leading into the next, Table 7.1 presents the environmental impact for six German/Swiss financial institutions using (in part) the VfU methodology in their environmental report: three regional German banks – Landesbank Berlin (LBB), Landesgirokasse (LG) and Bayerisches Landesbank (BLB), a major German insurance company (Allianz), and two Swiss banks, the Credit Suisse Group (CSG) and UBS.[6] The figures make it possible to compare the eco-efficiency of banks. In the case of CSG and UBS, data was available for two years; both years are presented, enabling a comparison across time within one institution. Unfortunately, the VfU tool fails to take into account the size, specific operations of banks or system changes (for example mergers), which leads to some anomalies. For example, smaller banks are obviously likely to use less paper while multinational banks will incur a much larger score for transcontinental business travel. The reference value may be ideal for ecological benchmarking, but is not ideal for comparing business; to take account of size, operations and system changes, a financial reference value might be more appropriate. Moreover, the

Box 7.1 Social and environmental indicators for the distribution policy of The Co-operative Bank, UK

The Co-operative Bank uses seven partner groups in its work and policy: shareholders, customers, employees and their families, suppliers, local communities, society (national and international), and past and future generations. In order to determine the social and ecological impact of each of the distribution channels, it is necessary to define precisely the infrastructure and activities per distribution channel. The Co-operative Bank has developed indicators to determine the social and ecological impact of distribution channels in general. The bank has identified three distribution channels:

1 Physical channels (bank offices, post offices, financial service centres, 'Handybanks' within Co-op stores, kiosks, ATMs).
2 'Remote' channels (post, call centres, interactive voice response, PC banking).
3 Virtual channels (mobile phone, TV and internet banking, plastic cards).

Of course, there is some overlap between the channels, which prompts the definition of a limited and well-defined set of activities and points of departure for each channel. In the use of post offices, for example, the ecological and social impact of the post offices themselves is not considered, rather the manner in which the customers use this channel. In total 15 areas have been defined against which the social and ecological impact can be measured. These are:

1 The building/location (production, use and waste).
2 Electronics (production, use and waste of computers and phones, etc).
3 Paper consumption (use of resources and energy as well as environmental contamination by chemicals).
4 Plastic cards (production, use and waste).
5 Transport (storage space, use of resources and energy by employees and customers).
6 'Financial inclusion'(access to banking services for various population groups).
7 User convenience (quality and loyalty).
8 Personal contact (participation).
9 Safety and privacy (quality and integrity).
10 Quality of the service.
11 Job security (quality of life, education and training).
12 Working conditions (quality of life).
13 Local economic development (quality of life).
14 Fair contracting (human rights, trade union freedom).
15 Collaboration between companies (cooperation).

The Co-operative Bank ultimately ended up with 19 indicators for its distribution policy:

1 Surface area of and space used by its bank premises (in m^2).
2 Percentage of its electronic equipment re-used, repaired or recycled.
3 Total energy consumption of the bank (in kWh).
4 Percentage of energy used from renewable resources.
5 Emissions from customer transport to bank distribution channels (CO_2 and NO_x).
6 Emissions from transporting bank paper supplies (CO_2 and NO_x).
7 Quantity (kg) and type of paper used.
8 Quantity of accounts offered to low-income households.

9 Quantity of accounts offered to small companies.
10 Customer (private and business) satisfaction with convenience.
11 Customer (private and business) satisfaction with the quality of the service.
12 Employee perception of job security.
13 Employee satisfaction with the quality of their job.
14 Satisfaction of family members of employees regarding working times.
15 Number of employees resident in deprived areas.
16 Supplier satisfaction in respect of fair prices and contracts.
17 Payment of a 'reasonable' wage to people working in IT system assembly.
18 Satisfaction with the bank as a full partner in the cooperative movement by members of the cooperative world.
19 Satisfaction with the bank as a good example of collaboration between members of the cooperative movement.

One of the bank's objectives for 2000 was to gain insight into performance against these indicators for at least 75 per cent of the aspects related to them. To facilitate this, in its 1999 Partnership Report the bank presented data to establish baselines against which it would measure (The Co-operative Bank, 2000, p61).

Sources: The Co-operative Bank (2000); Street and Monaghan (2001).

Table 7.1 *Internal environmental impact of banks with headquarters in Germany or Switzerland*[7]

Parameter	*LBB*	*LG*	*BLB*	*Allianz*	*CSG96*	*CSG98*	*UBS97*	*UBS99*
Electricity consumption (kWh/employee/year)	2886	4627	4816	4110	7500	7600	7300	6900
Heat consumption (kWh/m^2)	n/a	n/a	n/a	n/a	96	85	104	109
Water consumption (cubic metre/employee/year)	145	98	100	76	119	111	94	69
Total paper consumption (kg/employee/year)	120	113	120	258	240	270	252	263
Copier paper consumption (pages/employee/year)	3997	3895	5200	7200	8850	6790	10,600	12,900
Waste disposal (kg/employee/year)	n/a	n/a	n/a	n/a	290	286	285	281
Business travel (km/employee/year)	321	1040	1500	3700	1700	1300	3000	2700
CO_2 emissions (kg CO_2/employee)	n/a	n/a	n/a	n/a	2850	3000	2800	1800

Sources: European Commission DG XI (1998) for LBB, LG, BLB and Allianz (figures are for 1996); CSG (1998; 1999); UBS (1999; 2000).

tool doesn't address the quality of the monitoring system within a bank or the quality of the data reported. This standardization tool certainly needs to be improved upon, but is a positive development towards standardizing the measurement of internal environmental performance in the banking community. It has, in any event, made indicative benchmarking possible.

This chapter has been exclusively concerned with the environmental aspects of the production process, though in practical terms internal and external environmental care cannot be separated. If a bank wishes, for instance, to issue an eco credit card (with, for example, a donation per cardholder to the WWF), it is worthwhile having the card manufactured from biodegradable material. It is possible that environmental measures will require an investment that cannot be recovered in the short term, though such investments should be considered as essential for the company in the longer term – in this case, for example, for its influence on customers' perception of the company. There is, in short, a major strategic dimension to internal environmental care. Chapter 8 looks more closely at the interaction between internal and external environmental care and the more systematic approach of integral environmental care.

8

Organization and Communication About Sustainability

INTRODUCTION

The road to sustainable banking involves organizational change. This can be achieved in an autonomous, ad hoc or structured way. The approach chosen will depend on the bank's current phase of sustainability or the stance it adopts with regard to sustainability (ie defensive, preventive, offensive or sustainable). An offensive or sustainable stance demands a completely different organizational structure than, for example, a preventive stance. An important question is whether sustainable behaviour is centrally or decentrally organized and thus where authorization and responsibilities lie. What is the role of employees and what is the role of the management or the board of directors? What should the approach be, 'top–down' or 'bottom–up'? Should there be a separate environmental department or should responsibility for sustainability and the environment be integrated into various departments or parts of the organization? Such things do, of course, depend on the structure and culture of the organization and the markets in which it operates, both geographical and functional. What is more, all sorts of specific strategic questions play a role, such as should the position the organization has adopted towards sustainability be changed and if so, when is the best moment for an organizational change and how will support for this be generated?

Behavioural change is a prerequisite for organizational change and while it can simply occur autonomously, it is usually organized. Communication plays an extremely important role in engendering such a change: an essential aspect of internal environmental care is that the environmental programme and the environmental policy are communicated to all sections of the company. This is also true for the successes and the visible contribution to the immediate surroundings. But besides this, internal communication is also very important for external activities such as those discussed in Chapters 5 and 6. Together with internal communication, external communication can also contribute to organizational change; employees see the change bearing fruit, for example in terms of an enhanced company image, when the surroundings show signs of valuing the company's initiatives. What is more, the employees should be able to answer questions about sustainability posed by stakeholders. Both forms of communication develop employees' readiness to seriously set to work with the organizational change. It will no doubt be clear that internal and external communication work hand in hand and are mutually reinforcing; which category a particular activity falls into is simply a question of approach. External communication can, of course, entail commercial advantages. External communication activities that spring to mind are PR activities, an environmental annual report and signing up to international treaties, but they also include the expression of community involvement by, for example, lending support to various initiatives in society that encourage a sustainable development.

Communication about social responsibility and involvement (signing treaties, sponsorship and the like) is important, but not without risk. Before embarking on this, a company should consider that its internal environmental care may be called to account once it becomes visibly active. If an environmental care programme is set up, carried out and recorded in accordance with industry standards, developments in performance can be analysed and benchmarking between banks facilitated. A rule of thumb for companies communicating about their involvement in sustainability is to be on the lookout for company policies or actions which may be perceived by others as 'window dressing'. The ever faster dissemination of information to members of the public as a result of information technology makes it possible to assemble a group quickly and efficiently to protest or take action, making issues of external accountability still more important. The question is: what form should a bank's communications about its contribution to the pursuance of sustainable development take? And should it communicate them at all? What consequences can this have for its business dealings? Will they be purely positive or also negative, such as a loss of credibility? Does signing a treaty, for example, mean interest groups will be able to call the bank to account for financing a heavily polluting sector?

'The organization of sustainability within banks', below, will address questions about organization and a structural approach to organizing environmental care. The section beginning on page 169, deals with the closely related theme of internal communication. The role of external communication is handled on pages 172–8, and pages 178–80 examine its inherent risks and pitfalls. In conclusion, page 181 draws together these aspects of organization and communication.

The organization of sustainability within banks

Introduction

The internal aspects of banking activities are dealt with in Chapter 7. In order to structurally reduce the environmental impact of these activities, some banks have introduced an environmental management system (EMS). A systematic approach such as this is also appropriate for external environmental aspects. As the position of banks, as shown in Chapter 4, moves towards sustainable banking, these external aspects gain a more explicit role in an EMS. Banks with a preventive approach have been systematically tackling internal aspects since the start of the 1990s; external aspects were handled in a more ad hoc way and tended to arise from commercial activities rather than result from a specific environmental policy. The EMSs within banks with an offensive or sustainable approach are more complete, consisting of both internal and external aspects and – in the ultimate phase of sustainable banking – being an integral part of corporate policy and activities. Since the end of the 1990s a few banks have developed and introduced an EMS for both their internal and external environmental impact. Within an EMS a number of aspects play a role, such as management responsibilities, controlling and monitoring and communication. For process and communication reasons it can be interesting to get external certification for an EMS. The communication aspects will be discussed in the following sections.

Environmental management systems

The form an EMS takes and how sustainability issues are organized within a bank will depend on how the bank wishes to position itself in the world with respect to sustainability. One strategic decision, whether ad hoc or not, precedes the introduction of an EMS. An environmental care strategy consists not only of the setting of long-term environmental objectives, which may take the form of a statement of intent or an environmental policy statement, but also the adaptation of business activities in the long term to the company's internal capabilities and the market potential that becomes apparent for achieving the policy objective.[1] An EMS is a coherent whole consisting of policy, organization and administrative measures with the objective of controlling and reducing the strain that the company places on the environment. Thus it is an instrument to implement and control the environmental strategy, just as the financial administration is an instrument to keep costs low. An EMS usually consists of the following basic components, (KPMG/IVA, 1996):

- an environmental policy statement;
- an environmental programme;
- integration of environmental care into company processes;
- the allocation of responsibilities and tasks to people;
- measurements and registrations according to internal and external rules (permits);
- internal information and education;
- internal and external reporting; and
- examination of the environmental care system (environmental audit).

Not only production processes but also products are parts of an EMS. The setting up of an EMS within a bank usually involves a number of phases.

The first phase is essential top–down and has two objectives: firstly, to determine the environmental care strategy, and secondly to obtain management commitment. Company specifics such as the company's mission and the available budget will be significant at this stage. Once commitment has been obtained the strategy can be organized and communicated throughout the company.

The second phase involves the development of the policy and the programme to be implemented. An environmental policy statement is both functional and informative. It establishes the direction the organization will take and lets first employees, and later various external groups, know what is being planned. The purpose of the environmental programme is to establish tangible resolutions and objectives for the coming one to two years.

In the third phase the implementation of the programme takes place. For banks new to this area, a step by step plan and/or the setting of priorities will be necessary. Training and internal communication are an essential part of this phase.

In the fourth phase performance and the EMS are measured, reported and evaluated. This creates the basis for making adjustments to the environmental programme, or even adapting the policy. The model of the 'balanced scorecard' can be used for monitoring. This model offers management three perspectives to complement the usual financial perspective: customers, business processes and organizational learning. The adoption of this model is very suitable for use in issues of sustainability (Kaplan and Norton, 1993).

The above four phases have sub-phases: 'plan', 'do', 'check/act' and 'learn', in that order. Together they comprise the 'wheel of sustainability', as shown in Figure 8.1. It consists of a continuous loop. For example, after completing the four sub-phases in phase 2, the second round will start in sub-phase 1 of phase 2 where the definitive environmental plan and programme are established. Simultaneously phase 3 will start, in which the implementation plan will be developed and then, in the following sub-phases will be set up, implemented,

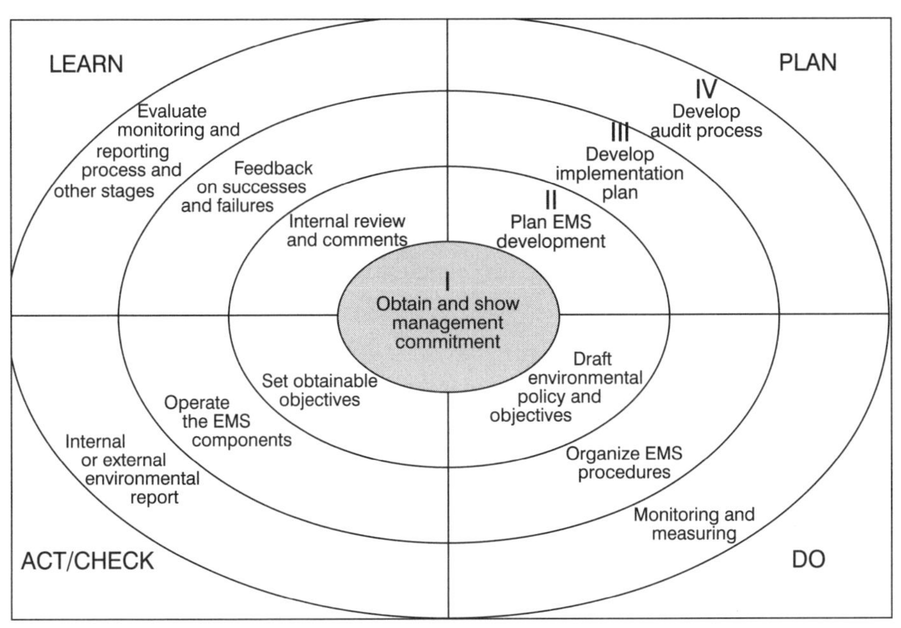

Source: adapted from Wempe and Kaptein (2000, p47).[2]

Figure 8.1 *Environmental management as a continuous process*

and adjusted. Movement is not only towards a higher phase but may also be towards a lower phase. Adjustments to the environmental policy, for example, will always require new management commitment. For the monitoring in the fourth phase use can be made of environmental management accounting models (see eg Bennett and James, 1998), so that in addition to the necessary monitoring of physical aspects the financial costs and benefits of the EMS can be measured and assessed.

In order to meet the specific requirements of banks, a number of British financial institutions, including Barclays, Abbey National and Prudential, joined forces and brought out the FORGE report.[3] The report is modular and very practical. It offers banks big and small the building blocks for setting up an EMS. The report identifies nine phases that can be clustered into the four phases shown in Figure 8.1. FORGE pays a little more attention to the obtaining of management commitment (phase 1), the first stage of which it describes as being the collection and development of evidence of the need for an environmental policy and strategy. This is indeed vital for obtaining management commitment. One bank will focus on commercial reasons (products) to achieve this, whereas another draws on ideological motives. Evidence can be looked for and developed internally or by an external agency. For some banks the first step has been to invite an external environmental guru to discuss with the board social developments related to the environment and the implications and challenges these could pose for the bank; Rabobank, for example, did this in the early 1990s. At the outset, a member of the board will, as a rule, be appointed as the owner of the problem and it will be that person's job to prepare the company for new developments.

Gradually banks will choose to implement a formal EMS. By integrating environment care into all business processes and procedures, this type of systematic approach brings with it numerous advantages. It can result in being treated more flexibly by government, cost savings due to, among other things, a reduction in waste and energy use, lower risks, better return or an increased market share as a consequence of new or improved products, or an enhanced public image. A formal EMS has, moreover, the advantage that it opens the way to certification, which brings with it external recognition of the company's efforts.

Certification of EMSs

Certification of an EMS involves a company receiving an internationally recognized certificate, once independent external investigation has shown that the EMS system meets certain requirements. Since the certification clarifies the quality of environmental care practised by the bank, it brings with it various advantages in relation to customers, suppliers and governments. If a bank wants to be considered for certification, this should be taken into account in the early stages of planning since it places special demands on the form and content of the EMS. A bank can do this by using internationally recognized standards. Even if there are no concrete plans to aim for certification, these standards are still useful given the expertise and experience they offer. Were a bank to later choose to aim for certification, little extra effort would be required.

An internally recognized standard is the International Organization for Standardisation (ISO) 14000 series. Within Europe the European Eco-Management and Audit Scheme (EMAS) is another popular certification, especially with German companies.[4] Companies which meet the EMAS requirements are added to a register and can use an EU logo; no EMAS logo exists as yet. The logo is a sort of quality mark that offers advantages in business-to-business dealings. EMAS is more demanding than ISO 14001 because it requires the independent verification of the annual environmental report. ISO 14001's main focus is on the EMS, while EMAS is primarily concerned with environmental performance. The certificates are otherwise fairly similar. EMAS is under revision. The most important proposals concern the expansion of the scheme to cover all sectors of the economy (thus also banking),[5] the integration of ISO 14001 in EMAS and the development of an official EMAS logo. In no country at present is certification to either of these standards mandatory. Box 8.1 briefly sketches interest in both standards and includes Figure 8.2, which shows the popularity of ISO and EMAS standards in developed countries among businesses in general. Box 8.2 concentrates on the fundamentals of ISO 14000 series.

For banks the most important advantages of certification are recognition that the bank's own position on the environment is sound. The impact of certification on business-to-business relationships is marginal, since the core of a bank's activities concern attracting and lending money and not acting as an intermediary in the handling of goods. Certification may, however, make it easier to communicate product demands to suppliers. Furthermore certification can provide PR with a useful tool, indicative of a progressive environmental policy.[6] This last factor can be extremely important for banks. USB admits that its ISO certification has been beneficial to the marketing of its sustainability funds to customers.[7] As can be seen from Table 8.1, various banks now have ISO or EMAS certification. Others, including ING Group, Swedbank and Nikko Securities[8] have already indicated that they aim to achieve ISO certification in

Box 8.1 ISO 14001 and EMAS in practice

By the end of 2000 around 3000 European companies had EMAS certification and by July 2000 around 15,000 companies in developed countries had ISO 14001 certification (81 per cent of the worldwide total of companies certified to this standard). ISO 14001 is, in absolute figures, most popular in Japan. As a percentage in relation to the population some northern European countries and Switzerland score highly. These are primarily countries with relatively small populations, which may account for their high scores. ISO 14001 certification in relation to a country's GDP gives perhaps a better benchmark. Sweden and Finland remain higher scorers. The US's low score stands out. In Europe, Germany is the frontrunner in absolute numbers for EMAS certification (70 per cent of all EMAS certificates); Austria has the highest score relative to population size. In one year (1999–2000) the number of ISO certificates worldwide has risen by 70 per cent; the increase for EMAS was less sensational and totalled 8 per cent over one and a half years (1998–2000).

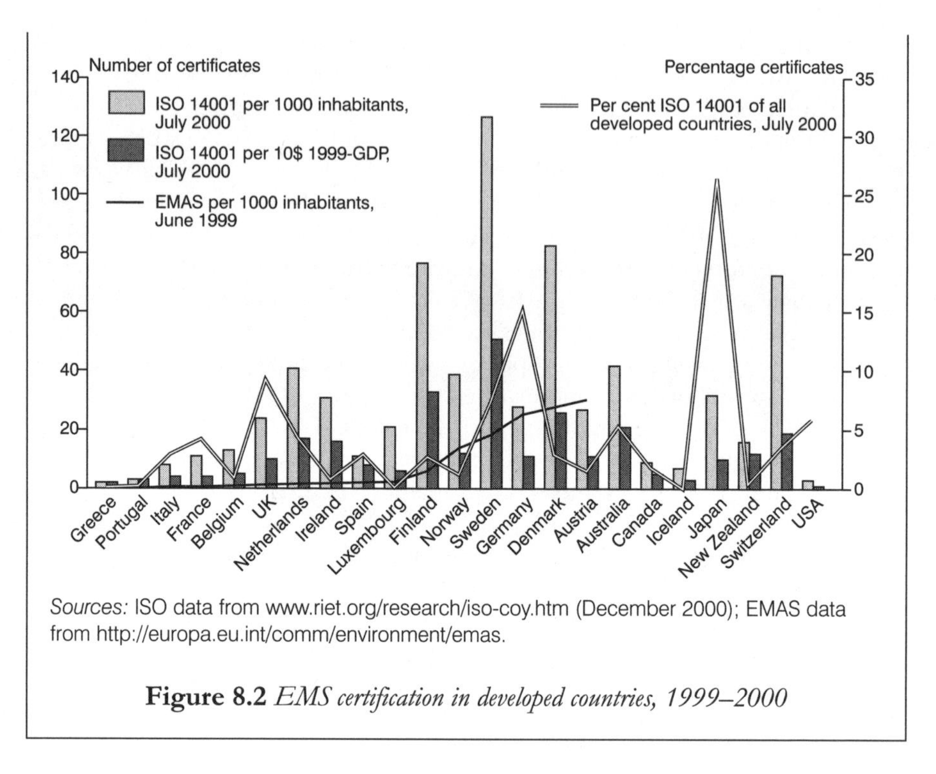

Sources: ISO data from www.riet.org/research/iso-coy.htm (December 2000); EMAS data from http://europa.eu.int/comm/environment/emas.

Figure 8.2 *EMS certification in developed countries, 1999–2000*

the near future. For European banks with worldwide operations ISO certificates will have a greater added value than EMAS. Once a European bank has achieved ISO certification, it can always aim for EMAS certification. Gaining an EMAS certificate before an ISO certificate appears to be less interesting to European banks. The banks in Table 8.1 with only an EMAS certificate are the local and regional banks. All financial service providers in the table with both an ISO and EMAS certificate, first achieved ISO certification.[9]

EMSs in perspective

A bank with a defensive stance will not have an EMS. Nevertheless it should be possible to identify some minimal agreement between strategic, tactical and operational levels. The activities geared to environmental improvement are in principle ad hoc and are usually a direct result of pressure applied by the government. This defensive phase, given its ad hoc and lagging nature, will not be further dealt with here.

A bank with a preventive stance will appreciate the advantages and necessity of environmental care. This will find expression in a partial EMS, a systematic approach to organization with the aim of integrating the internal and external environmental objectives into policy. It consists in any case of an environmental policy statement and an environmental programme. In contrast to the defensive phase at least one person will be responsible for such tasks as coordination of environmental activities. This is usually an operational function.

Box 8.2 The ISO 14000 series

The ISO 14000 series gives guidelines for the content of an EMS and subjects related to this. Among others, it consists of:

- ISO 14001 Guidelines for implementing an EMS
- ISO 14010–14012 Guidelines for environmental auditing
- ISO 14020–14024 Environmental labelling
- ISO 14031/32 Guidelines on environmental performance evaluation (EPE)
- ISO 14040/43 Lifecycle assessment (LCA)
- ISO 14060 Guide for inclusion of environmental aspects in product standards

Within the series, ISO 14001 is the only standard intended for certification by third parties. Within the methodology of ISO, an EMS is a management system in which the processes and procedures (on paper or software) focus on the continued improvement, monitoring, auditing and reporting of environmental performance.[10] Briefly, ISO 14001 concentrates only on processes, not on products. Banks may be interested in ISO 14020–14024 for the development and marketing of, for example, sustainable investment funds, but since this (currently) does not involve a standard for certification, these guidelines' only added value is the experience and knowledge they contain. Moreover, these guidelines can be usefully referred to as a source in a bank's own reporting.

In addition, ISO 140001 does not specify any emissions or savings objectives for a company. In short, it does not provide a benchmark for the environmental performance at any given moment. So, company A scores four on a sliding ten-point scale for environmental friendliness and has an ISO certificate, while company B (of a comparable size and nature) has a score of eight and no certificate. While this reflects the fact that participation is voluntary, it could also indicate that company B does not satisfy the ISO criteria.

ISO 14031–14032 standards can offer a solution since they do now focus on environmental performance. Within this standard, two indicators have been identified: environmental perfomance indicators (EPIs) and environmental condition indicators (ECIs). The latter group offers a company insight into the general condition of the environment and can help a company to determine its influence and impact on the environment. EPIs function internally as a management tool, offering planning, monitoring and control. Externally they become information tools for specific target groups and benchmarking tools to enable performance comparisons between companies. The EPIs are subdivided into operational performance indicators (OPIs) and management performance indicators (MPIs). OPIs focus on the environmental impact of inputs (such as energy and purchasing), outputs (such as products, services, waste and emissions) and throughput (such as the installation and functioning of company offices). MPIs offer support for the evaluation of company performance in its objective of reducing environmental impact. MPIs concentrate on, among other things, how people function, and decision making and policy in relation to environmental care.[11] A group of Swiss and German financial institutions have developed guidelines for the product side of banking MPIs and OPIs based on ISO 14031.[12] Page 175 will return to this initiative. In Chapter 7 the VfU initiative for internal environmental care EPIs was discussed.

In addition, environmental tasks will be fulfilled by departments such as purchasing and credit risk control. An environmental commission, representing various disciplines and departments, will generally be set up.[13] It will be the task of such a commission to make an inventory of environmental issues and

Table 8.1 *ISO 14001 and EMAS certification among financial institutions, 2000*[14]

Financial institution	*Country*	*ISO 14001*	*EMAS*
Deutsche Bank*	Germany	✗	
UBS*	Switzerland	✗	
Credit Suisse*	Switzerland	✗	
The Sakura Bank	Japan	✗	
BBVA*	Spain	✗ (4 sites)	
Swedbank* (Föreningssparbanken)	Sweden	✗ (1 site)	
Raiffeisenlandesbank NÖ-Wien	Austria	✗	✗
Österreichische Kommunalkredit	Austria	✗	✗
Deutsche Ausgleichsbank	Germany		✗ (2 sites)
LBS Bayerische Landesbausparkasse	Germany		✗
Landesbank Sachsen Girozentrale	Germany	✗	✗
Landesgirokasse Stuttgart	Germany	✗	✗
Sparkasse Heidelberg	Germany		✗
Frankfurter Sparkasse	Germany		✗
Triodos Bank*	Netherlands	✗	
Gerling (insurance)	Germany	✗	✗
Victoria Versicherungen (insurance)	Germany		✗ (5 sites)
Tokio Marine* (insurance)	Japan	✗ (head office)	n/a
Yasuda Fire and Marine Insurance* (insurance)	Japan	✗ (head office)	n/a

Sources: * environmental annual reports and financial annual reports from the individual banks; all other institutions from www.bmu.de/sachthemen/oeko/emas_e.htm (December 2000).

advise different levels of the organization. The emphasis will lie on internal environmental care. Knowledge of environmental aspects will be developed internally with help from external experts, instructions, checklists and appropriate training.

Allocating budgets to environmental care will be essential in this phase for giving it credibility in the organization. Employees and management are often not aware of the environmental aspects of their own activities. Raising consciousness about aspects of sustainability is therefore very important. Developing support and a sense of involvement among staff and management demands time, attention and energy.[15] A clear and consistent emphasis on the importance of sustainability by top management will help foster support. Internal communication is particularly important in this phase. External communication about the objectives and the environmental stance will usually be limited.

In the offensive phase environmental care goes so far as to include external reporting and the setting of clear environmental objectives. A system is implemented to enable measurements and recordings on the basis of internal and external guidelines. These measurements can, for example, be used for internal control of progress and as the basis for a possible environmental report. A separate department is often set up to determine internal and external care objectives, but it must be clear that sustainability should be regarded throughout the whole organization as being in the company interest. This 'sustainability' or 'environmental' department usually reports directly to top management. It is responsible for the environmental policy and is a source of expertise for all departments, but especially those with key roles for the offensive stance. External expertise is only engaged when necessary for specific tasks, for example an agency might be approached for advice about the decontamination of ground; most knowledge is available internally. The department is usually a staff department and determines policy. Very offensive banks will allot management of environmental care to one person on the board of directors (Schmidheiny, 1992). Operational tasks will be delegated to the line organization and the sustainability or environmental department will offer support to line departments. Various committees exist, though they are now more horizontally (thematically) than vertically organized, for example a platform for the development of climate products. The EMS is regularly examined so that the system and activities can be continuously improved (ie by an environmental audit).[16]

In the sustainable phase environmental care and sustainability form an integral part of the business of the company. The sustainable position is adopted by the whole organization, from the top to the bottom. Employees are explicitly encouraged to make the approach their own and will be rewarded for doing so, for example in the form of bonuses dependent on a person's contribution to the sustainability of the company instead of, as is usual, a bonus for contributing to financial profitability. Responsibilities and authorization will spread throughout the whole organization. A separate department may well take care of coordination, but in principle the system is so complete that the internal communication between line departments is optimal. All departments meanwhile possess the knowledge to make sustainable behaviour possible. However, while it started as a top–down process, the bank's environmental care has become a bottom–up process. Responsibility and authorization for environmental care are both decentralized and central. Strategic, tactical and operational tasks are increasingly separated from one another and take place at levels that reflect their significance to the company's core activities. It is appropriate to talk of an integrated sustainable business operation.

In each of the above three phases (preventive, offensive and sustainable), environmental achievements will be communicated internally, together with good financial results, for example savings, and commercial successes. Top management's readiness to allow the environmental ambitions to assume a key role in decision making is a prerequisite for the successful implementation of an environmental care strategy. Since environmental achievements often affect the whole company while their costs lie with one department, an additional general budget (determined by management) is a requirement for the implementation of environmental care in the preventive and offensive phases. Support by staff and management is another precondition for a successful environmental strategy. Communication is one important tool to achieve this.

THE ROLE OF INTERNAL COMMUNICATION

Internal communication has three sequential objectives:

1 to increase staff and management's knowledge;
2 to contribute to a positive attitude towards sustainability; and
3 to foster sustainable behaviour by staff and management.

Raising the level of staff and management knowledge is of great importance in the defensive, preventive and offensive phases (and diminishes in importance in every higher-level phase).[17] An environmental policy statement with the organization's long-term objectives can be a useful tool, as can a declaration of intent. Environmental annual reports can make explicit to staff the environmental impact of various, apparently clean, banking activities. Information about sustainable development, the importance of environmental care, the results achieved by competitors and so on can be disseminated via the intranet or in newsletters. An easily accessible knowledge centre is another constructive tool. No one activity will suffice, a range is required.

An environmental policy statement, signed by the board of directors, establishes long-term objectives. It is usually for external audiences and is qualitative. The environmental programme, by contrast, sets short-term internal qualitative and quantitative objectives and activities. It may cover a period of, for example, one year. The programme must include concrete and appealing objectives if it is to succeed in influencing staff and management attitude and behaviour. One objective may be to reduce paper use by 25 per cent over four years. Supporting information and education will require heavy investments of time, energy and money. This may include internal seminars, conferences, theme days and participation in broad external networks. In a preventive strategy, giving creditor assessors extra training will be something which is particularly time-consuming. Supporting information would cover checklists and instructions for insuring against environmental risks and for lending. This could also take the form of databanks with environmental information by sector. Many banks already have such tools. A logical place to store them is the intranet because it enables changes to be easily made and quickly communicated. It is an ideal way of keeping up with changes in environmental legislation that affect the activities of clients.

Companies employing a preventative or an offensive strategy will be primarily concerned with the second and third objectives: attitude and behavioural change. Sustainability will have to be a company-wide concern before creative innovations take place at product level. To help employees take ownership of the sustainability initiative, ABN AMRO introduced the so-called environmental yardstick, as reported in its 1998 environmental annual report. This is used by employees to gain an impression of the environmental impact of their own activities.[18] This practical tool may have a positive influence on attitude and behaviour and may increase employees' knowledge.

Attitude and behaviour can also be influenced by discussion and training sessions in which metaphors are used to break down employees' preconceived ideas about environmental care and to alert them to opportunities. Focused training is also necessary to equip account managers with the skills and knowledge they will need when dealing with clients, as mentioned above for credit assessors. Bottom–up initiatives usually fail because they encounter an uncomprehending middle management: 'the environment just costs money', 'the customer's not interested...!'. Consciousness-raising training among middle management can address this problem and thus accelerate the change process. The commitment of top management is clearly important here to send a message to middle management.

Some banks use 'green teams', in which colleagues work together to think up tangible savings or to develop new products. Such teams can be supported by internal or external professionals but the idea is that the changes in employees' attitude and behaviour will be lasting precisely because they themselves took the initiative. Two major Dutch banks (ABN AMRO and Rabobank) use so-called eco teams. This is a programme developed by Global Action Plan (GAP) in which a small group of colleagues in a department collaborate to work out how the environmental impact of their own department can be reduced. The programme has proved in practice that it does lead to behavioural change. Employees realize that they can solve many environmental issues simply and without great expense. Their solutions may even save money. The tangible results of their efforts increase employees' support.

Communicating environmental successes, and particularly successes achieved by employees, will further encourage support throughout the company. This is true for both internal and external environmental care. The ever present danger is that attention drifts towards products because they generate revenue and it is always more fun to generate revenue than make savings. Internal environmental care is too easily regarded as the business of a few departments. It is the job of communication to correct this view: each employee is responsible and has the opportunity to contribute to sustainability in the internal business operations. As stated in Chapter 7, internal environmental care in banks is primarily a question of behavioural change, not of technological change. Communication should emphasize this. For external care, a technical approach, using new and adapted products and services, may well be a tactic, but it must complement behavioural change, not be a substitute for it. Visionary declarations from the top, whether formalized in written declarations by the bank or not, are an important influence on the attitude of staff and management. Communication 'from the outside to the inside' can also be an important impulse for a more sustainable attitude. For example, a relatively short media announcement, like 'bank x signs sustainability declaration' can lead to positive attention among stakeholders that subsequently raises staff interest or results in their being asked about sustainability.

Sustainable behaviour by employees can also be promoted by financial reward. At its most extreme this would mean bonuses that were dependent upon individual contribution to the bank's sustainability targets, either for internal environmental care, for new environmental products or environmental investments and projects. Such a scheme would be interesting because sustainable projects do not get much chance in current bonus systems. Many bank initiatives related to external care deal with risk reduction or niches for new products. The bulk of activities, however, concern regular financing and the environment is paid scant attention. The normal bonus structure within banks encourages the employee to focus on large-scale projects. Sustainable projects on the other hand are usually relatively small scale. To enable a shift to sustainable projects a bonus system better focused on sustainability is required. Moreover, indicators would need to be developed to assess a project's level of sustainability, and not simply the environmental risk it poses. These indicators would thereafter be applied to the new bonus system. Such a discussion is fundamental to achieving a phase of sustainable banking and can have far-reaching consequences.[19] Within, for example, the Rabobank and the IFC the first tentative steps are being taken to make this a topic for discussion and to bring about change.

THE ROLE OF EXTERNAL COMMUNICATION

Introduction

The offensive or sustainable phases may entail various external communication activities that aim to improve the bank's image in a broad sense. This will give credibility to the claims already made internally. The term 'in a broad sense' is used because image can include on the one hand immediate commercial activities (such as sustainable investment funds) and on the other hand involvement and responsibility (although image in this case is obviously more a consequence than an aim). Among the activities discussed are codes of conduct, environmental reporting, participating in networks and sponsoring.

Codes of conduct

The publication of a statement of intent, code of conduct or an environmental policy statement can be important examples of external communication. A general statement of intent puts into words the bank's direction, the key objectives and the key values. It forms, as it were, the compass for everyone in the organization, thus fulfilling an important function as internal communication. But it is also of importance to the outside world: the statement of intent shows what the organization stands for and what it can be called to account for.[20] Box 8.3 contains a number of quotes taken from the statements of intent and policy statements of various banks. There are clear differences between the statements. Some banks are very progressive and use financial performance as their means and sustainability as their target. Other banks opt for a more defensive or preventive approach. Not all banks are equally clear about what sustainability or the environment means for their own priorities.

In addition to a statement of intent or an environmental policy statement, signing an international environmental declaration can be an important communication device. Dozens of banks have signed the ICC charter for sustainable development in business (see Appendix V).[21] The ICC declaration is a general declaration in which signatories commit to striving for sustainability. It has characteristics similar to those of ISO 14001, but carries no obligations. The more far-reaching UNEP declaration for banks on environment and sustainable development has since been signed by 179 banks (June 2001; see Appendices VI and VII). The most important commitment of the UNEP declaration is that besides the precautionary principle, a bank will integrate environmental consid-

Box 8.3 Quotes from banks' statements of intent and environmental policy statements

The goal of the policy is to fully integrate environmental aspects into all company decisions, so that ABN AMRO is able to contribute to a sustainable society. (ABN AMRO, 2001, p.5)

Enhancing the sustainability of society is ASN Bank's key objective and leading principle in all its economic activities. Enhancing the sustainability of society means contributing to changes which aim at ending processes in which the adverse effects are pushed to the future, or passed on to the environment, nature or poorer sections of the community. Economic activity means recognizing that a return has to be made in order to secure a healthy future for ASN Bank over the long term and also recognizing the necessity of managing the funds entrusted to ASN Bank in a manner which justifies our clients' expectations in this respect. (ASN Bank, 1999, p1)

At Bank of America, we believe environmental protection is an integral component of doing business. We are mindful that the company's actions and leadership can make a difference in our own business practices around the world, in the communities where we do business and in our customer and client relationships. We are committed to making that difference. (Bank of America, 2000, p3)

Barclays believes that a responsible environmental policy makes sound business sense, as well as being an integral part of good corporate citizenship. We are therefore keen to demonstrate good environmental practice, both through our own activities and in raising the awareness of others. (Barclays, 1999, p2)

Citigroup believes that working to conserve and enhance the environment is a good business practice. As a concerned global corporate citizen, Citigroup analyses the potential environmental impacts of its business activities and takes action to either reduce environmental risk or promote benefits. (Citigroup, 8 October 2000, www.citigroup.com/citigroup/homepage/environment/mid.htm)

We, The Co-operative Bank, will continue to develop our business taking into account the impact our activities have on the environment and society at large. The nature of our activities are such that our indirect impact, by being selective in terms of the provision of finance and banking arrangements, is more ecologically significant than the direct impact of our trading operations. (The Co-operative Bank, 2000, p36)

We strive to be among the most progressive companies in terms of environmental management and we maintain an environmental management system in compliance with the relevant standards. (CSG, 1998, p8; 2000, p68)

Acting responsibly when it comes to the environment is part of the way that we at Deutsche Bank view ourselves. In addition to observing environmental legislation, we strive to protect natural resources like our air, water and soil. (Deutsche Bank, 2000, p10)

> *Fortis companies' everyday concern for the environment is evident in many ways. They demonstrate their willingness to reduce their share of the burden imposed on the environment through numerous practical initiatives. Examples include the recycling of plastic cups, the sorting of waste in the workplace and the responsible disposal of used office supplies.* (Fortis, 2000, p29)
>
> *In everything we do we are committed to achieving best practice relating to the environment. This is a priority for us.* (NatWest, 1998, p1)
>
> *The Rabobank Group believes sustainable growth in prosperity and well being requires careful nurturing of natural resources and the living environment. Our activities will contribute to this development.* (Rabobank, 2000, p3)
>
> *Our commitment to the environment is woven into all aspects of our operations. In recycling programmes ... in how we factor environmental stewardship into our purchasing decisions ... even in our products like the Royal Bank World Wildlife Fund Visa Affinity Card.* (Royal Bank of Canada, 2000, p39)
>
> *UBS regards sustainable development as a fundamental aspect of sound business management and ... is committed to continuing the integration of environmental aspects into business activities.* (UBS, 2000, p14)

erations into its internal processes and commercial decisions. Although reference is made only to environmental legislation, it has been left to the banks' own discretion whether or not to extend credit (or more expensive credit) to companies lacking a balanced and strong environmental policy.[22] Research in 1995 showed that there is no statistical difference between the activities of signatories and those who have not signed.[23] This means that a signatory is not necessarily a preventive or offensive bank, but may have a defensive stance. It may also mean that some proactive banks simply don't want to communicate about their environmental policy and activities or dislike public codes, like that of UNEP.

Environmental reporting

In Denmark and The Netherlands the publication of an environmental report is mandatory for companies. In Sweden and Norway companies must report environmental data in the financial annual report. In the US, Canada, Spain, Australia and South Korea it is required that financial risks and polluting emissions are reported to a certain government or market-regulating body. This usually applies to highly polluting industries but can also apply to big listed companies, in which case big banks are affected, as happens in Canada. No country requires banks to publish a separate environmental report. Within Dutch law however room exists for this to happen in time.

Most banks reaching an offensive phase will start to publish an annual environmental report. External reporting has three functions:

1 to justify (to give an account to stakeholders about implemented policy);
2 to inform (to give stakeholders information to enable them to take decisions regarding their relationship to the company); and
3 to publicize (to create good PR).

The stakeholders can, of course, be members of the organization itself. A bank's openness about its relationship to the environment creates confidence, an important commodity for banks. Many environmental reports have a primarily informative function but a shift towards justification is already evident.

Environmental reports differ widely in their form and content. The ING Group has been producing an environmental annual report since 1995 (ING Group, 2000). Their reports take a quantitative approach to internal activities and a qualitative approach to products. Various environmental annual reports are separately checked and approved by an accountant or environmental expert. NatWest's environmental report (1998) opens with a lengthy environmental policy statement and contains the results, both positive and negative, of an external environmental audit; it gives an overview of the results achieved by the bank since 1990. Bank of America (2000) has embraced the far-reaching CERES principles and used them as the basis for its report.[24] VanCity Savings has based its sustainability report largely on the GRI standard (see pages 48–50; VanCity Savings, 2000). Recent years have seen initiatives by several bodies to develop standards for environmental reporting by the financial sector: the FORGE group (FORGE, 2000; see page 164); VfU for internal aspects (VfU, 1998; see pages 156–8); EPI Finance for external aspects of environmental care (EPI Finance, 2000). Those of the VfU and EPI are the most concrete. EPI Finance proposes for commercial banking, investment banking, asset management and insurance three management performance indicators (MPIs; 1–3) and two operational performance indicators (OPIs; 4–5) (see Box 8.2):

1 environmentally relevant staff positions and available knowledge in the form of environmental departments;
2 environmental management training;
3 environmental management audits;
4 integration of environmental aspects into core activities; and
5 specific financial environmental products and services.

UBS and HypoVereinsbank, both participants in EPI Finance, show the percentage of lending checked against environmental aspects as being 54 per cent at UBS and 100 per cent at HypoVereinsbank. Over the same reporting period 939 employees at UBS received environmental training (UBS, 1999, p4). HypoVereinsbank invested 0.25 per cent of its total assets in sustainable funds (HypoVereinsbank, 2000, p2). EPI Finance is, just like VfU, a good and interesting standard for financial institutions. The question is to what extent these standards will be adopted by other banks.

A more extreme form of reporting is ethical accounting (EtAc). Developed at the end of the 1980s, this is primarily concerned with structurally informing and satisfying a wide range of stakeholders. In this method a balance is sought between short-termism and long-term wishes or necessities. This concerns not only profit figures or environmental responsibility, but also social aspects such as employment and emancipation. The objectives and values of the organization are determined by continual communication with stakeholders. One of the great advantages of this method is that it enables the organization to react quickly to changes in its environment. Box 8.4 explains the method and the process behind it and the Danish Spar Nord Bank's experience of EtAc is briefly presented.

A good environmental report has benefits both internally and externally. Internally, a good report forces a thorough analysis of the environmental effects of the product(s) and/or production process. This raises employee awareness and develops their sense of involvement. It addition, it fosters conscientious care for the environment which in turn leads to policy making. Externally, a good report fosters the credibility of and trust in the company and initiates a dialogue with stakeholders. Just as a tendency towards integration of environment and the economy exists, so the environmental report will eventually be integrated into the financial annual report. Environment policy after all is a part of overall company policy.

Other external communication activities

Some banks have chosen to publish informative brochures for their business partners, NatWest and ING, for example. Their brochures about environmental management are for employers (see Chapter 5). Most banks disseminate information about their activities and position on environment issues via the company website. UBS no longer publishes its full environmental report in hard copy, but simply places it on its website, which also makes it easily updatable. An abbreviated form is published as a brochure. Other information on the website can include inspiring examples of environmental success that should encourage business partners to aim for sustainable development, and in so doing reduce the bank's risks as a lender. A bank can also organize conferences, seminars and workshops to encourage customers to work in a sustainable way.

Sponsorship is another external communication activity. Various banks sponsor nature conservation organizations, for example ING and Bank of America. This is more neutral than sponsoring an environmental organization. The bank's borrowers may dislike the position of a particular environmental organization and associate the sponsoring bank with the views of this NGO. This type of sponsorship is more likely to be at project level or to be low profile. Via its Project Funds, the Rabobank, for example, sponsors various environmental projects, including some in developing countries. The Swiss Bankers' Association sponsored an internet site set up in 1998 by UNEP that is devoted to environmental management by banks.

Finally, participation in external networks. This too works both internally and externally. It can reinforce internal knowledge, attitude and behaviour. Externally it is a way of exercizing influence and showing involvement. Examples of such networks are the UNEP workgroup for sustainable development and banks, the WBCSD and the Environmental Bankers' Association in

Box 8.4 Ethical accounting and the Spar Nord Bank

The process of ethical accounting consists of eight steps which the company completes each year. A continual process of communication with stakeholders forms the basis. The process can be presented as a circle, as below in Figure 8.3, in which step 1 in year x follows step 8 in year x – 1.

Establishing who the stakeholders are in step 1 is important. The number of groups should be limited so that the process is manageable. A discussion is held with stakeholders about the business objectives (economic, ecological and social). This should lead to a certain degree of consensus among all stakeholders including, of course, the company itself. In step 2 these objectives are made concrete for each group (resulting in propositions). In step 3 a survey is used to establish how successful a sample of all stakeholders regards the business as being in working to achieve these objectives.[25] In step 4 an EtAc report is published based on the survey results, even if the results are not good. In step 5 the report is evaluated with the stakeholders, who themselves establish the agenda. In step 6 proposals for improvement are made on the basis of the evaluations in step 5, and these are presented to the management throughout the company. The decisions reached are reported in the EtAc report for year x + 1. In step 7 the accepted proposals for improvement for each section of the company are worked out in detail and budgeted. Step 8 involves carrying out the plans that were made and budgeted in step 7.

One important advantage of this EtAc process is the continual dialogue with the stakeholders, which enables the business to react quickly to changes in their perception of what is valuable. This has a favourable influence on the image and financial performance of all sections of the company. It enables, for example, new products to be quickly developed. Another advantage is the continual and systematic drive for improvement within the company. There are, however, a number of disadvantages: securing agreement among stakeholders is not a natural part of the process; most of the above-mentioned advantages will be publicly available in the EtAc report and therefore available to competitors; while the revenue is not immediately visible, the costs are.

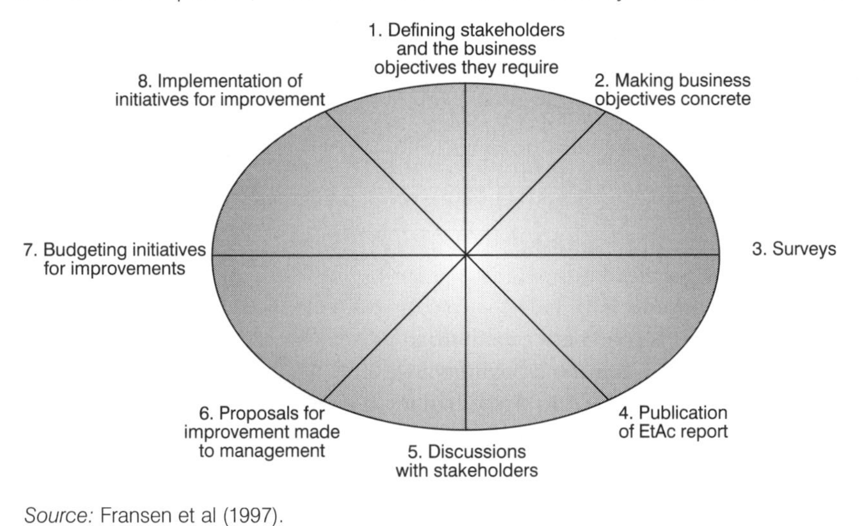

Source: Fransen et al (1997).

Figure 8.3 *The ethical accounting model*

Furthermore, the extent to which the scale of the organization influences the method is not clear and the method does not enable benchmarking against other companies since the objectivity of the EtAc report cannot be guaranteed. Whereas an environmental annual report includes situational (or numerical) indicators, the EtAc report contains only indicators of perception. All that is possible is the analysis of a company's performance over time.

The Danish Spar Nord Bank (SNB) is the seventh biggest bank in Denmark in terms of assets. Having published an EtAc report annually between 1989 and 1994, it is experienced in this field. In 1994 it published 'The Ethical Accounting Statement'. The SNB has identified four groups of stakeholders: employees, customers, local community and shareholders. For the SNB one of the most important advantages of EtAc was that it enabled the bank to quickly find out about changes in the needs and values held by its stakeholders (SBN, 1994). Since 1997 the EtAc report has been replaced by the bank's Quality Accounting Statement (SNB, 2000) and a QA report which forms part of the annual financial report. This uses the EtAc methodology but is directed at two main groups: customers and employees.

the US. As part of UNEP six network meetings have now been held at which banks have openly discussed both developments and their experiences.[26] National bankers' associations can fulfil a similar networking function for environmental activities. Thus, the bankers' associations in the UK and Canada have published brochures for businesses about banks and the environment.[27] Furthermore, the value of contacts with the environmental movement and central and local government is obvious. In 1999 Dutch banks and the central government set up an environmental dialogue between banks and governments. The aim of this dialogue is to encourage the further exploration of sustainable development within banks and to jointly solve problems encountered. In Germany, the US, Sweden and Switzerland there are similar initiatives, also usually set up by national banking associations.[28] Some initiatives have lead to banking sector guidelines. Others have focused on research into new products and breaking down barriers to effective environmental performance.

External accountability and the risks of external communication

While communicating a sense of social responsibility and involvement by, for example, sponsorship or signing sustainable declarations, is important to realize that it is not without risk. Before embarking on this, banks should consider that once they become active externally, their internal organization may be called into question and must be able to withstand scrutiny. A rule of thumb about communication about social involvement is that a bank should be alert to 'green washing'.[29] The ever faster dissemination of information to members of the public as a result of information technology makes it possible to assemble a group quickly and efficiently to protest or take action; consider the dissatisfaction Shell shareholders felt about the proposed dumping of the Brent Spar oil platform and the discussion about human rights in Nigeria arising from the pollution being caused there. In just a few hours the whole world was aware of Shell's plans. The consumer backlash quickly followed (Van Riemsdijk, 1994).

In another case, a Dutch environmental organization protested in 1998 against ABN AMRO because of its environmental code and the fact that it was financing Freeport-McMorran Copper and Gold, an American company that was planning to exploit an extremely polluting copper and gold mine in West Papua New Guinea. The NGO published a brochure entitled 'Do you know what ABN AMRO is doing with your money?' (Environmental Defence, 1998). Many clients sent ABN AMRO standardized letters which stated their opposition to this kind of financing. At the end of 1998 the bank agreed to require Freeport to pay for an independent investigation. Freeport had to agree the scope of the investigation with the local population and the NGOs involved (ABN AMRO NewsNet, 17 December 1999). The environmental organization considered the action a success and it enjoyed worldwide attention.[30] NGOs had already called governments and polluting companies to account, but incidents like the case of ABN AMRO are establishing an important precedent: that NGOs will also put pressure on financiers and investors. This precedent was followed in 2000 by an action by Greenpeace and Friends of the Earth in The Netherlands and other developed countries. As a result of research by AIDEnvironment, these NGOs called all major Dutch banks to account for their financial links with palm oil plantation projects in Indonesia.[31] As a result of negligent trading behaviour there were worries about funding forest destruction. Each bank reacted in its own way. The Rabobank denied involvement in such scandalous projects in its sustainability report (Rabobank, 2000, p20). As a rule, the environmental aspects of a company or project should be investigated before financing is granted, especially when it is in a developing country. The investigation should extend beyond financial considerations to the ecological and social risks and the risk to reputation. When this does not happen environmental or ethical issues sooner or later surface, to bad effect. This is especially important for banking because it requires the public's trust. What is especially concerning for banks is that an interest group's accusation need not be valid to damage public perception. This was partly the case in the issue of Shell's proposed dumping of the Brent Spar oil platform in the North Sea.[32]

Negative publicity has direct consequences for the sale of products and profitability. Share values fall by an average of 2 per cent during the period of dispute; greater decreases have occurred.[33] Not surprisingly, shareholders respond by demanding a sound environmental policy. Negative publicity also has consequences for staffing. A positive image on the environment and sustainable development usually attract good workers. At the end of 1997 it appeared abundantly clear: incidents such as Brent Spar and Nigeria had caused Shell to fall sharply in the estimation of jobseekers, including top managers. In the list of most popular European companies Shell slipped from third to eleventh place. The apparently environmentally friendly BP, on the other hand, jumped to second place (*Algemeen Dagblad*, 25 September 1997).

The signing of the UNEP declaration raises the question of whether stakeholders will harass a bank with actions or press charges if they regard the bank as having contravened the principles of the declaration. This applies to the bank's larger and smaller customers. However, it is harder for a bank to find out about a small company's approach to sustainability. Some politically acceptable activities are not socially acceptable. This appeared to be the case with Brent Spar. When a bank adopts an open, honest approach, it can limit consumer actions. By making itself vulnerable, a bank indicates that it has a positive approach to sustainability, but is nevertheless unable to entirely prevent environmentally damaging credit activities.

Banking differs from other business sectors in how it is affected by external accountability. While issues of responsibility and legal accountability are in principle the same, there is a far lesser risk of consumer action. An example illustrates this. The consumer actions against Shell at the time of the proposed dumping of the Brent Spar oil platform caused turnover to fall by 30 per cent in Germany. In The Netherlands it was just 1 per cent and the reason lay to some degree in the AirMiles programme: Shell petrol stations were just too good a source of AirMiles for many Dutch motorists.[34] Worry about personal financial loss apparently acts as a brake on the public's readiness to take action. A similar mechanism operates in banking: consumers unsatisfied with the policy of bank A do not quickly cancel their mortgage and move to bank B. There exists a sort of 'zone of tolerance'. It does vary per consumer but in general is fairly wide and is created by the transaction cost to the consumer of change. Customers do act in other ways, for example by demonstrating, writing letters, and collecting signatures. And indeed an economic action can very much be felt in terms of potential new clients and business partners who can simply change banks.

Moreover, the possibilities for taking quick, directed action are ever greater. Various organizations use the internet to keep an eye on companies in all sorts of fields and incite consumer action when they consider it necessary. Morgan Stanley Dean Witter and Credit Suisse were held accountable for their considerable participation in a government bond issue to finance the Three Gorges Dam in China in 2000. In the same year Sumitomo was branded in a similar way for its co-financing of the Sardar Sarovar Dam in India.[35] The National Wildlife Foundation has developed a Finance and Environment programme to mobilize the public and clients should one or more banks finance a project that poses a threat to nature or the environment. The website carries a standard letter together with the addresses and the CEOs for all banks in the US.[36]

A more potent reason why banks are chosen as targets is that in principle the public sides with the weaker party, which is never the bank; banks are regarded as having deep pockets. When powerful businesses are involved, such as Shell and major banks, the drive to take action increases sharply. The time when polluters themselves were the focus of attention is slowly passing: the 'mighty' financiers behind such businesses will be increasingly held accountable. An open dialogue with the public can reinforce the mutual realization that striving for sustainability holds many possibilities. Banks can even prevent action against them by engaging in a stakeholder dialogue to find out what the public considers acceptable. A continuous open dialogue and demonstrably good behaviour are therefore the only ways to keep the possibility of future unforeseen collisions under any kind of control.

CONCLUSION

To integrate sustainability into a bank it is necessary to integrate it into the whole business operation, from auxiliary services, premises, the granting of credit, investments, and insurances, to project financing and marketing. The allocation of tasks, authorization, responsibilities and means must take place via a clear organizational and consultative structure for environmental and sustainability aspects. This structure must, of course, be appropriate to the bank's structure, size, culture and ambitions. Attention and support are best guaranteed by there being one or more people at the top who are given responsibility. The challenge is to encourage staff and management to integrate environmental care into their activities. This can be done by use of directives or can be 'bottom–up'. In such a bottom–up scenario, a broad programme of communications influences the knowledge, attitude and behaviour of staff and management to such an extent that employees themselves initiate change. In particular, giving an incentive to people who live in a reality in which there seems little or no room for sustainability is important (as eg in financial market activities).

Ideas stemming from the 'learning organization' can be of great importance (De Geus et al, 1997). What is important is the individual's freedom to work towards environmental care or sustainability in his or her own way. Organizational learning can take place in various ways: through the systematic application of experience, through continual experimentation (and thus by making mistakes), through comparison with other banks (the use of benchmarking) and through the transmission of knowledge and experience (Buiter et al, 1995). Organizational learning also requires a change in the management process and a solution for the problem of the gulf between the individual's personal interest and that of the company. In a dynamic environment learning is essential for survival and can be of great value in the change process which environmental care brings with it (Collins and Porras, 1994).

Another method is that a dominant leader introduces changes by using directives. This method in particular requires a separate department at staff department level. Eventually the activities should be embedded in the line organization. Responsibility and authority for the implementation should also lie with line managers. In this 'top–down' method it is probable that less use will be made of the knowledge, experience and skills present lower down the organization.

During the integration of environmental care or sustainability into the total company concept, the elements summarized below should be realized (De Groene, 1995). The form and weight of each element is, of course, dependent on the size, structure, culture and objectives of the bank concerned.

1 Organization structure:
   - line responsibilities;
   - total responsibility at the top;
   - advice, coordination and control among staff-level employees;
   - possible (specific) knowledge from external agencies.

2 Tasks, responsibilities and authorization:
   - written for staff-level functions (eg legal affairs and a separate staff department for the environment and sustainable development);
   - written for line functions with specific tasks (eg credit risk management, general and technical services and purchasing).
3 Consultative structure:
   - at management level, staff level and the work floor;
   - horizontal and vertical;
   - possibly thematically (eg about financing sustainable energy).
4 Internal company instructions (where possible written):
   - measurement and registration of internal environmental aspects;
   - internal controls;
   - education and providing information;
   - investigation of the environmental care system;
   - internal and external reporting.

Part III

In Reflection

'The world we have created today
as a result of our thinking thus far,
has created problems which cannot be solved
by thinking the way we thought when we created them.'
Einstein

9

Sustainable Banking in Perspective: The Cases of 34 International Banks

INTRODUCTION

In earlier chapters the activities of a range of banks were explored. Various examples were given and many were taken from smaller banks such as Triodos Bank and The Co-operative Bank. Within mainstream banking, the same few banks were often referred to, such as UBS, ING, Deutsche Bank and Bank of America. These are clearly the banks most active and accessible in the field of sustainable banking (accessible in that information about them exists because they themselves publish environmental reports or are covered by other studies). Since the preceding chapters focused on the actions of individual banks in particular cases, an incomplete picture has been created in which the context in which environmental activities within a bank take place is not clear, nor is it clear to what extent environmental care has permeated throughout the whole banking sector. This chapter rectifies that by looking closely at the environmental activities of a number of banks selected for an a priori unrelated criterion: the scale of their banking activities. A set of environmental activities has been defined and each bank in the study is assessed for its performance in those activities.

Therefore a sample of 34 mainstream banks is taken. Box 9.1 describes the geographical distribution of the banks in the survey, standardization issues and the selection criteria that were used.

The most valuable and accessible source of information about each bank's road to sustainability is its most recent environmental report. Where the bank does not publish one, its most recent annual community report has been used, or failing that its most recent annual financial report; these reports cover the period 1998 to 2000. In addition, the banks' official websites were used. Evidence in other publications of a bank's activity in one of the listed areas that is not corroborated by the bank's own official, public channels including press releases, has been ignored.[1] Although interviews or questionnaires are appropriate methods for information gathering as well, the scope of this survey prevented the use of them. In short, banks have not been approached directly; not publicly communicating clearly about certain environmental aspects has been taken as an indication that the bank attaches insufficient priority to them,

Box 9.1 Sample survey of mainstream banks

The selection of banks was made on the basis of three criteria:

1 Total assets: This factor has a direct bearing on the banking sector's direct and indirect environmental impact. A bank must have assets of at least €100 billion.
2 Location: Only banks from developed countries are considered and a maximum of three banks have been used per country. This is to prevent an over-representation of a few limited countries (50 per cent of the top 50 banks in terms of assets have headquarters in Germany, Japan or the US).
3 Number: a maximum of 50 banks may be included in the sample.

In total these selection criteria were fulfilled by 34 banks. All of these banks are in the top 78 banks of 1999 in terms of assets.[2] The use of the second criterion considerably broadened the scope of the survey to 15 countries. Their geographical distribution is shown in Figure 9.1.

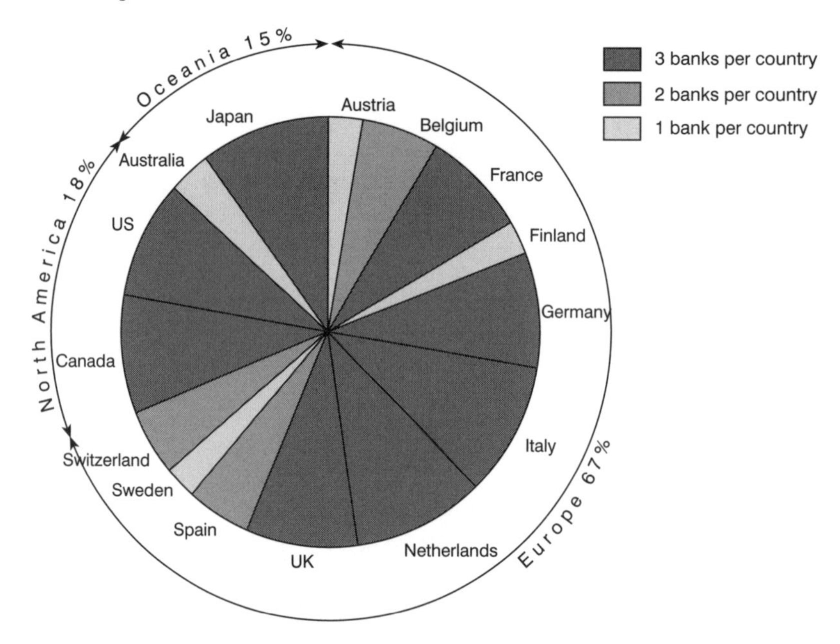

Figure 9.1 *Geographical distribution of the 34 selected banks*

The selection criteria account for the high proportion of European banks in the selection. Still, this presence can be justified by the fact that sufficient legislative and cultural differences exist between European countries to give them separate identities. Moreover, the European banking market is important relative to other world regions (see Table 4.1). In the analyses in the following sections frequent comparison is made between the three world regions Europe, North America and Oceania (Japan and Australia). Where the analyses focuses on numbers of banks in a world region, standardization for the region is applied. This means that for the issue in question the share of banks scoring in the total banks in one region will be compared with the share of banks scoring in the total banks for another region. For example, if four banks in Oceania (out of four) and three banks in North America (out of six) have signed the UNEP declaration, the standardized score for Oceania is 100 per cent and for North America 50 per cent. This facilitates comparison. Obviously, the results for Europe will be stronger.

The accumulated assets of the 34 banks are worth €13 trillion; that is twice the accumulated 1999 GDPs of all low and middle income countries (World Bank classification). The top ten selected banks are, in fact, the world's top ten in terms of assets. Their assets alone roughly amount to the accumulated GDP for all 108 'developing countries' in 1999 (World Bank, 2000, p274). The ratio of the assets of these 34 banks to the GDP of the 23 developed countries (as defined in this book, see Box 1.1) is 57 per cent. Clearly, the banks selected have a significant stake in the development of the global economy.[3]

Appendix VIII shows the following features for each of the 34 banks selected:

- name and origin: full name, and world region and country of origin or headquarters;
- size: assets, number of offices, number of employees, number of countries with offices;
- performance: cost–benefit ratios and profitability; and
- strategy: the banking markets in which they are active and their geographical reach.

and belonging to the defensive or at the most the preventive phase. The accuracy of the information sources or quality of the reported data has not been challenged.

The analysis has involved scrutinizing for each bank the following set of activities:[4]

1 *Codes of conduct, environmental reporting, and EMSs (pages 188–90):*
   - environmental statements and codes of conduct;
   - environmental reports;
   - standards for and certification of EMSs.
2 *Environmental care in practice: reported policy, objectives and data (pages 190–2):*
   - quantitative and qualitative internal and external environmental data;
   - internal and external objectives for the near future;
   - environmental policy.
3 *Environmental care in practice: products and risk management (pages 192–8):*
   - environmental risk analyses, sector exclusions and use of World Bank or OECD guidelines;
   - investment products, loans, leases, savings accounts, credit cards and insurance products with an environmental feature;
   - micro-credits, debt-for-nature-swaps and climate products;
   - environmental advisorial services or activities.
4 *Socioeconomic activities and sponsoring (pages 197–8):*
   - community building, sponsoring and internal social objectives or activities.

All these aspects are integrated on pages 199–200 in an integral sustainability score. Pages 200–3 sums up and relates the integral scores to general features of banks, like size, profitability and efficiency.

Codes of conduct, environmental reporting and environmental management systems (EMSs)

This section gives a picture of how banks communicate about sustainability and their interest in the standardization of environmental activities. In Figure 9.2 an overview of such communication is given.

Codes of conduct, such as those of the UNEP and ICC, are popular among the 34 major international banks studied. More than half of them have signed the sector specific UNEP declaration and half the more general ICC declaration. Although in absolute numbers (in and outside the sample) mostly European banks have signed up to the UNEP declaration on sustainability, this is not supported in the standardized analysis of the 34 large international banks in relative numbers. Standardized for differences in numbers of banks per region, the North American banks score a little better than European banks. What is striking is the absence of Oceanian banks in the UNEP declaration.[5] The ICC declaration enjoys popularity in all regions, whereas European banks show a clear preference for the UNEP declaration. Canadian and German banks have simply signed both declarations, which may be evidence of peer pressure: when one or two of a country's big banks take a particular stance on sustainability, other banks quickly follow. The same trend seems to apply to membership of the WBCSD. Only four banks are members, but they are all European and mainly Dutch. Early in 2000 the Rabobank in The Netherlands announced its membership of the WBCSD and hereby its support for sustainability. At the end of that year ABN AMRO followed, and in 2001 ING is deciding whether to join as well.[6]

Environmental reporting by banks appears to be very much a European phenomenon. Banks in other regions lag behind. Cultural differences are a determining factor. Transparency is not highly regarded in Japan and is sometimes even considered a weakness.[7] While European banks focus more on environmental aspects, North American and Oceanian banks concentrate more on community involvement. For this reason, a distinction is made in the figure between banks publishing an environmental report (32 per cent), banks which devote at least one column[8] to the environment in the annual financial report or mention it on their website (24 per cent), and banks which mention only community involvement in the annual financial report or on their website (15 per cent). Accumulated, sustainability in some form is then reported on by 71 per cent of the selection.[9] French and Italian banks stand out in that they report on neither the environment nor their community involvement in the above sense.

Six banks publish a 'sustainability' report covering both environmental and social aspects (Bank of America, NatWest, Rabobank and the three German banks). Barclays covers environment and social and/or community aspects in two separate reports. The two Swiss banks and the two remaining Dutch banks publish 'pure' environmental reports. This leaves 23 out of 34 banks publishing no environmental or sustainability report at all (68 per cent). Once again, conformity within a country is striking; all banks selected in Germany, Switzerland and The Netherlands publish an environmental or sustainability report. Some peer

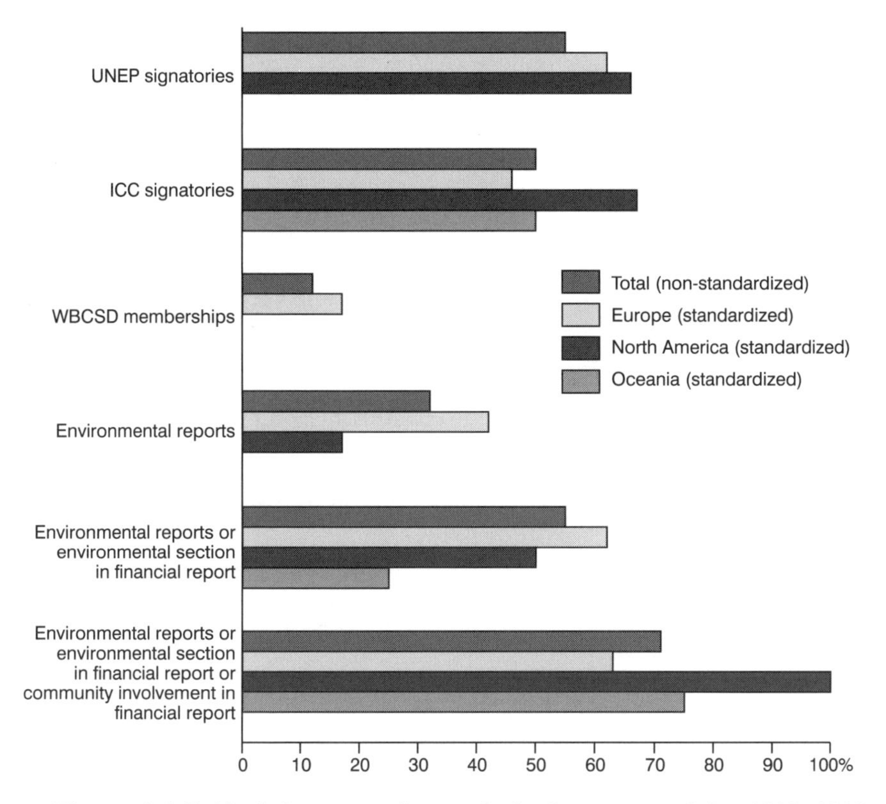

Figure 9.2 *Public declarations and reports by banks on sustainability, 1998–2000*

pressure also seems to exist in the UK, where in addition to NatWest and Barclays, Lloyds TSB for example publishes an environmental report. Remarkably, while the economy is global and most banks are as well, peer pressure appears confined within national or regional borders. In North America, Bank of America is alone in publishing an environmental report, and has been doing so since 1993, when along with NatWest it pioneered this among mainstream banks. Next to Bank of America and NatWest, the two Swiss banks and ING are notable for the frequency with which they publish; all have now published five to six annual or multi-annual environmental reports. The other six banks have just started publishing an environmental report (one or two reports up till now). Not a single bank has yet based an environmental report on the GRI guidelines.

The most proactive banks attempt to systematically reduce the environmental impact of their internal processes. A formal EMS can achieve this. No bank has achieved or wished to achieve EMAS or BS-7750 certification for its internal or external communication or its environmental processes. Four European banks do have ISO certification: BBVA (some locations), Deutsche Bank (national office network) and UBS and Credit Suisse (for their worldwide EMSs). ING's 1998 environmental report mentioned the company's aim of achieving ISO certification but the 1999 report makes no mention of this anymore. It is not out of the question that some European banks will opt for EMAS II certification, despite ISO 14001 being more widely recognized. There is no sign that banks in North America or Oceania will acquire similar certification in the near future.

In the mid-1990s, UBS, Credit Suisse and HypoVereinsbank were involved in the development of the VfU standard for internal environmental care within the banking sector. In 2000 these same banks, now together with Deutsche Bank, worked on the development of the EPI standard for external environmental care within banks. In that same year, two British banks, Barclays and NatWest, were involved in the development of an internal and external management standard for the banking sector, the modular FORGE method. While other banks in the sample have received these industry standards with interest, they have not yet been adopted.

Environmental care in practice: Reported policy, objectives and data

Figure 9.3 shows the internal environmental activities of the 34 selected banks. The following conclusions can be drawn. Close to 60 per cent of the banks worldwide have an environmental policy statement: looking at the regional differentiation this applies to 67 per cent of the European banks, 50 per cent of the banks in North America and 25 per cent of the banks in Oceania (standardized). A good 60 per cent of the banks report the environmental impact of their internal processes in qualitative terms (eg 'we reduce the use of paper'). This breaks down into more than 70 per cent of the banks in Europe, 50 per cent in North America, and 25 per cent in Oceania (standardized). Transparency in quantitative terms (eg 'we have reduced the use of paper by 20 per cent') is, however, a lot less prevalent: 38 per cent of banks in total, none at all in Oceania and only 17 per cent in North America and 50 per cent in Europe. The European banks also offer the most transparency about their objectives related to the reduction of the environmental impact of their internal processes for the near future (42 per cent). HypoVereinsbank and Rabobank mention, for example, a better monitoring of this impact.

Differences within regions also occur. The Bank of America is the only North American bank to score on all aspects covered in Figure 9.3. Most other banks in North America make no mention of any of these aspects. Exceptions to this are Citigroup and the Royal Bank of Canada (environmental policy and qualitative targets). Moreover, this Canadian bank reports a future objective: ISO certification for its EMS. In Oceania only National Australia Bank (NAB) has an environmental policy, and Sumitomo is the only bank that reports on some of its internal environmental impacts qualitatively. Not a single bank in this region reports its activities quantitatively, nor reports having any objectives for the near future. Within Europe, too, there is great variation. The Italian and French banks score negatively on all facets in Figure 9.3; Société Générale is the exception, which makes limited mention of the environmental impact of its

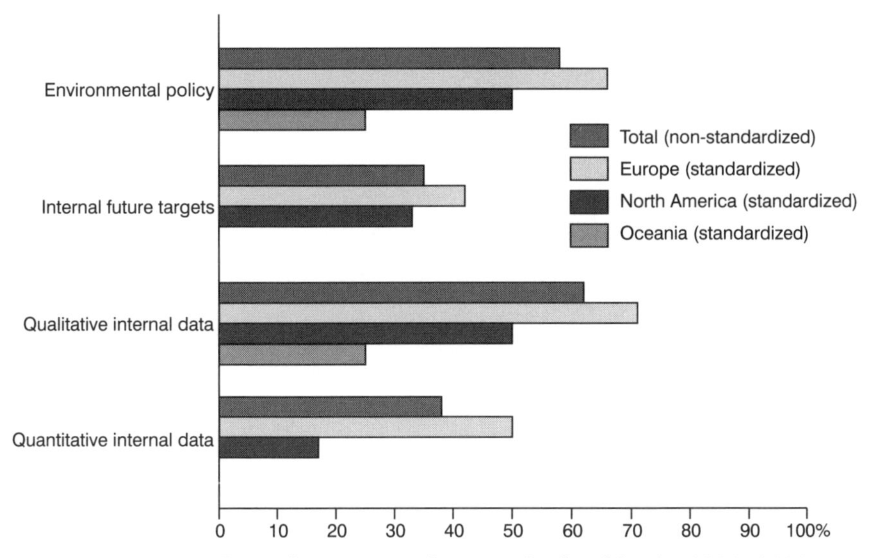

Figure 9.3 *Internal environmental aspects of selected banks, 1998–2000*

internal processes in qualitative form. In contrast to this, the Dutch, German, Swiss and English banks (except HSBC) all score positively on all aspects. BBVA and Fortis score slightly less (all aspects except objectives) and Bank Austria, MeritaNordbanken, Svenska Handelsbanken and HSBC score on two facets (environmental policy and qualitative care).

As for environmental policy, a distinction can be made between banks with a comprehensive policy statement of at least one page, published separately or within an environmental or financial report, and banks with a limited expression of environmental policy (ie only a few policy sentences on the subject in the financial report), and banks with no policy statements on environmental issues at all. In Figure 9.3 the first two groups have been aggregated; in Figure 9.4 the differences are shown.

It is striking that it is principally the North American banks which have a comprehensive environmental policy (50 per cent); a limited expression of policy objectives does not occur here at all. 46 per cent of the European banks have a comprehensive environmental policy, while for 21 per cent a few words in the financial report suffice. Another striking thing is that UniCredito, BSCH, Royal Bank of Canada and Bank of Montreal are signatories to the UNEP declaration, but don't report any environmental policy objectives. Apparently signing this declaration is no guarantee of a bank's having its own environmental policy. What it may indicate is that these banks regard the statement as their own environmental policy (as eg Dresdner does in its environmental report), albeit that there is no mention of this in their annual reports. On the other hand, that a bank pursues an environmental policy is no indication that it subscribes to the tenets of the UNEP declaration. Bank of America is widely regarded to be among the front runners in terms of environmental care but has nevertheless not wished to sign the UNEP declaration.

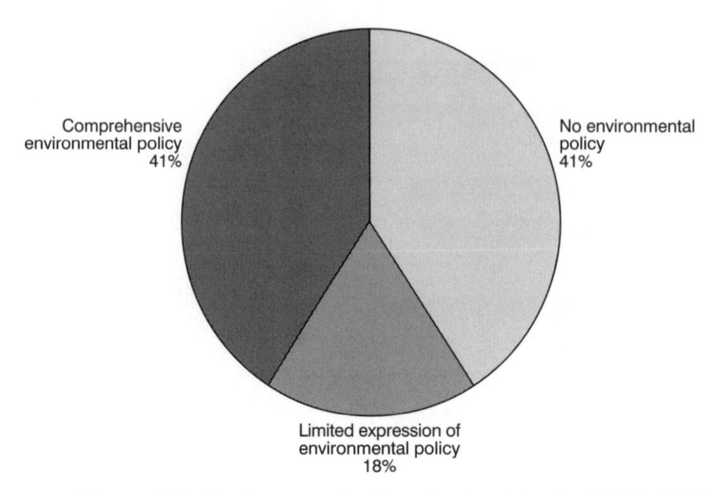

Figure 9.4 *Environmental policy of selected banks, 1998–2000*

Environmental care in practice: Products and risk management

Introduction

In one way or another 85 per cent of the banks express concern for external environmental care in qualitative terms, such as environmental risk assessment. Only four banks clearly regard environmental issues as leveraging no client opportunities or threats. These are: Bank of Montreal, BNP Paribas, SanPaolo IMI and Bank of Tokyo-Mitsubishi (BTM). Only 29 per cent of the banks offer some quantitative insight into their external environmental activities, such as the financial volumes of environmental loans. Switzerland and The Netherlands are the only countries in which all banks do this is. Similarly, around 25 per cent of all banks explicitly state their objectives for external environmental activities. Of these, 78 per cent see possibilities for, and have ambitions regarding the development of, climate products. These banks are Royal Bank of Canada, Bank of America, Deutsche Bank, HypoVereinsbank, ING, UBS and Credit Suisse. Rabobank does not appear in this list but given its participation in the World Bank's Prototype Carbon Fund (it and Deutsche Bank being the only participants from the financial sector), it does see possibilities. Both Barclays and Credit Suisse have formulated objectives regarding social accounting. ING mentions the possibility of signing the UNEP declaration and Rabobank has set itself the goal of reporting in a better and more structured way. Furthermore, the two Swiss banks and HypoVereinsbank and ING explicitly state that they would like to develop and market further environmental products.

In the following sub-sections banks' external environmental care is looked at in detail. The section beginning on page 193 looks at the integration of environmental risk analyses into lending to and financing customers and projects, the extent to which this involves the use of international guidelines

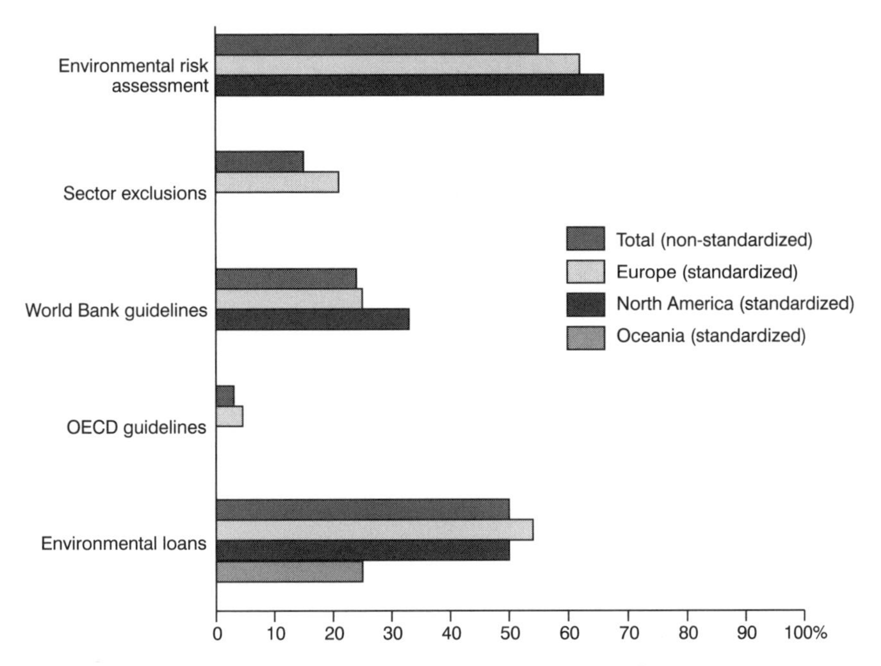

Figure 9.5 *Environmental aspects of granting credit, 1998–2000*

and the extent to which banks exclude sectors on the basis of environmental concerns. In anticipation of the section on pages 195–7 the extent to which banks offer specific environmental loans to their customers is also shown. The section beginning on page 195 deals with the interest the 34 sample banks have shown in specific financial products related to sustainability. These include investment funds, leases and insurances. And finally, the section beginning on page 197 offers an overview of the degree to which individual banks use the products and services discussed.

Environmental risk assessment and guidelines

Figure 9.5 gives an insight into how international banks deal with the environmental risks inherent in the granting of credit.

Fifty-six per cent of the banks pay close attention to environmental aspects when setting up credit and financing agreements. This percentage is low; banks have been fully aware of environmental risks since the beginning of the 1990s. In the US, CERCLA even caused some banks to go bankrupt. What is more striking therefore is that not all North American banks conduct risk analyses. No evidence of this was to be found at either Chase Manhattan or Bank of Montreal. Citigroup is the most transparent of all banks in the sample and offers complete insight into the losses suffered and funds set aside for environmental risks (though only regarding the environmental risks inherent in its insurance activities). No bank in Oceania, Belgium, France or Italy conducts environmental risk analyses. All other banks (except for BSCH) have an environmental risk policy for their credit facilities.

Only within the group of banks who indicate that they explicitly take into account the existence of environmental risks, do offensive banks exist which make sector choices or use international guidelines for financing; when financing projects or companies in developing countries or countries in transition, some of these banks explicitly adhere to the guidelines of the World Bank (24 per cent) and/or the OECD (3 per cent) and/or set one or more sectors aside from financing (15 per cent). Only HSBC adheres to the generic OECD guidelines while not following the sector specific World Bank guidelines. Banks who do take account of the World Bank guidelines with regard to their international activities are Citigroup, Bank of America, Deutsche Bank, HypoVereinsbank, ABN AMRO, Rabobank, UBS and Credit Suisse. North American banks are more eager to use this guideline than European banks (standardized).

By contrast, only European banks explicitly state sectors or activities which they will not finance. In this, the Rabobank stands out with its code of conduct for the financing of GMOs. Deutsche Bank reports that it does not wish to finance projects or companies which 'it believes pose serious threats to the environment' (Deutsche Bank, 2000, p22). HypoVereinsbank has a similar phrase. HSBC will not work on defence equipment and landmines, and NatWest will not finance projects that could lead to the destruction of the tropical rainforests. Such exclusion of sectors or activities is for banks still a delicate subject. The banks with something to say on this subject are quick to note in their annual financial or environmental reports that strict adherence to this principle is not always possible in the financial world. In many cases exclusion by one bank simply means a project is financed by another bank, whereas were the bank with an active interest in the environment to be involved, it could exercise some influence over the environmental consequences. It is in this respect interesting to see that Bank of America reports that it wishes to play no part in the building of the Three Gorges Dam in China, while Société Générale reports (and proudly it seems) that it is involved in this project.

The question is whether it is actually sensible to aim for the exclusion of sectors. On the one hand, there are perhaps certain sectors or projects that the entire reputable financial world will refuse to engage in, such as the drugs trade or projects involving the destruction of the rainforests. On the other hand, many projects are so specific that it would be better if more general criteria for financing on a project-by-project basis were applied (instead of excluding entire sectors). Where, for example, a bank does not wish to finance weapons producers, how should the bank react to a request for credit from a company that derives 2 per cent of its turnover from supplying parts to weapon producers? To overcome this issue, some banks apply generic codes of conduct when assessing financing. ING, for example, has a comprehensive set of 'Business Principles' and HSBC has published a brochure on its position on responsible financing.[10] Transparency on policy and investment decisions could enhance the financial sector's image.

Half of all banks have specific environmental loans in their portfolio of services. Such products are quite popular in Europe and North America. In every country except Australia and Switzerland, at least one bank offers an environmental loan. Within Europe, ten banks offer an environmental loan guaranteed by the EIF. In total nine banks have an environmental loan without fiscal or other support from the government, of which five are in Europe. NatWest has set aside €33 million of its profit to subsidize attractive interest rates on business loans for environmental projects (NatWest, 1998, p13).

Financial products for environmental care

Besides environmental loans, a bank may offer its clients a whole range of other products and services related to sustainability. Figure 9.6 shows a broad scope of such financial products and services offered by the selected banks.

As shown in Chapter 5, a whole range of different investment funds related to sustainability are offered by banks nowadays. Examples include funds aimed at environmental pioneers, ie aimed at a select group of companies with a distinctive environmental or social profile, or aimed at the best companies within a sector from an environmental perspective (best-in-class approach). Of the selected banks, ten offer customers the opportunity to invest in such funds (in total 29 per cent of the selected banks); European banks are the most active in this field. Many differences exist between banks, however. All Swiss, German

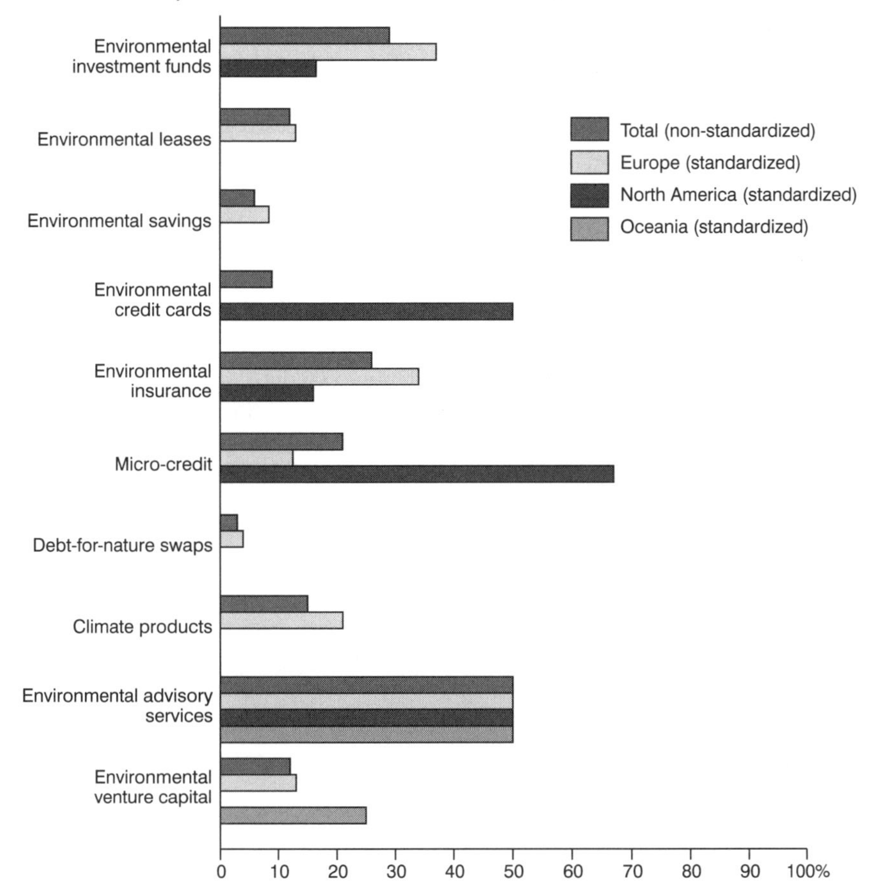

Figure 9.6 *Financial products geared to the environment, 1998–2000*

and Dutch banks are active in this area. The Dutch banks are supported by governmental fiscal facilities. Within the remit of the Fiscal Green Ruling both ING and Rabobank have developed a green savings product, and ABN AMRO has an investment fund that is quoted on the stock exchange. These savings products are the only environmental savings products offered by any of the 34 banks. Peer pressure is clearly evident in The Netherlands in both the Fiscal Green Ruling and the introduction of best-in-class investment funds. Rabobank assumed the lead in both areas; the two others quickly followed with comparable products. Swiss banks are also very active in the area of sustainable investment funds. UBS now has four funds: an environmental pioneers' fund, an eco-performance fund, a new best-in-class fund aimed exclusively at the Japanese market (Dr Eco), and a fuel-cell basket for its investment banking activities. UBS is clearly setting the tone. Dresdner, Citigroup and ING offer wealthy customers the opportunity to compile their own sustainable portfolios. Société Générale has launched its Bull Ethical Certificate, HypoVereinsbank offers (in addition to an equity fund) index certificates for the Dow Jones Sustainability Group World Index and Deutsche Bank is the custody bank for the WWF Panda Return Fund.

Environmental leases are offered only by European banks and environmental credit cards only by North American banks. Together with Deutsche Bank, all Dutch banks are active in the market for environmental leases (a total of 12 per cent of the banks selected). In The Netherlands this is again complemented by governmental facilities. Only three banks (9 per cent) have an environmental affinity credit card (or similar): Royal Bank of Canada has a 'WWF Visa Affinity Card' and Citigroup a 'Environmental Defense Platium Master Card'. In both cases, per purchase the bank makes a donation to a charity of the user's choice (at no cost to the client). Bank of America offers its clients check boxes, whereby the consumer pays a little extra per cheque and this is donated to a good cause of his or her choice.

One activity of banks that requires little from the core financial services is the provision of environmental advisorial services to industrial customers. Throughout the world's regions, the percentage of banks which offer this is the same: 50 per cent, though what it involves varies from publishing brochures about the realization of energy saving to offering customized consultancy and on-site consultancy. Unicredito and NatWest are the front runners in this area.

Just as investment funds, insurance is a product for which an environmental version is more likely to exist. Many banks give their customers the option of taking out an insurance policy for environmental damage that they have either suffered or caused to third parties (26 per cent). No banks in Oceania have such a product and in North America only Citigroup is active in this area. Within Europe, The Netherlands is notable: every Dutch bank in the selection offers its customers environmental damage insurance. Insurers in The Netherlands jointly developed an innovative insurance product, which was subsequently offered by all banks (although Fortis lags behind on this).[11] The Dutch banks are also All Finance institutions for which insurance activities form an important part of their portfolio of financial services. It is important when considering insurance products to see whether they really do belong to the core activity of each bank in the sample.

Adjusted for those which don't, 69 per cent of the relevant institutions offer environmental insurance to their customers. See Table VIII.2 in Appendix VIII for a list of banks which do see insurance as an integral part of their services. In other words, when it comes to banks where insurance is a relevant business activity, the majority of these banks also have an environmental insurance product.

In conclusion

The external products and services of active banks towards sustainability has been discussed in this section. Table 9.1 shows for each bank whether a particular product or service is available. The list is not exclusive. Individual banks offering a product which is included in Figure 9.6 but not in Table 9.1 have already been mentioned in the previous section.

Most of the above-mentioned products and services will be found at banks taking an offensive stance on sustainability. Banks which want to conduct their business in a sustainable way will go one step further. They will stimulate innovations and apply these to their own activities in an effort to find a solution to the diverse aspects of the issue of sustainability. Activities appropriate to this category include: environmental venture capital activities, micro-credits (in particular in developing countries), climate products and debt-for-nature swaps. With regard to the latter, only Deutsche Bank has to date played a role and this initiative really has not been widely adopted among mainstream banks. New financial innovations are perhaps necessary to offer solutions to both the question of debts and the dire need for nature conservation in developing countries. Micro-credits are a very interesting financial instrument offering an economic way out of poverty to people who normally cannot obtain regular bank financing , by supplying them with very small-scale finance. Seven banks (21 per cent) offer micro-credits (four banks in North America, three banks in Europe and none in Oceania) although only four banks actually use this financial product in developing countries: Royal Bank of Canada, Citigroup, Deutsche Bank and Rabobank. A whole new area is financial products and innovations related to international climate policy. This area is still very new and only the progressive banks are as yet active in it (15 per cent, all European). The section beginning on page 192 showed that many banks believe that interesting possibilities lie ahead in this area.

SOCIOECONOMIC ACTIVITIES AND SPONSORING

That the social component has played a role in the pursuit of sustainability alongside environmental care over recent years for many banks is apparent from the scores of banks on a number of social components. This is usually in the form of charity and sponsorship For this reason, sponsorship benefiting nature and the environment has been included in Figure 9.7. This is an area in which nearly 60 per cent of all banks are active. The differences between the regions are not very great. From the figure it appears that for North America and Oceania, community involvement is a key issue. In both regions all banks are

Table 9.1 *Financial products and services offered by individual banks, 1998–2000*

	Environ-mental assessment	*Environ-mental loans*	*Environ-mental funds*	*Environ-mental leasing*	*Environ-mental insurance*	*Environ-mental advisory services*	*Environ-mental venture capital*
Royal Bank Canada	✗	✗				✗	
CIBC	✗						
Bank of Montreal							
Citigroup	✗	✗	✗		✗	✗	
Bank of America	✗	✗				✗	
Chase Manhattan							
Bank Austria	✗	✗				✗	
Fortis Bank							
KBC Bank		✗			✗		
MeritaNordbanken	✗	✗				✗	
BNP Paribas							
Crédit Agricole		✗					
Société Générale			✗				
Deutsche Bank	✗	✗	✗	✗	✗	✗	✗
HypoVereinsbank	✗	✗	✗			✗	
Dresdner Bank	✗		✗			✗	
Banca Intesa		✗					
UniCredito Italiano		✗			✗	✗	
SanPaolo IMI							
ABN AMRO	✗		✗	✗	✗	✗	
ING Group	✗	✗	✗	✗	✗	✗	
Rabobank Group	✗	✗	✗	✗	✗	✗	✗
BSCH							
BBVA	✗	✗					
Svenska Handelsbanken	✗						
UBS	✗		✗			✗	✗
Crédit Suisse	✗		✗		✗	✗	
HSBC Holdings	✗						
Barclays Bank	✗	✗					
NatWest Bank	✗	✗			✗	✗	
NAB						✗	
BTM							
Fuji Bank							✗
Sumitomo		✗				✗	
Number of banks	19	17	10	4	9	17	4

active in this area; for Europe the figure is 71 per cent. Community involvement includes voluntary projects (whether by employees or not), and investments in and sponsorship of social activities in local communities. Banks active in the internal US market are even encouraged in this direction by a legislative incentive. Lastly, this figure shows the internal socioeconomic policy of the 34 selected banks. Internal socioeconomic activities include focused training, equal career opportunities for men and women and arrangements for shares for staff. The total percentage of the banks in one way or another active in this area is 82 per cent. North America once again scores 100 per cent.

INTEGRAL SCORE OF SUSTAINABLE BANKING

This chapter has highlighted important differences between regions, countries and banks with regard to sustainable banking. The current position of banks on a broad range of issues has been analysed.[12] The picture created is mixed, both between banks and for each bank individually. To achieve a better picture an integrated score based on the model in Chapter 4 can be useful.[13] This involves creating one final score from the results in the previous sections, on the basis of which it can be calculated whether a bank can be judged as engaging in defensive, preventive, offensive or sustainable banking. The methodology for this integral score is set out in Appendix IX. However, the index does not give the relative environmental impacts of individual banks. That is, the environmental impact of a bank gaining a relatively low integral score might be better than that of a bank gaining a high integral score. In short, this section creates a picture of how active banks are, not what their environmental impact is.

The results of this approach are shown in Figure 9.8.

It appears that the majority of the banks adopt a defensive position towards the environment (53 per cent). The overall picture shows a group of ten front runners (30 per cent) who are very proactive, a group of six followers (18 per cent) and a group of 18 stragglers (53 per cent). As explained in Appendix IX this last group has subsequently been divided into two groups (26 per cent each) labelled 'highly' and 'lightly' defensive. Among the 'lightly' defensive banks, indications can be found for a move towards a preventive approach. Within each of the three main groups peer pressure occurs, most strongly among the front runners who quickly react when one of their peers releases a new product onto the market. Peer pressure is also visible within communication activities.

There are regional differences. The average score in Oceania is defensive, whereas in North America and Europe it is preventive. Figure 9.9 shows the differences that exist between countries. Note that for most countries the selected banks contain aggregated a major market share in their home country. This is especially true for The Netherlands (80 per cent).

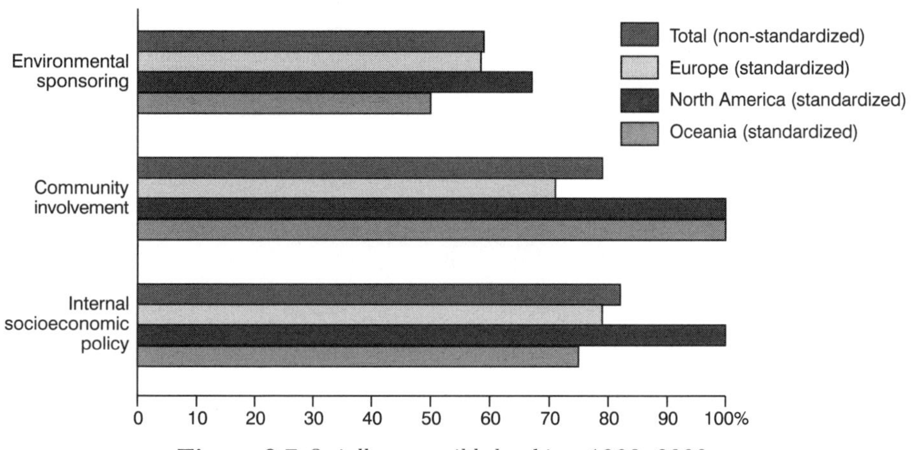

Figure 9.7 *Socially responsible banking, 1998–2000*

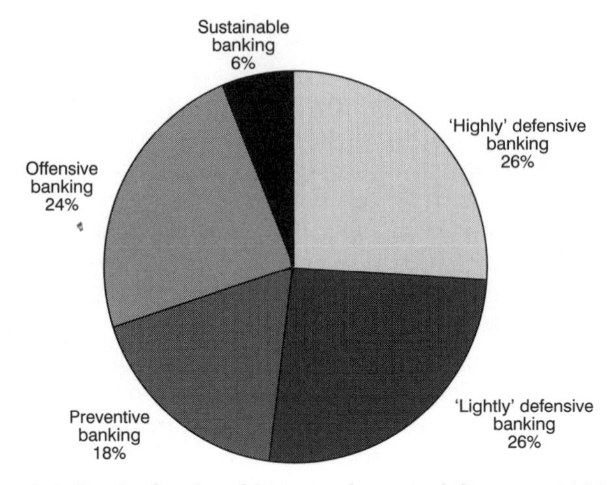

Figure 9.8 *Banks distributed by integral sustainability score, 1998–2000*

Figure 9.9 shows that all the front runners are European banks. The greater weight accorded to sustainability and the environment within corporations in that region can be accounted for by culture, market forces and legislation. These factors also account for some of the differences between countries within Europe. One cultural factor is the greater weight accorded to the environment within Europe, especially north western Europe: in North America and Oceania the emphasis is on community involvement. Moreover, in North America (and to a lesser extent Oceania) environmental concern is more closely linked to charity than in Europe. This is especially visible in the social activities (Figure 9.7) of American banks. It is also apparent in these banks' products since these banks only score better for those products containing an element of charity, such as credit cards.[14] As for legislation, governments have traditionally taken a greater role in social affairs in continental Europe than in Anglo-Saxon oriented countries. Social issues therefore have a higher priority at Anglo-Saxon banks compared to their continental European competitors. Moreover, European governments frequently make use of market instruments to encourage banks to produce financial products benefiting the environment. The Netherlands is a prime example of this. Paradoxically, the US and UK governments are increasingly using regulation as an instrument to enhance banks' contribution to sustainability. Table 9.2 shows the total score of each bank.[15]

Conclusion

To conclude, a connection is sought between the profile and the performance of a bank and its progress towards sustainable banking. A two-sided Pearson correlation test has been carried out on the basis of the individual data per bank (see Appendix VIII); Table 9.3 shows average figures per banking phase. A connection has been sought between the banking phase and the efficiency and size of the bank (in terms of assets) and the degree to which the bank is inter-

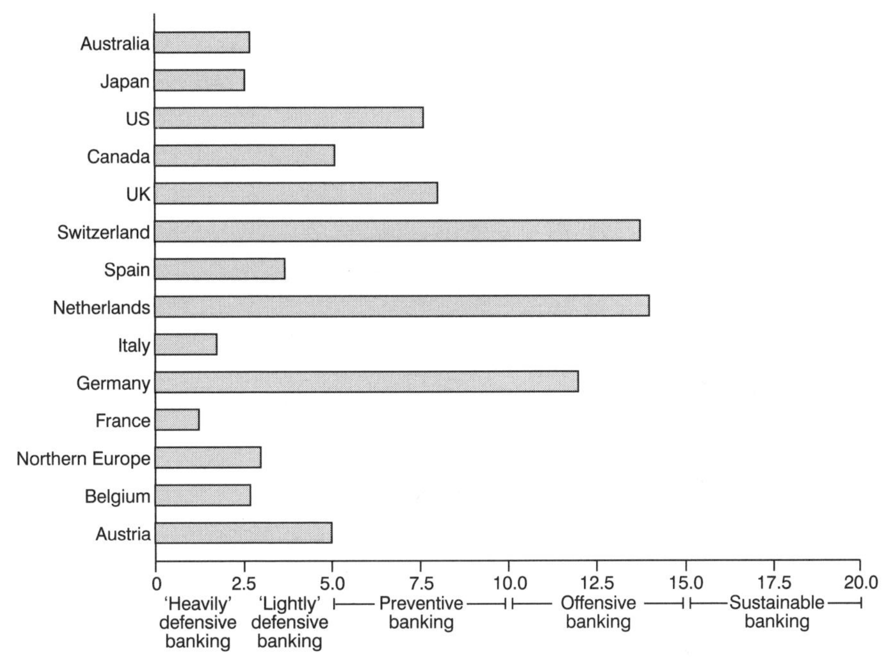

Figure 9.9 *Differences between countries aiming for sustainable banking, 1998–2000*

national. Although the correlations are weak, each is significant (ie significant at the 5 per cent level for efficiency and assets and at the 1 per cent level for international focus). In short, the figures in Table 9.3 give a tentative idea of causality, but should be interpreted cautiously.

With these limitations taken into account, it is striking that sustainable banking is not practised by the more efficient banks. Obviously, the small sample of sustainable banks plays an important role here. On the whole, there seems evidence of a negative trend from defensive towards offensive banking. Apparently environmental care goes hand in hand with net costs. Where proactive banks want to know from businesses what the costs and benefits are of environmental care, they could lead by example and themselves provide better data. None of the 34 banks offered great insight into this area. Barclays was the only bank that offered information about its investment costs and the operational benefits of further energy savings. Were banks to make such information available, the identified trend could either be dismissed or confirmed.

Especially big and/or very international banks will be critically followed by the media and NGOs for their role and contribution to sustainability. They are the most well known and fulfil a symbolic function for big money. For this reason a connection has been sought between the size (in terms of assets), the extent to which the bank is international and the phase of sustainability in which the bank currently operates. In Table 9.3 an upwards trend towards sustainable

Table 9.2 *Integral sustainability scores per bank, 1998–2000*

Defensive banking	*Total score*	*Preventive-sustainable banking*	*Total score*
SanPaolo IMI	0.0	Bank Austria	5.0
BNP Paribas	0.6	BBVA	5.1
Bank of Tokyo-Mitsubishi	0.9	HSBC Holdings	5.4
Société Générale	1.4	Barclays Bank	8.0
Banca Intesa	1.7	Dresdner Bank	8.0
Bank of Montreal	1.9	Citigroup	9.4
Crédit Agricole	2.1	Royal Bank of Canada	10.1
BSCH	2.2	NatWest Bank	11.2
Chase Manhattan	2.3	ABN AMRO	11.3
Fortis Bank	2.8	Bank of America	11.6
UniCredito Italiano	3.0	HypoVereinsbank	11.8
National Australia Bank	3.0	ING Group	13.4
Fuji Bank	3.0	Crédit Suisse Group	13.6
KBC Bank	3.1	UBS	14.1
Svenska Handelsbanken	3.1	Deutsche Bank	17.0
CIBC	3.6	Rabobank Group	17.9
MeritaNordbanken	3.8		
Sumitomo	4.0	MAXIMUM	20.0

banking is evident for both of these; that is, the higher the level of assets and the higher the degree of internalization, the higher the integral sustainability score. The banks which are 'highly' defensive are an exception to this rule. It is noteworthy that the average size and the degree of internationalization of the banks in this group is greater than for those in the group engaging in 'lightly' defensive banking. It is precisely the big and internationally operating stragglers who run an increasing risk of damaging their reputations.

Further research should look into this correlation. A bigger survey sample and a higher number of banks per country and per phase is necessary to do so. Further research might also look into the quality of the data reported and add interviews to the desk research to the methodology of this chapter. This would also reveal potential gaps in the presented activities of individual banks; the order in Table 9.2 might change due to such gaps or quality issues. Moreover, it

Table 9.3 *Relationship between bank profiles and phases, 1998–2000*[16]

Phase (number of banks)	*Average efficiency (%)*	*Average assets, (EUR, billion)*	*Average degree of international focus (%)*
'Highly' defensive (9)	63.2	389.6	26
'Lightly' defensive (9)	59.7	260.5	21
Preventive (6)	65.4	411.0	37
Offensive/sustainable (10)	69.0	474.3	41
Offensive (8)	66.7	452.7	41
Sustainable (2)	78.0	560.5	43
Correlation coefficient (significance level)	0.41 (1.6%)	0.35 (4.2%)	0.45 (0.8%)

would be interesting to look at the activities of banks from developing countries and smaller banks and compare their activities with those of, respectively, the banks from developed countries and the bigger banks. Finally, the empirical study made it necessary to add an additional stance or phase. Seemingly the defensive category can be divided into those banks that do not act towards sustainability at all ('truly' or 'heavily' defensive) and those that act but hardly see the cost and efficiency advantages of the preventive phase. More theoretical and empirical research should clarify this distinction.

10

Sustainable Development: A Paradigm Shift

Introduction

Many people live from day to day, confronting and dealing with problems as they arise. They accomplish this within the existing context and the current paradigm, rarely stopping to consider this context and paradigm. While the previous chapters have been written on the basis of the current paradigm, but with a new perspective, this chapter addresses the possibility of a new paradigm.

A paradigm (from the Greek words *para* and *deigma*, meaning 'pattern') reflects the structure upon which a concept is based. It creates a framework through which we view the world and a way to give our observations meaning. Examples of paradigms are worldviews, norms and values, and management philosophies. What characterizes a paradigm is that the relevant authorities no longer discuss whether or not this structure behind the concept is valid; they simply accept it.[1] Furthermore, if a current paradigm is contested, the dominant coalition will always exert great effort (often backed by considerable emotional drive) to defend it, regardless of signals suggesting that the paradigm has serious shortcomings or limitations.[2] The paradigm determines – usually at a subconscious level – what information we accept (as well as what we ignore), the kinds of questions we ask and how we look for answers. If the approaches suggested by the paradigm for finding answers frequently fail to lead to answers, however, a situation arises in which a new paradigm can develop. What happens then can be called a paradigm shift (Kuhn, 1970). A change in paradigm is a completely new way of thinking about old problems. Old concepts no longer apply or are given a new meaning, while new concepts develop. To distinguish between the meanings of concepts when a paradigm shift occurs, the terms 'first-order processes of change' and 'second-order processes of change' will be employed. First-order processes of change are changes within an existing paradigm, while second-order processes of change imply a paradigm shift or a change in the system at a higher level of complexity (the change of a system; Watzlawick et al, 1973).

Down through human history, certain major far-reaching paradigm shifts have occurred. Examples such as the Agricultural Revolution and the Industrial

Revolution were changes involving mainly our material level of existence. The Enlightenment is a good example of a paradigm shift on a more intellectual level. In addition to these major paradigm shifts, however, whole series of smaller changes have taken place. From a historic and evolutionary viewpoint, this is an organic process governed by the principle of selection. Ideas that at that moment are mature and considered most valuable are accepted. Often, minor events at a micro level lead to major consequences at the macro level.[3] Changing perceptions or perspectives (ie first-order processes of change) can thus lead to a paradigm shift (ie a second-order process of change). In the study of systems, this is known as the 'development' of a system (In 't Veld, 1987).

Systems can 'develop' or 'grow'. 'Growth' indicates a quantitative change in an open system without a change in its internal structure.[4] An example is crystal: although its size can increase, its internal structure remains the same. 'Growth' is usually possible within systems, but only up to a certain point. At this point, a system can no longer manage a larger quantity; further structuring is impossible. This crisis is accompanied either by disintegration or a leap to a higher level of structuring. In the study of systems, this is spoken of as 'development'. Cell division within the human body is a good example of this. A cell grows to a certain size and then divides to become two cells. At that point, further growth at another level of structuring is possible. Second-order processes of change involve the development of a system. For this transformation, a crisis (in concepts, in the system) may be required. The growth of a system, on the other hand, is a first-order process of change (Voogt, 1995).

Changing perceptions within banks have led them to bring green investment funds onto the market. This occurred after research and pioneering work had shown that not only could they earn an ecological return but also a good financial return. This is a case of the growth of a socioeconomic system (first-order process of change). In fact, the system has since grown (specific environmental products or services are now possible) but the internal structure remains unchanged (the dominance of the financial return). In a phase of sustainable banking, sustainable businesses will be financed at relatively favourable tariffs; not from a financial risk standpoint, but from the standpoint of sustainability. This can only happen if a change in the internal structure of the economic system occurs, or rather if the system develops (second-order process of change). Such a change can be a new paradigm in which financial return is replaced by sustainable return as the dominant factor in human actions.

As long as a narrowly defined financial return remains dominant, sustainability will not be attained. Only if sustainability is fully internalized within the financial return can the attainment of sustainable development be possible through striving for financial return. In which case, it is better to talk of a sustainable return, in order to indicate that a system transformation is involved. Sustainable development therefore implies a second-order process of change. This is not a destructive, but a constructive process. Such a paradigm shift is usually not gradual.[5] The new paradigm often contains the old one in the form of a partial truth. According to various authors, the coming decades will involve change accompanied by a brief period of chaos (ie unpredictability: a period in which the old paradigm no longer works but the new one has not yet been devised; Harman, 1979, Russell, 1992, Stikker, 1992 and Meadows et al, 1991). Afterward, another system (ie another paradigm) will become dominant.

Sustainable development entails a different philosophy of life, another way of seeing things, and other ways of valuing things. The organic (complex, many interdependencies) character of this process, as well as preconceptions derived from the old paradigm, make expressing a new paradigm unfeasible and unproductive. This involves an open process. What is possible, however, is suggesting elements that can play a role in a sustainable society during a paradigm shift. Although this chapter does not focus on providing answers per se, it is intended to stimulate thinking in this direction by addressing the foundations of current thought and other ways of looking at things.

The section below deals with the question of whether changes of the first order are sufficient and sketches possible futures. Since the answer is negative, developments in Western thought (both philosophical and economic) over the past millenia are the subject of the section beginning on page 207. The section beginning on page 211 then focuses in on the deficiencies in current economic paradigms and in the section beginning on page 213 the manifestation of a paradigm shift is discussed. The section beginning on page 216 concludes with some non-Western worldviews, offering them as possible alternatives, not as solutions.

ARE INCREMENTAL STEPS SUFFICIENT?

All activities aimed at sustainability involve meeting certain economic conditions related to solvency and corporate continuity. Within the current paradigm, banks can go no further than the current socioeconomic system allows. All incremental (gradual) steps to sustainable development discussed in the preceding chapters indicate the right direction and should definitely be taken to encourage sustainable development. Such steps to sustainable banking, while being included in offensive banking, can actually be summarized as working toward eco-efficiency (Schmidheiny and Zorraquin, 1996). Are such steps (first-order processes of change), however, sufficient to attain a really sustainable society? The following parable (Meadows et al, 1972, p29) illustrates this question:

> *The Earth can be seen as a closed ecological system. Accordingly, a parallel can be drawn with a pond and its water lilies. In this pond of a certain size, the water lilies double in number every day and the pond will be filled after 30 days. When this happens, other life in the pond becomes physically impossible. After how many days will the pond be half full?*

The answer, of course, is 29 days. Eco-efficiency may reduce environmental impact per product, but volume effects such as economic growth and population growth will usually cancel out this effect. The fact is that millions of people (mainly living in the Southern hemisphere) are now dying due to environmental problems. According to a World Bank report, rapidly increasing air pollution will be resulting in 850,000 premature deaths and 7.5 million new cases of bronchitis a year (World Bank, 1997). In terms of the parable, then, how many days have passed?

Naturally, the exact size and quality of the ecological system (the pond) is not really known. Furthermore, the causality question in today's complex world with its multitude of interdependencies, is difficult to answer. Moreover, it is hard to say if humanity has reached day 29, let alone if it is now one o'clock in the morning or five to midnight. The lesson of the parable is that it is the volume effects that count while eco-efficiency is aimed only at relative effects. Zero per cent pollution is never possible (this follows from the Second Law of Thermodynamics; see Appendix X). If the entire world would consume at a lower rate of environmental impact per product by means of increased eco-efficiency, and all the previously produced goods would be replaced, an increase in economic and population growth may be possible without immediate short-term volume effects. Even so, this would mean a gain of no more than a few days in terms of the parable. Day 29 would soon be staring us in the face again. The parable seems to imply a doomsday scenario, but this is not at all the case. It is precisely the drawing of this conclusion and then giving up that is a doomsday scenario and an underestimation of human capacities to arrive at a solution. The parable indicates that a problem exists and requires a solution to guarantee the continuity of human life on Earth.

If the developments of population size, affluence and the use of natural resources are placed within an historical perspective, what results is a picture indicated in Figure 10.1.

On the basis of Figure 10.1 it cannot be concluded that day 29 has dawned. What can be concluded is that humanity has arrived at an exceptional period in its history that involves major, rapidly occurring changes. The urgency of taking measures is increasing. The vertical axis can represent several different factors such as population size, degree of affluence or the use of natural resources, the scale changing according to the factor. Figure 10.1 is not intended to indicate an exact value for the Y-variable, but is supposed to illustrate this special period in human history. It is evident that the upward quantitative trend characteristic of the 20th century cannot continue unabated; after all, the Earth is a closed ecological system with physical limitations (also see Appendix X). Nevertheless, the entire Western system is based on material growth. Are alternatives conceivable? Does the economic system have an internal mechanism for stabilization or change? Before such questions relating to second-order processes of change can be discussed, however, the existing paradigm and economic system should be described.

THE EXISTING ECONOMIC SYSTEM

In describing a paradigm, place and time are important factors. In addition to an historic component, describing the broad worldview is necessary. This section provides a brief description of the development of Western thought in terms of philosophy.[7] Next, a connection is made with the development of economic

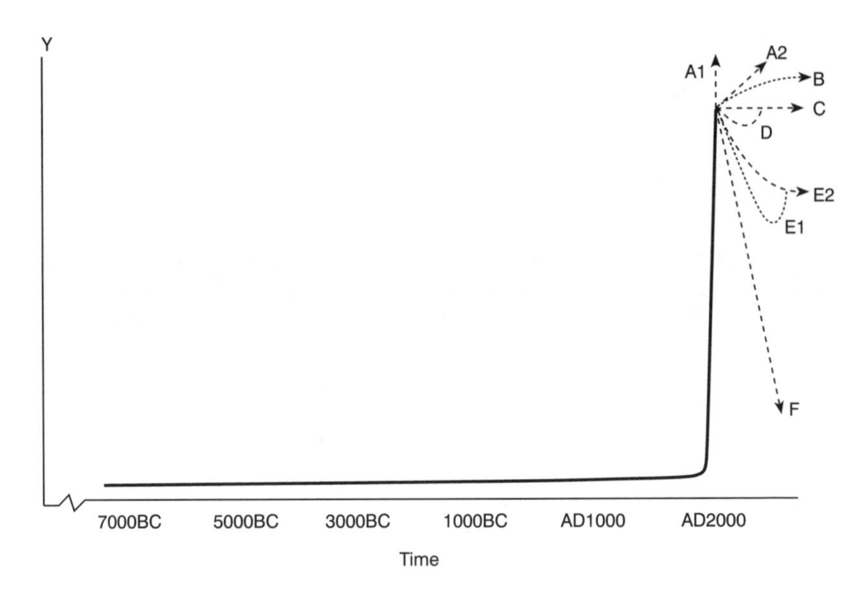

Figure 10.1 *The 20th century*[6]

thought and, in turn, the current economic system (Kastelein, 1984; Ekelund and Hébert, 1990; Hoelen, 1995 and Pearce, 1992).

Between 800BC and 200BC, a transformation of thinking occurred in various societies throughout the entire world (Greece, Palestine, India, China and Persia). At the individual level, a critical reflection of man originated in which two basic ways of thinking were categorized: one related to a physical world and one related to a spiritual world. The Ancient Greeks were the first to give logical thought priority. Emotions and logical thought were seemingly torn apart. During the Middle Ages (usually regarded as the period from around AD500 to AD1500), the spiritual thinking of the church predominated. Medieval philosophy is generally characterized as the desire to create a fundamental harmony between belief and reason. Empirical facts that opposed this harmony were seen as optical delusions to be ignored. The Renaissance (14th, 15th and 16th centuries) returned logic, independent of spiritual ideas, to the forefront.

From the 16th century, attention was increasingly devoted to the material world and the 'part' (as opposed to the 'whole' of the Renaissance). With rationalism (Descartes, 1596–1650), a sharp line was drawn between the material and the immaterial (ie body and soul). Furthermore, the individual (Descartes: 'I think, therefore I am') became the key focus. All kinds of questions could be analysed by means of logic. The view was that by analysing natural phenomena, questions of a more spiritual nature could be answered. This was done by means of abstraction and causality. This mechanistic approach has dominated Western thought ever since the scientific revolution of the Enlightenment (17th and 18th centuries). In general, this approach states that everything in the world (or even the universe) can be reduced to matter and movement while excluding any other factors. Knowledge of the parts should lead, through synthesis, to knowledge of reality. Such abstractions have led to discrepancies between the micro and macro levels: the whole is not equal to the sum of its parts. Major problems developed in regard to causality, especially when it came to reciprocal influence among factors. Hume (1711–1776) talked about the 'induction problem': if in reality there is a necessary connection between cause and effect, is it then possible to extrapolate to the future as based on the discovery of a causal connection in the past? According to Hume, such a necessity exists only in the expectations of the observer.[8] Natural laws, therefore, could not be extrapolated into the future. Hume was seen as a sceptic. However, a broad philosophical discussion about the causality question is rarely conducted these days.

The revolution in natural sciences has been accompanied by the introduction of linear thinking (the idea that one phenomenon can be derived from another phenomenon and will occur independent of place or time) into Western thought. During the Enlightenment, another important step in Western thought was taken. By means of science and technology, 'man came to be perceived as the master and owner of the natural environment' (Bor and Petersma, 1996, p238). The idea that people could control the natural environment and are obliged to use it for their own ends (the intrinsic value of nature had, in fact, eroded) was generally accepted. This idea has been an important factor in the estrangement between people and the natural environment. This philosophical optimism then became realized with the Industrial Revolution.[9] Notions such as efficiency, production, usefulness, utility, labour and value made their entrance along with the Industrial Revolution and upcoming economic thought.

The development of philosophical thought is reflected in economic thought.[10] The classic economist and moral philosopher Adam Smith (1723–1790) contributed greatly to these ideas. As a matter of fact, the development of our current economic paradigm is largely traceable to this economist. In the time of the mercantilists (16th and 17th centuries), the government held great power, while its view related to the operation of economic processes was limited. A powerful state and strict regulation of the national economy were the basis for this paradigm. Adam Smith attacked this paradigm because it benefited the producer over the consumer while this economic system was contending with inefficiencies.[11] With his theory of the 'invisible hand', Smith lay the basis for economic liberalism and for the transition from a static feudal system to a dynamic industrial capitalism in which everything centred on individual freedom. This was encouraged by private property, the basis for capitalism.

Smith was heavily influenced by deism, a belief that saw human socioeconomic life as a cosmos controlled by natural laws – a cosmos that is completely accessible to human analysis.[12] Smith recognized a natural order in the chaos of things. The central theme of his principal work published in 1776, *An Inquiry into the Nature and Causes of the Wealth of Nations*, is how production develops without being guided from above. It was in this regard that he spoke of the 'invisible hand' (currently known as the price mechanism or market forces): the pursuing of one's own interests by individual consumers and producers (at the micro level) will maximize the affluence of all individuals at the macro level. The assumptions of our current economic paradigm can be traced back to the abstraction that companies are attempting to maximize their profits and consumers are attempting to maximize their utility and in which prices provide the best possible allocation of these benefits. Such a maximizing behaviour at the micro level will then lead to the best possible level of affluence at the macro level. The neoclassical and classical school is thus based on a mechanistic worldview.[13]

By assuming the existence of a natural order and harmony, no attention needed to be devoted to ethical questions (mechanistic worldview). As long as all individuals maximized their own benefits, the benefits of each individual and the group would automatically be optimized. On the other hand, a liberal economist by the name of Jeremy Bentham (1748–1832) suggested that crime, for example, might be in the individual's interests but not in the collective interests. Smith, too, drew this conclusion in 1759. But by means of a strong judicial system (accompanied by heavy sanctions) and a strong government, the function of the 'invisible hand' could be ensured (the result being 'utilitarianism').

Alfred Marshall (1842–1924) pointed to the existence of external effects[14] such as environmental problems, ie entities to which no price is attached and that for this reason remain outside the economic system. He also saw the economic system as a self-regulating system. John Maynard Keynes (1883–1946) called this into question and focused on macroeconomic phenomena. According to Keynes, it was possible to have economic balance in the event of insufficient employment (meaning that the system is not self-regulating). Furthermore, the economic system was unstable, according to Keynes, because individuals were confronted with insecurity (and not having access to sufficient information which made them unable to act fully rationally, unlike what the neoclassical economists presumed). Since the system itself was unable to arrive at the best possible outcome and was unstable as well, the government should play a corrective role in the economy. Due to Keynes' ideas and the crises early in the 20th century, the government received an active role in regard to employment and also in regard to incomes. Like Smith's ideas, Keynes' ideas have persisted.

Economic thinking has continued to develop as based on such foundations as Smith's systematic framework and Keynes' ideas. Today, we can distinguish three systems – three kinds of economic orders. The economic order is the economic organization of society and concerns the decision making in regard to the disposition of scarce resources. At the core of economics is the analysis of supply and demand. Economic order therefore involves how the connection takes place in society between the preferences relating to supply and demand. In simple terms, those requesting goods or services are asking for sufficient products of good quality for low prices, and those offering these goods or services are looking for the highest possible profit or turnover. When this connection occurs by means of the market mechanism (the requested and offered quantity of goods/services adjusts itself in such a way that an equilibrium price results), what we have is a *free market economy*. If this connection is determined by the government, what we have is a *centrally planned economy*. The key concept here is constraint. A mixed form is known as the *directed economy* (Broekman, 1992 or Hoelen, 1995). For a long time, the US was seen as an example of the first category; the former Soviet Union, China and Cuba as examples of the second category; and Germany as an example of the third category. The distinctions, however, are no longer easy to make. With the destruction of the Iron Curtain, the globalization and liberalization of the economy, and deregulation by national governments (and within Europe by the establishment of EMU), the boundaries have become so vague that the prevailing economic order can be described by a single name: a *mixed market economy*.[15] Central to this is the price mechanism as a method for organizing society. The government establishes the limits within which economic activity takes place (eg environmental permits). The government also plays a role in regard to public goods, employment and social security.[16]

When seen as a paradigm, current Western philosophical thought can be characterized by such concepts as liberalism, individualism, mechanistic thinking, a distinction between the material and immaterial, linearity, causality and abstraction. Current economic thought can now be characterized as a mixed market economy within which material progress dominates. And in addition to Western philosophical thought, other key concepts that apply would be markets, the price mechanism, private property, rational subjects (ie armed with full information), efficiency (ie without waste) and the idea that the economic system is not self-regulating.

ARE THERE SHORTCOMINGS ASSOCIATED WITH THE CURRENT ECONOMIC PARADIGM?

The current economic system has provided capitalistic society with much affluence, but the system fails to consider various abstract or non-monetary qualities such as the natural environment, leisure time and do-it-yourself activities. Maximizing utility has turned into maximizing consumption; in our utilitarian and materialistic society, 'better' has turned into 'more'. The current economic system is primarily based on quantitative growth. Up to a certain point, qualitative growth (in which immaterial values in particular play a role) can run parallel to quantitative growth. But when the system considers only values that can be expressed in prices, there inevitably comes a time when the two forms start to diverge. Daly and Cobb (1989) have calculated that welfare in a broad sense (ie economic development corrected by such factors as environmental quality and health) has declined in the US since the 1960s while an economic, frequently used indicator of welfare such as the GDP has shown a continuous rise. This is why many assert that zero growth (see Appendix XI)[17] is needed. This is questionable, however; an economic system aimed at qualitative growth is probably more sustainable, especially from a social standpoint. A comparison with human development illustrates the limited character of the current economic system. Up until adulthood, a person grows quantitatively (height) and qualitatively (knowledge).[18] A stage is reached when quantitative growth is limited, adulthood. After this the qualitative aspect takes precedence. The economic system appears unable to make this shift on its own and quantitative growth continues to play the dominant role.

The current economic paradigm – and its many disciples[19] – display an almost total confidence in the self-regulation of the economic system. Due to the price mechanism, 'the shore will turn the course of the ship around'. In other words, there may be some disasters ahead ('the shore' in the form of ecological catastrophes such as the forest fires in Indonesia or a major climate change due to the increased greenhouse effect), but these will eventually be corrected ('turned around' by substitution or technological innovations induced by price incentives).[20] At this time, however, too many goods are still freely available; ie they are not considered scarce and thus play no role in the economic mechanism ('the ship' itself).

Large numbers of people are embracing organizations in the Western world working toward environmental conservation. The desire to find solutions for environmental problems evidently exists, but people feel that they have little if any chance to contribute to these solutions in their normal activities (consuming and producing). What we have is a 'prisoner's dilemma', as explained in more detail in Appendix XII. An appropriate example of this dilemma is the reaction (of companies, employees, government agencies, etc) to new environmental legislation. To conceptualize this, Table 10.1 is included.[21] Although inhabitants of country A believe that things cannot continue as they are, the assumed damage to their competitive position makes it 'impossible' to do anything for the environment as long as country B (the rest of the world) does nothing.[22] The greatest possible global affluence (including the environmental factor) would be attained if both parties would combat environmental pollution: both country A and country B would reach an affluence level of 5. The idea that country B implements a stringent environmental policy means that country A should have a better competitive position and therefore a higher affluence level (8) by doing nothing. As a result, country B has an affluence level of only 1. Country B will display the same behaviour so that neither implement any environmental policy. The total affluence level of country A and country B will therefore be 3. This is not the best possible solution. After all, if both countries would implement a stringent environmental policy, the total affluence level for each country would be 5 for each. This result can be attained by means of agreements or cooperation.

The reliance on the price mechanism is only partially justified because various goods or values are not being included in the price mechanism. Furthermore, the mechanism does not operate perfectly – at least not in such a way that the outcome per se is what society wants. The reason for this can be traced to vested interests, existing institutions, incomplete information, a lack of trust, greed, physical limitations (the economic system assumes the existence of relative scarcity while in reality absolute scarcity exists) and the time factor.

Table 10.1 *An application of the 'prisoner's dilemma'*

	Party B: no	*Party B: yes*
Party A: no	3, 3	1, 8
Party A: yes	8, 1	5, 5

'The ship' (the economic system) needs to be seen as a gigantic oil tanker. At full speed, an oil tanker will not be able to react once the shore comes into sight; it is too cumbersome and difficult to manoeuvre. The desired changes take too long to put into effect. This time factor also plays a major role in the current economic system.[23] Essentially, vested interests will be against a change of course and try to postpone or even prevent such a change of course by means of various institutions or a direct application of power that originates from economic scale (Rowell, 1996). When this occurs, the previously mentioned economic stimuli will fail to have any effect.

In short, the existing economic system is simply unable to regulate itself, ie the economic system is unable to make sustainable development possible. The corrective role of government – establishing environmental requirements and regulating producer and consumer behaviour – is necessary. Corrections on the government's part, however, have proven insufficient in practice. The problem is that it is the vested interests in a democracy that hinder this process: companies, because it can affect their continuity, accomplish this by direct lobbying, while political parties, because it can reduce the purchasing power of their voters, will be threatened a reduction in votes. Individual citizens, companies and banks can accomplish a great deal by means of eco-efficiency, but there can be no sustainability because of ultimately running up against the limitations of the current economic system that is based on material economic progress.

The demand for a new course is therefore justified. Another system, another way of thinking – a second-order process of change – is required. After all, if one realizes halfway through day 29 that it's day 29, it is most likely too late to turn the tide. Waiting until the shore turns the ship is not an acceptable alternative. Before the shore is really in sight is when the change of course has to begin. What this means is that the ship is going to have to learn to see better. Awareness of the shore, the clumsiness of the ship and the fact that everyone aboard the ship can affect the course of it are part of this lesson. Individuals and companies are a part of an economic cycle, and they introduce effects that adversely affect the environment. The challenge is to interrupt that cycle. As a reaction to the Exxon Valdez oil disaster, certain people in the US, for example, purchased electric cars since they realized that the oil disaster was a result of their own need for petrol. Such examples illustrate that the causes for environmental problems are actually very close to home and that this is where the solutions can be found as well.

SUSTAINABLE DEVELOPMENT AS A PARADIGM SHIFT

According to the previous section, the current economic system has acquired its own raison d'être and no longer has any connection to the objectives for which the system was originally established. A system is always a means to achieve something. Before people realize it, however, they can become imprisoned by the system itself: the system will have become the objective. As an example, country A's weapons system was established essentially to scare off country B from attacking country A. When country B follows suit, an arms race results.

Arming oneself has then become the objective and no longer a means. A second-order change (ie a dramatic change in thinking that sooner or later is reflected in the internal rules of a system) is necessary to interrupt this paradox. Such a change could be a decision by country A not to use weapons as a scare tactic any more but to try to use political cooperation with country B to rebuild trust in one another and thus bring the arms race to an end.

The original objective of economic growth was to achieve a more affluent world (or nation). Economic growth is therefore a means to this end. Society indicates what economic progress is, and not vice versa.[24] The order in which the questions are asked should thus be turned around in order to arrive at any real change. In the current economic paradigm, quantity is of first importance instead of quality. This leaves little room for qualitative or immaterial matters; they are not expressed in the price mechanism. This can also be seen in the Brundtland definition of sustainable development: people are conceived as beings that want to satisfy only material needs.[25] This fails to consider human tendencies toward intellectual and ethical development, working toward developing personal relationships, creative expression, charity, altruism (whether or not directed to one's own descendants), etc. The current paradigm sees people primarily as consuming creatures; this, then leads to our unbalanced behaviour. Sustainable development would take on an entirely different meaning if people were seen, for example, to be constructive, producing beings. If so, the system could be designed in such a way that the development of our constructive abilities could be maximized. This insight is just one example of a possible second-order process of change. Watzlawick, Weakland and Fisch provide a good description of second-order processes of change by means of a puzzle. Box 10.1 goes into this puzzle in more detail (Watzlawick et al, 1973, pp 44–48).

Evidently, the solution for all kinds of dilemma-like problems can be found by revising the perception of reality, ie taking a closer look at the foundations of the underlying system – which, in this book, is the socioeconomic system. In regard to the problem of sustainable development, first-order processes of change are an example of 'treating the symptom', while a second-order process of change would go to the heart of the problem and therefore escape from the predominating ideas determined by the dominant paradigm.

A common misconception about the nature of change is that a problem can be solved by replacing what has to be changed by its opposite. In spite of the criticism that the fundamentals of capitalism (or the market economy) have received, communism can by no means be considered the solution for its ailments – especially considering the impoverished situation of the environment under Soviet communism. Actually, throwing the market mechanism overboard would be like throwing out the baby with the bath water. Instead, what is needed is to consider the underlying assumptions and to devise new ideas and outlooks.

Yet another misconception is that the solution could simply be 'more of the same'. A commonly held idea is that more economic growth will provide the means to solve the environmental problem. Although this is basically true, it disregards the fact that economic growth is also one of the very causes of the environmental problem. A second-order change involves transformation,

Box 10.1 An illustration of a second-order process of change

This thought puzzle can clarify a second-order process of change. The reader is invited to solve the following puzzle before reading further. The assignment is to connect all of the nine dots by drawing four continuous straight lines (ie not removing the pen from the paper). The solution is given in writing under the diagram so as not to give it away visually beforehand.

•   •   •
•   •   •
•   •   •

To explain the solution in words, the columns are given numbers (1, 2 and 3) and the rows letters (A, B and C). The solution is as follows: draw the first line from C3 (you can actually start with any corner dot) through B2 to A1. Then draw the second line from A1 through B1 and C1 to a new point with the imaginary position of D1. Draw the third line from D1 through C2 and B3 to another new imaginary point, A4. Then draw the last line from A4 through A3 to A2 (or A1).

Many readers, after trying diligently to solve the puzzle, will conclude that there is no solution possible. This is because they themselves have limited how a solution could be found by implying that the perimeter formed by the dots is also a limitation. The problem seems unsolvable because one wants to solve it in a certain way. This approach to a solution can be compared to changes of the first order. Indeed, within this self-defined framework or system, there is no solution. The solution lies beyond the framework. Only by means of a change in the system (a second-order process of change) can the puzzle be solved; by taking a mental leap outside of the framework of dots, the solution is possible. Then, the solution is surprisingly simple. The analogy with many daily problems is obvious. Watzlawick et al conclude: 'The solution results not from a vision of the dots themselves but from a revision of what one presumes in regard to these dots.'

discontinuity (ie changes that interrupt routines) or a logical leap to another structural level.

Many second-order solutions appear to be paradoxical or illogical in practice.[26] They take on this character when viewed from the dominant framework of thought. As Einstein once said, 'The world we have created today as a result of our thinking thus far, has created problems which cannot be solved by thinking the way we thought when we created them'. Usually, we interpret today's problems by means of yesterday's frame of reference and wind up selecting outdated answers for tomorrow so that we at most have then created new problems (the environmental problem is a good example of this; Peters and Wetzels, 1997). Instead, we should focus on stepping beyond the dominant paradigm and being open to other conceptions of reality, with reflection being the key in this process. Seeing beyond our blind spots – by recognizing the existence of 'impossible worlds' (comparable to the puzzle in Box 10.1) – leads to an expansion of our awareness. Learning about other cultures and other ideologies can lead to throwing implicit preconceived ideas overboard. By then integrating the experiences gained during this learning process with principles of the dominant philosophy (but not starting from the dominant philosophy), the basis for a transformation to a new paradigm is laid.[27]

Sustainable development: Another way of seeing

A second-order process of change involves another way of seeing. To explain this, it might be useful to compare Western and Eastern ways of thinking. Instead of a mechanistic view, it is also possible to consider an organic or holistic view.[28] Today's society is structured in such a way that the total of everything no longer counts (or counts only as a derivative). So why would you assume a holistic attitude? From a holistic or organic viewpoint, the problem of sustainability looks entirely different. Box 10.2 attempts to illustrate this. Box 10.3 then touches on certain points related to an alternative Western social movement while Box 10.4 refers to non-Western cultures. These boxes are given not with the intention of felling value judgements but only to show alternative lines of thought. The idea is that being open to fundamental alternative views of society and a willingness to learn from them can pave the way to the necessary second-order changes toward sustainability.[29]

Box 10.2 Another view

Mechanistic thinking is all-pervasive, even in healthcare. Using the human body as an analogy, we can make the following comparison. If someone has cystic fibrosis, the entire body suffers from this disease (eg because one's general condition declines). Whereas earlier, when doctors made this diagnosis, they would have focused attention on the lungs and the pancreas, medical insights have progressed to the point that doctors now focus on such aspects as nutrition to encourage recovery. Evidently, the whole functions more poorly if part of the body is neglected, while a part functions poorly if the whole is neglected. For some time now, it has been possible to observe a trend in society – and also in healthcare – in which the body and mind are no longer seen as being strictly separate. A positive attitude to life seems to benefit survival.

There exists an analogy between the nature of man, the economy, the environment and sustainable development. The first situation in which the doctor focuses attention only on the lungs and pancreas is comparable to the economic system (ie the 'invisible hand'). The progressive view that other parts of the body as well – and maybe even the whole body – might be involved, can be compared to the attention to the environment and ecology. Attention to the interactive effect of body and mind can be compared to sustainable development.

Attempts to work towards sustainable development create room for ethical questions again. If the world could be seen as a whole, extreme riches in certain parts of the world existing next to systematic malnutrition and hunger in other parts of the world would never be permitted. Sustainability problems would never become so serious because people would tend to act on them sooner. Allegations that human life in the developing world is of less value than human life in the Western world can be accepted within the narrow mechanistic viewpoint but not within a holistic perspective of sustainability in which the entire society is considered and in which a balance exists between material goods and immaterial values.

Box 10.3 A new era?

In the late 1960s, several schools of thought such as the esoteric and occult movements, the ecological movements and the Eastern religions more or less merged and a mutual recognition of values occurred. Ferguson, an American writer, interpreted all these movements as expressions of a new age in her book *The Aquarian Conspiracy*. What these movements have in common is that they want to offer an alternative to the rationalistic, materialistic views of man, the world and the cosmos that are commonly held in Western thought. The New Age movement does this by emphasizing the unity of man, the natural environment and the cosmos. Accompanying this is the placement of great emphasis on an intuitive, instinctive approach to problems and other matters (Ferguson, 1981).

According to astrology, the Earth entered a new sign of the zodiac (Aquarius) in the year 2000. This marks a new era, the Age of Aquarius, with possibilities for the development of a new human consciousness that operates holistically instead of dualistically. To initiate this process, people should devote their attention to the spirituality of life, ie to the spiritual dimensions of existence, the Earth and the cosmos. People can develop this new consciousness themselves by means of any of several meditation techniques. This should result in individual persons and ultimately all humanity arriving at the self-realization that consists of a unity with the 'source of existence'. In this conception, man is seen in the middle of an evolutionary process in which he is developing (whether or not by means of several lives, ie reincarnation) onward to higher levels. Important in this entire process is a change in paradigm, ie the change in the common pattern of thought.

A holistic sustainable philosophy is based on an integrated unity. At the core of the mechanistic economic paradigm is the freedom of the individual. As a rule, this means the right to self-determination at the micro level (the part). When coupled with striving for self-interest, this right can lead to adverse consequences at the macro level (the whole). After all, if there are no limitations exerted externally, considering the advantages and disadvantages of a certain decision will only take place as based on the consequences for the part. Individual decisions should therefore be brought back into balance with the whole. There needs to be an internal mechanism that ensures that physical limitations at the macro level are not exceeded. This mechanism could be other norms and values. Central to this are two elements: responsibility and commitment. Related to these are trust[30] and altruism (also for future generations). This still leaves room for self-interest, but not in the narrow sense in which the idea of the 'invisible hand' is taken at its extreme to mean 'everyone for himself and God for all of us'.

Taking responsibility for the environment – for the whole – is ultimately part of the solution for the sustainability problem.[31] This applies to everyone: producers, consumers, citizens, governments and banks. And this taking of responsibility may be the real paradigm shift. After all, companies or citizens are not islands. What is involved is behaving in such a way as you as a person would behave or would want to be treated. Shifting pollution onto the environment or engaging in antisocial behaviour thus become impossible – not so much as based on externally controlling motivations but as based on an internal motive. The

Box 10.4 Other cultures

It is impossible here to discuss all cultures and non-Western philosophies. Only a few will be described very briefly in an attempt to provide a comparison with modern Western thought. The box is intended only for purposes of illustration and cannot possibly describe the rich development and elements in these cultures and ways of thinking.[32]

Native peoples (such as Aboriginals, African Bushmen and Native Americans)

These groups often assume a sense of public spirit, respect for the environment, and man as a part of nature (and the cosmos) instead of as master and user of the natural environment. Spirituality is very important. Where private property rights are a condition for the functioning of the capitalistic system, most native peoples (such as those who live in the tropical rainforests) see private property as a contradiction in terms. All natural resources, including land, belong to the community. In practice, this philosophy works very well and is implicitly sustainable. It is doubtful, however, whether such a system can also succeed on a large scale and accompanied by continuous population growth.

Buddhism

In Buddhism, the inner self and happiness are important. Happiness cannot be attained by power, prestige or material objects. Everything hinges on an understanding of oneself and the world around us. This can be reached by concentration, patience, generosity and altruism. The ultimate objective is enlightenment. This can be reached by means of meditation. In Buddhism, the spiritual, the qualitative or immaterial, plays a central role.

Taoism

Taoism assumes a complementary way of looking at things (yin and yang) instead of conflicting opposites (as in Western thought;[33] see Figure 10.2). Among Taoists, there is no antithesis between the material and the immaterial (while Western philosophy sees a clear and solid distinction between the two). Reality is usually seen as consisting of complementary parts that are symbolized in the Chinese yin and yang: good is not possible without evil; there is no male without female; we experience no sun if there are no clouds or rain. Taoism assumes a harmony (there is a universal unity) between man, the Earth and the universe. The attitude toward the natural environment is respectful and is in harmony with human life. Taoists think in terms of the whole (holism) and in dynamism (evolution[34]).

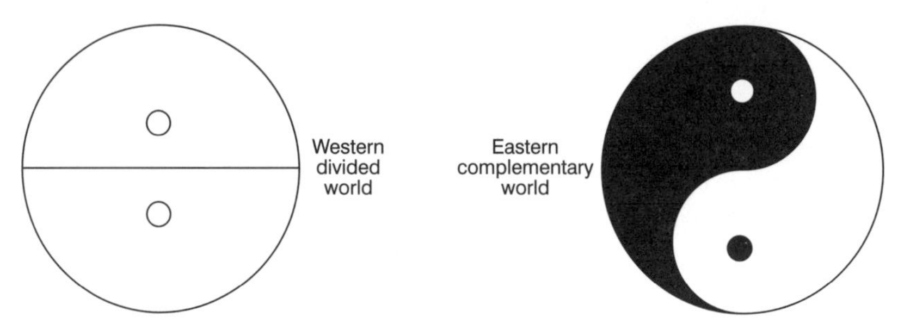

Figure 10.2 *Yin and yang versus Western thought*

Other examples of an entirely different organization of social life include the Amish living in North America who reject every form of modern life and material possessions. Another example is the Jain religion in India. Jainists see every living creature as an independent soul that must be treated with respect. This goes to such an extreme that Jainist monks carefully sweep the street as they walk along so as not to kill any insects. Every tie with material life is also broken. And a simple Hawaiian saying goes, 'There are two ways to be rich: make more or desire less'.

government's role will then be entirely different; on a large scale, companies will no longer look to the government to determine how far their responsibility goes. One's own norms and values will determine what activities we engage in and what we do not engage in.

11

Towards Sustainable Banking

INTRODUCTION

The effort directed towards sustainable development is an irreversible process that will ultimately result in a second-order process of change. This different approach to the economy will emphasize the importance of a symbiosis between the economic system and ecology as the source of life – and not seeing the ecological dimension simply as a limiting condition for economic activities. After all, a limiting condition will always be seen as a restriction. By applying a different way of thinking in which the economy and ecology are seen as complementary (as yin and yang as opposed to conflicting opposites), it will be possible to develop another way of living and working. This will involve an open process with a link to current reality. This final chapter will attempt to create a connection between the more philosophical discussion in Chapter 10 and current reality as discussed in Parts I and II. This will be accomplished by using today's reality as a foundation for a view to the future and from that point, distilling activities or developments for the future. Naturally, the question of what this means for banks, and how can they contribute to it will be addressed.

Sustainable development, however, usually has a broader context than that employed in this book. In addition to an ecological component, it often involves social and economic components. In our current societal orientation the economic component is already a leading factor, and the social component is more generally accepted (this does not mean, of course, that these components do not require extra attention). Furthermore, these components can conflict with one another and threaten to make the concept of sustainable development so broad as to become all-inclusive. Such a situation can easily lead to a lot of discussion but little action. Considering that the Earth can be 'used' only once, it would be advisable to change the ecological and economic courses as quickly as possible to a course in which the stewardship of the ecological system is more important than the 'use' of it. Naturally, how this is accomplished will ultimately have to be to some extent compatible with the social and economic dimensions. By devoting attention first to the ecological dimension, it will be seen that much more is possible without people having to lose their jobs en masse and without having a country fall into decline. This may mean that paradoxical steps may need to be taken towards the ecological dimension initially which can then be followed by looking into a new system in which sustainability is integral.[1]

The trends: A summary

People have been confronted for centuries with the impact their activities have had on the natural environment. Even the Ancient Greeks believed environmental issues to be important. Since the Industrial Revolution, the human impact on the ecological system has become so great that the negative effects of industrial progress have become apparent. Economic development has brought much prosperity but has also led to social and ecological problems. In recent decades, environmental protection has been legislated and businesses are gradually setting a new course. A dominant trend in this regard is the striving for a delinking of economic growth and environmental degradation, chiefly by means of eco-efficiency. Among businesses, efforts aimed at enhancing sustainability are increasingly becoming an integrated part of their operations – not based just on a direct or indirect commercial viewpoint (indirect in that natural resources are a condition for the long-term continuity of operations), but also – and often – as based on social responsibility. Consumers and citizens are seeing a course in which efforts aimed at sustainable development are more and more self-evident. Sustainable development is a trend and not just hype; it is inescapable: a condition for the continued life of people on the planet.

Another trend is socially responsible entrepreneurship. Sceptics say that this is hype and that as soon as the economy takes a turn for the worse, it will be shown to be just a passing fad. However, focusing solely on the ecological dimension, it has been shown that doing business based on preventing harm to the environment is not just a passing fad. Instead, it is very much based on cold, commercial reasoning. The same applies to offensive business, although in this case, companies depend more on decisions being made by customers. Since sustainable development is inescapable, sustainable business will also be inescapable. Due to the intermediary position of banks, this also applies to sustainable banking. This process will proceed either gradually or abruptly depending on the position taken now, with winners emerging and losers falling by the wayside.

It is also possible to observe a trend in the character of environmental policy in which the responsibility for environmental policy is increasingly being turned over to private parties. The required publication of an environmental report in Denmark, for example, increases the knowledge and insight into the environmental problems within a company, and this can lead to a more preventive form of environmental protection or an offensive stance in regard to sustainability. This environmental information can also allow third parties such as other companies or consumers to compel, or third parties such as environmental organizations or citizens to request companies to do more to protect the environment. Other examples are the emergence of liability lawsuits and market-based environmental instruments, such as the flexible instruments of the Kyoto Mechanism. This trend in particular implies either threats or challenges for banks.

Companies and banks react in different ways to the challenge of sustainability. Only a few companies and banks – mostly niche players – are now operating in a semi-sustainable manner. Others feel themselves closely tied to the limitations set by the current economic system. Nevertheless, important steps have been made within the dominant economic paradigm, such as an internalizing of the external environmental impacts in prices which has become a necessary condition for the encouragement of sustainable development. After all, the 'environment problem' is based on the fact that the current paradigm endows the environment with little if any economic value. This internalizing is developing as a result of government requirements and instruments (eg levies and subsidies) and also more often by demands from the market (buyers, suppliers, employees, regulatory institutions and competitors).[2]

Banks occupy an exceptional role in efforts devoted to sustainable development. Due to their intermediary position in the current economic system, they are playing both quantitative and qualitative roles. There are various ways in which banks can make good commercial use of their efforts toward sustainable development and can fulfil a social role in attaining sustainable development. What follows is a brief summary of the activities that banks can undertake at this time.

An important task accomplished by banks is the assessing of risks; the increasing responsibility of companies and the farther-reaching internalizing of external environmental impacts on market prices will mean added risks for banks in comparison to the current situation. If banks adjust their extension of credit to accommodate this, this will be a way that they can internalize environmental needs into their prices. Attaching value to the environment can also take place by means of certain products such as fiscal green investment funds, environmental funds and environmental leasing. Securitization (providing insight into the future gains of certain investments and then making them available as investment properties) is another possibility for coming closer to a sustainable society (eg by the securitization of the expected yields from solar energy). Various technological developments that should be able to reduce the pressures on the environment by factors of 10 to 20 are still associated with high levels of innovation and risk. Banks can fulfil a special role in this regard by financing these activities by such means as specific innovation funds. Banks can also establish various institutions such as micro-credit in developing countries, donating to exceptional activities in regard to sustainability, and setting up products with regard to combatting climate change. Obviously, most banks implement an internal environmental protection policy to reduce their own direct environmental impact.

As part of the regular process of extending credit, banks will employ risk-related premium differentiation (as a reward for lower environmental risks) to encourage sustainability among businesses. Banks can also opt to encourage sustainability by means of non-risk related premium differentiation; regardless of the risks, the sustainable activities of businesses or the activities of 'sustainable businesses' can be financed at lower costs to the customer.[3] In practice, however, this will lead to a decrease in the bank's financial creditworthiness and rate of return, and its continuity can even become endangered. Due to the innovative nature of many sustainable activities and sustainable companies, the risks will in principle be higher than those of more conventional companies. To allow sustainability in banking activities regardless of these higher risks, a bank can set up a separate fund or reserve part of its assets for such activities. The government can also stimulate this by tax rulings such as the green tax ruling in The Netherlands or the European EIF guarantee for bank loans to SMEs.

Completely sustainable banking, without external assistance, is not yet possible on a large scale. Although steps can be taken in this direction, a large-scaled integral sustainable balance sheet seems likely within the framework of the current paradigm only if the costs of a sustainable bank are considerably lower than those of non-sustainable banks. After all, such a case allows a lower debit interest without endangering the interest margin and the continuity of operations, while a buffer is developed for credit risks. This is possible only if the risks of sustainable businesses are considerably lower than for those of other businesses or if the government makes more resources or facilities available. Since this is not yet the case and there are few borrowing companies that are operating even partially sustainably, large-scaled offensive banking is the most feasible solution currently available. From the standpoint of sustainability, offensive banking does not imply complete sustainability. By learning and innovating, however, sustainable banking will become more attainable step by step.

A LINK TO THE FUTURE

To achieve sustainable development, a reduction in the total environmental impact is necessary. At an aggregate level, this can be achieved in three ways: a reduction in per capita affluence, a reduction in population growth, and a reduction in the environmental impact per product (see the formula in Box 2.4).

A reduction in population growth (while it may be desirable and which plays a major role in the Southern hemisphere) is a very ethical question, but it is one that is difficult for private parties to influence. For this reason, it is not addressed further here.

A reduction in economic growth is found undesirable by many (this applies especially to developing countries). Others, however, see it as a necessary condition for sustainable development. However, such a difference in opinion is not at all necessary. In regard to the reduction of the total environmental impact, the rate of growth is not as important as the nature of that growth.[4] Economic growth in sustainable activities is quite possible without an increase in the total environmental impact; if a different economic course is taken, it can even decrease.

A more sustainable economic growth can be achieved by means of reducing the environmental impact per product. In general, this can be accomplished by five means: changing a country's sector structure, behaviour changes among consumers, behaviour changes among producers, technological developments and miscellaneous means.

A change in the sector structure would manifest itself differently in each country. In general terms, this might be accomplished in many Western countries by having fewer energy-intensive sectors and a less intensive system of agriculture (see Appendix II for an elaboration of a possible future sector structure in The Netherlands in 2030).

A change in behaviour is not easy to direct but will have to result from an individual and collective wish to achieve a sustainable development.[5] The fact that there are many members of nature and environmental organizations in developed countries illustrates this wish, but this has yet to be expressed in terms of consumer behaviour. Although the demand for environmentally friendly products is rising, it is still relatively small. For one thing, this is related to the availability and accessibility of these products. Offering sustainably produced food products in supermarkets instead of only in special stores is an important step forward in this regard. For another thing, the price of environmentally friendly products is in most cases still too high. Actually, this is only to be expected (*ceteris paribus*, ie without cost-reducing innovation): environmentally friendly products do not shift any environmental damage off onto the environment or society at large but internalize this into the product (by such means as taking preventive measures) and thus into the price. The government can play an important role in this regard; after all, this is a case of unfair competition by environmentally unfriendly products that are actually dumping environmental damage onto the environment at no cost (also known as ecological dumping). Such products should therefore be taxed at a higher rate. This is already being initiated in the greening of tax systems in certain European countries such as Germany, Denmark and The Netherlands. Another possibility is the granting of subsidies to relatively environmentally friendly products. Furthermore, the abolition or restructuring of 'environmentally harmful' subsidies is being reconsidered at the international, European and national levels. Examples are the energy subsidies for large-scale and energy-intensive sectors (like aluminium smelting operations). In addition to influencing changes in behaviour among consumers the government also has a great deal of influence on the change of behaviour among producers. The above-mentioned greening of the tax systems and abolishing of environmentally harmful subsidies are examples of this. Changes in producer behaviour are obviously linked to consumer behaviour, but may also result from commercially applicable technological developments.

Technological innovation is therefore a fourth factor in achieving sustainable development.[6] Technological innovation is also a means of having environmental friendly products at a lower price than environmental unfriendly products. Within the near future, much will be possible, even within the limiting conditions of the current economic system. It will be primarily offensive companies that will be taking on this challenge. Technological breakthroughs will not only be possible in the long run, but they will also be necessary. By doing so, environmental impact can be reduced by a factor of 4 to 20. Examples are dematerialization, closed ecological cycles, sustainable industrial parks, cars that operate on fuel cells and more efficient freight transport (eg by means of underground tubes). These would definitely seem to be feasible within the foreseeable future but these kinds of long-term decisions will have to be made now. After all, before such technologies can be applied, a great deal of knowledge will have to be accumulated and research and experiments will have to be conducted. Banks can contribute to these efforts by supporting and financing such innovations, such contributions being included in sustainable banking.

One interesting new development is the 'alternative fulfilment of needs' (AFN). This could be called a fifth factor which is a combination of the second, third and fourth factors (changes in behaviour and technological innovation). To reduce environmental impact, companies have developed various activities that have a certain chronological order. At first, companies shifted to end-of-pipe measures to prevent environmental pollution (eg placing filters on chimneys). These measures do little to change the actual production process or product itself. Next, the production process itself was examined ('process-integrated measures' such as co-generation or combined heat and power projects). Today, many companies are taking a closer look at the product itself (environmentally responsible product development, an example being a television produced by environmentally responsible methods). In the case of AFN – the newest development and one that can significantly contribute to sustainable development – the degree of freedom for reducing environmental impact is even higher: only the consumer's need is still important; the material requirements of the product or production process are no longer factors. Obviously, the various methods supplement one another. Real-life examples of AFN are email instead of handwritten letters, novel protein foods instead of meat, a shared car instead of a personally owned car, videoconferencing instead of physical mobility, and the use of bacteria and insects instead of chemical crop control agents. AFN thus implies an entirely different way of approaching consumer needs or a fundamentally different way of looking at products. What becomes apparent is that if such developments continue, certain products such as meat will suffer in terms of the demand for them. In terms of extending credit, banks should be aware of businesses that either carry a certain product that can be threatened by, for example, an AFN product or that may be able to develop such products themselves.

A seemingly similar development and one in which the consumer's need is also a major factor is the operationally sustainable leasing of products.[7] In this development, the producer remains responsible for the product throughout its entire lifecycle. A consumer can lease a carpet, for example. If the consumer wants another carpet after so many years, or would rather have wooden floors, the product need not be thrown away. The producer, who still owns the carpet, takes the carpet back, cleans it, and leases it to another customer. At the end of the carpet's lifecycle, the producer recycles the carpet by processing it into another carpet. Leasing can thus contribute significantly to encouraging closed cycles, and closed cycles are by definition sustainable in terms of their raw materials use.

These examples illustrate that achieving sustainability within the current economic reality is already partially possible by means of innovative inventions, new combinations and another way of looking at producing and consuming. Such developments are challenges for banks in the areas of risks (certain products will disappear), markets (new companies, sectors and products with enormous growth potential)[8] and financial products (eg innovation funds and leasing).

THE ROLE OF BANKS IN A SUSTAINABLE FUTURE

Banks and other businesses are tied to the limiting conditions that the current economic system imposes on their activities. Market players exhibiting new values, however, will make it possible to stretch these limitations. After all, the limits of the current economic system are not static; they only seem to lack flexibility as yet.

Now that Western society has attained a certain degree of affluence and the negative effects of uncontrolled economic growth are evident, an era is emerging in which material progress is no longer the most important factor in our lives but in which progress will be considered progress only if it involves an improvement in quality – not only in regard to products and production processes, but more importantly in regard to life itself. Within certain minimum conditions of continuity, both social, ecological and financial rewards will be determining factors. Sustainability will become a key concept. In this regard, financial rewards will not be seen as an objective but as a limiting condition for guaranteeing the continuity of business operations.

For decades, it has been proposed that a bank's role within the social process actually goes no further than its most basic function as a transformer of money in terms of place, term, size and risk. According to this concept, a bank is reactive (or a 'follower'). When society starts attaching value to certain things, a bank will have to anticipate this. In this way of thinking, a bank plays no prominent, proactive role in the forming of a sustainable society. The current view that banks are reactive might change in time; it's not an established fact. Change is possible. To encourage the process toward a sustainable society, banks may want to become active instead of reactive.[9] After all, it is particularly due to the intermediary position of banks that they can create conditions and can actively contribute to efforts that will make sustainable development possible. Sustainable development will imply a mind shift or cultural shift within banks in this regard – something that can and will not take place from one day to the next. The banks' own perception of themselves in regard to their role in social processes will have to be modified. More than it does today, a bank will determine what it finds to be socially or ecologically acceptable or unacceptable and will act accordingly to make a transformation to sustainability possible. After all, sustainability involves vicious circles (in which banks follow the changes in society while changes in society depend partially and sometimes to a great degree on the activities and position of banks) and banks can break these circles.

There are two criticisms of this viewpoint. The first is that any role other than a reactive one is essentially impossible. This linear concept, however, is limited by definition. It also implicitly presumes that one is a prisoner of the system – a rather fatalistic presumption. The second criticism emanates from the first: 'every man to his trade' (the importance, in other words, of core competency). Although this is essentially a generally accepted strategy, it requires insight into what the trade is. If Shell, for example, did not preselect on the basis of oil becoming economically exhausted (eg by establishing a factory for the production of solar energy), this organization would be expected to exist for only 20 to 30 years to come. This process is being accelerated by social pressure. Shell's trade is possibly not so much oil as energy. It is clear that what makes a world of difference is how core competency is defined. A bank's core competency might not be intermediating in a strict sense but providing knowledge, information and services relevant for financial decisions by economic actors. If so, this would open the way to a more active instead of passive role.

If banks assume a more active role, what kinds of choices and dilemmas will they be facing? Will active banks become full-scale extensions of the government for upholding and implementing environmental policy? This will not actually occur; banks will continue to make autonomous decisions and not become extensions of the government. Banks will take various decisions themselves, an example being whether or not to continue or restructure certain banking activities. After all, certain banking activities themselves are incentives for environmentally polluting activities. A bank will run considerable social risks in this that can become financial risks over time (eg the financing of today's intensive livestock industry).

Surely with investments involving timetables of 10 to 15 years, a bank should be asking itself what could happen in the interim and what its risks could be (both in the social and direct financial sense). A bank can also ask itself exactly what the social relevance is of circulating money in the financial markets without actually having any control over exactly where that money winds up (for that matter, for example, it might be financing the arms trade without being aware of it). The exclusion of social and long-term risks requires banks to accept a proactive role. Such a position can be seen as an intermediate category for a more active and guiding role for banks (as opposed to the traditional reactive one) and therefore as a step toward sustainable banking.

The question of whether this will require making strict sector-related choices arises. Chaos theory teaches that during a transformation, a symmetry break or 'splitter' (Peters and Wetzels, 1997) develops. At that point, more than one truth exists. Such a situation does not demand making choices. On the contrary, making choices excludes other possibilities. Plurality and heterogeneity are important; while one borrowing client is still living according to the old way of thinking, another will already have stepped into the new one. And it is precisely by being open to both of these realities that the best possible customer value is attained and a bank is able to undergo a transformation by internalizing its dynamic environment. Not by choosing but by guiding clients with a high environmental impact to sustainability and by supporting clients with sustainable intentions or activities will a bank shape the process of change towards sustainable banking.

Learning and innovating are necessary for banks on their way to sustainable banking. Setting up innovation funds and funds for subsidizing certain relatively sustainable activities or businesses are examples of this. Just as important, however, is participating in networks; accumulating knowledge about systems of sustainability; the drive to participate in experiments aimed at sustainability; supporting certain sustainable initiatives with knowledge, experience or money (especially in developing countries); participating in innovation projects; contributing to raising social consciousness concerning sustainability; applying chain analysis to investment decisions; donating money to certain projects; etc.

More than two-thirds of environmental problems are associated with local consequences. It is precisely at the local level that a connection can already be made between financial and social or ecological rewards. Certain investments in the local living environment have a direct or indirect effect at the local level. If a certain company wants to reduce its emissions or wastes that affect the local environment (eg waste water or offensive odours) and needs financing to accomplish this but its collateral, financial position or rate of return on its investments are insufficient for regular bank financing, a locally rooted bank might still be able to finance this (and sometimes even with favourable conditions). The decrease in the impact on the environment that results from such an investment can be felt locally and improves the living conditions of the bank's clients and employees. This can have positive effects on the bank's image and other activities. In this way, a proactive bank can internalize some external environmental effects in its investment decisions so that the improvement of the environment also demonstrates its profitability in economic terms. In the current paradigm, local goodwill can enable relatively favourable financing without suggesting that the bank is engaged in making charitable donations.

CONCLUSION

Care for the environment has been a major consideration among production companies for several decades. Banks started devoting attention to this matter only halfway into the 1990s. Within the banking sector, environmental concerns were long considered as having a soft image associated with idealistic reformers. By no longer seeing the environment as simply a cost item or risk, but understanding that banks themselves as well as their clients can earn money from care for the environment, the issue of how to deal with the environment suddenly takes on an entirely different character. By the year 2001, many banks are not only actively involved but are also actually trying to get the better of one another. Even so, image is not the only motivation behind these activities. Insights have changed as a result of pioneering in the field of sustainable banking at such smaller banks as the Umweltbank, the Triodos Bank and The Co-operative Bank. Now, the environment and sustainability are more often seen offering possibilities to generate additional profit or turnover through the development of specific products or services. This change in outlook occurred halfway into the 1990s among several banks in Europe, Oceania and North America.[10] An example of this is that various large banks such as ING and UBS have now set up sustainable investment funds. The environment as a selection criterion for investment funds thus made its appearance among large banks.

Surprisingly, the possibilities for such financial innovations had existed for a long time, but except for a few national and international niche players, these opportunities were either ignored or not taken seriously. Individuals in the large banks continued to limit their definitions of the environmental question just as most people limit their ability to solve the nine-dot puzzle in Box 10.1. Meanwhile the solution lay beyond these limitations. The examples listed in this book are opening up the way for a larger-scale application of premium differentiation in the extension of credit – the core banking activity. It is more or less accepted as being based on risk but not yet as based on ideological grounds. Individuals open to the implications of getting off the beaten track, as illustrated in the puzzle, will also make non-risk related premium differentiation acceptable within the foreseeable future. If this kind of development is embraced on a large scale, it will become standard practice. Without putting itself in the role of the entrepreneur or being an extension of government policy, sustainable banking will then be a generally accepted market practice and no longer out of the ordinary.

Sustainable development will go hand in hand with changes. These changes may be gradual and occur over a long period, similar to the gradual shifts in concepts within the banking sector, or they may occur suddenly within a short period of time (usually as the result of a crisis such as a serious environmental catastrophe). From their own perspective, banks usually deal with gradually changing views. Even so, banks can also be confronted with sudden changes. A good example of such a shock is the introduction of the direct liability of banks in the US for the environmental damage done by the banks' clients in the 1980s. Another shock in the coming 10–20 years might be the demise of an entire sector due to the effects of climate change or policy, sectors in which banks have occupied long and considerable positions.

Banks themselves will also have an impact on these changes and will be able to guide the path towards sustainable development. The changing view that a bank might not always be a follower but can also take the helm will first be very gradual but may shift radically at a certain point. Banks that follow this course and are prepared to take part in such a radical change will become the sustainable banks of the future – and this also refers to their own continuity of operations. Real sustainable banking does not imply that banks should write off clients that are currently not operating as sustainably as they might, but that they assist these clients along the difficult road to more sustainable business practices. At the same time, banks will have to support clients with sustainable investment ambitions (usually ones that are still ahead of their time and whose investment ambitions are highly innovative) with specific financial instruments to help them meet their objectives. For this reason, it is definitely inadvisable for banks to make choices (or to require them to make choices) from among main sectors when their long-term objective is sustainability. Only when clients remain stubbornly opposed to a more sustainable course (ie continually unwilling to introduce improvements that would lead to a more sustainable direction) would such issues arise. But then, it would make good business sense to withdraw in consideration of such a client's poor perspectives for the future. Dealing openly with this process of change, both internally and externally, will lay the foundations necessary for sustainable banking. This is not a process than can or will take place overnight. A key factor in this process of change is how a bank sees itself within the global setting.

Appendix I

Environmental Performances of Developed Countries

Figures I.1 and I.2 provide a snapshot of a development in the environmental impact and performances of the developed countries. The figures are from 1998/1999 (OECD, 1999b). The lines on the graph are simply a visual aid and are not representative of any connections. GDP and population trends are shown as a reference. The figures reveal significant differences among the developed countries.

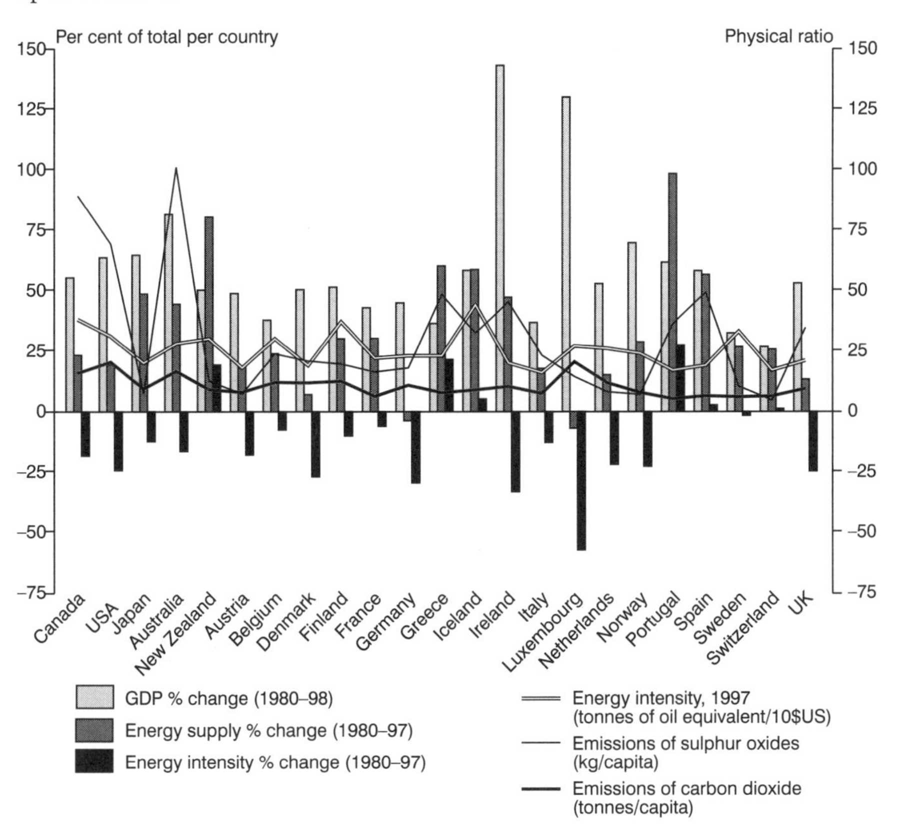

Figure I.1 *Energy consumption and related pollution of developed countries*

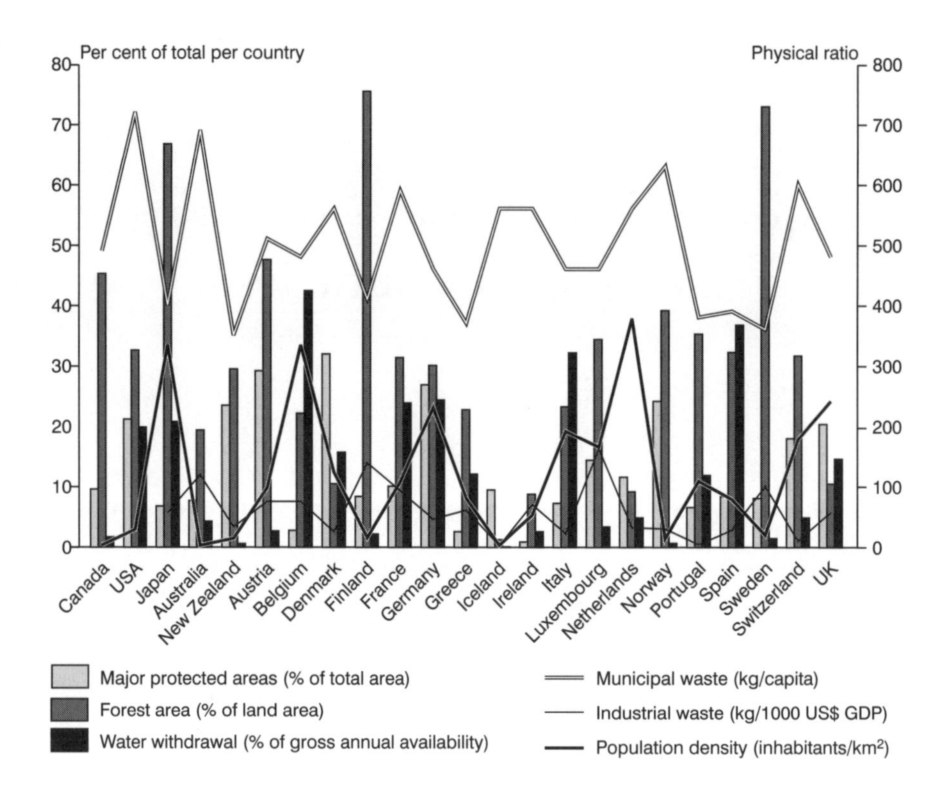

Figure I.2 *Protected forest, water and waste in developed countries*

Appendix II

Sectoral Changes in the Pursuit of Sustainability: A Dutch Scenario

In Table II.1 scenarios are constructed which reflect the effects of certain forms of sustainability policy on a number of macroeconomic variables in The Netherlands in the long term (2030). The results of this SEDS study are summarized briefly below.[1]

Environmental technological developments and economic structures are central to this study. Only the supply side of the economy is examined; a structural change in consumption patterns is not included. The central question is what a sustainable Netherlands will look like in 2030. Four scenarios, 'Strong Together' (strong sustainability policy with the rest of the world), 'Strong Alone', 'Weak Sustainability' and 'Balanced' were considered, the last being an in-between scenario, in which ecological systems are considered to be tolerant.

Irrespective of the outlook for sustainability, the food and confectionery industry, the chemical industry, public utilities, the transport sector and animal and crop farming will come under pressure. There will be a contraction of agriculture in all scenarios while horticulture will only experience contraction in the two strong sustainability scenarios. Growing sectors can be distinguished in addition to contracting sectors. The differences per scenario here are, however, greater. Both strong scenarios show growth opportunities in the service sector, with strong growth particularly in nature education and recreation. In the two weaker scenarios, construction and installation businesses and miscellaneous industries will grow strongest.

The macroeconomic data show that the development of sustainable economic structures does not have to be to the detriment of employment opportunities. A key point about this study is that consumption patterns, relative price changes and the instrumentalization of environmental policy do not play any explicit role. The scenarios were developed for the very long term. Moreover, the structural changes apply to the same degree in every business in a sector. This will not happen in practice. Within any one sector, growing and contracting businesses are likely to appear. The results should therefore be interpreted tentatively.

Table II.1 *Scenarios for a possible sector structure in 2030*

	1991	*Baseline*	*Strong together*	*Strong alone*	*Balanced*	*Weak*
Macroeconomic outcomes (in euro billions)						
Unemployment (x 1000)	202	N/A	113	236	113	113
GNP (euro billions)[2]	226	454	395/443	340	458	534/512
Production value (euro billions)	458	949	792	666	955	1166
Production structure (in euro billions of production value)						
Food/confectionery industry	35	60	23	23	29	34
Oil industry	11	12	37	3	5	7
Chemicals	18	40	13	11	18	23
Rubber/synthetics	4	10	29	3	29	29
Metal	43	89	66	25	41	152
Wood/furniture	3	5	2	16	3	16
Miscellaneous industry	20	37	17	16	111	111
Subtotal	135	253	187	97	235	371
Energy extraction	9	8	24	4	24	24
Public utilities	10	15	13	8	11	13
Construction/installation	35	93	34	30	137	186
Transport	27	67	38	36	43	49
Services	214	463	437	435	439	475
Subtotal	295	646	546	514	654	747
Livestock farming	10	15	6	6	7	9
Crop farming	1.3	2.3	0.9	0.9	1.4	1.4
Horticulture	6	12	4	4	15	9
Forestry	1.3	2.7	5.0	1.4	7.7	7.7
Fishing	0.4	0.9	0.5	0.5	0.5	0.5
Subtotal	19	34	16	12	32	27
Cleaning	1.3	2.3	1.4	1.4	6.4	7.3
Nature education/recreation	8	14	42	41	28	14
Subtotal	10	16	43	43	34	21
TOTAL	**458**	**949**	**792**	**666**	**955**	**1166**

Source: based on a study by the Institute for Environmental Issues.

Appendix III

IFC Project Classification for Environmental Assessments

Below, examples of projects which normally would fall in category A, B or C of the IFC are listed.[3]

Category A projects

- Large dams and reservoirs
- forestry (large scale);
- agro-industries (large scale);
- industrial plants (large scale);
- major new industrial estates;
- major oil and gas developments, including major pipelines;
- large ferrous and non-ferrous metal operations;
- large port and harbour developments;
- projects with large resettlement components and all projects with potentially major impacts on human populations;
- projects affecting indigenous or tribal populations;
- large thermal and hydropower development;
- projects that include the manufacture, use or disposal of environmentally significant quantities of pest control products;
- manufacture, transportation, and use of hazardous and/or toxic materials;
- domestic and hazardous waste disposal operations;
- projects which pose serious occupational or health risks; and
- projects which pose serious socioeconomic concerns.

Category B projects

- Agro-industries (small scale);
- electrical transmission;
- aquaculture and mariculture;
- renewable energy (except large hydroelectric power projects);
- tourism (including hotel projects);
- rural water supply and sanitation;

- rehabilitation, maintenance, and modernization projects (small scale);
- manufacture of construction materials;
- general manufacturing;
- textile plants;
- telecommunications; and
- greenfield projects in existing industrial estates.

Category C projects

- Advisory assignments;
- factoring companies;
- life insurance companies;
- mortgage securitization;
- securities underwriters and broker/dealers;
- technical assistance; and
- rights issues.

Appendix IV

Example of an Environmental Risk Checklist

The following list is a specimen 'checklist' for potential environmental risk. It is not an existing control list but one that is derived from the environmental risk aspects discussed in Chapter 6 and includes elements of existing checklists.

1 Sector:
  - 1.1 Environmentally sensitive?
  - 1.2 Specific market circumstances?
  - 1.3 Potential changes to government policy?
  - 1.4 Environmental issues in the past?

2 Strategic aspects:
  - 2.1 Product lifecycle?
  - 2.2 Pressure from the chain?

3 Environmental policy:
  - 3.1 Is an environmental policy in place?
  - 3.2 Environmental audit?
  - 3.3 Environmental management system?
  - 3.4 Certification?

4 Management:
  - 4.1 Executive involvement?
  - 4.2 Stance with respect to the environment (defensive, preventive, offensive, sustainable)?
  - 4.3 Employee involvement?

5 Legal requirements:
  - 5.1 Permits required/present?
  - 5.2 Other government requirements?

6 Legal position:
  - 6.1 Administrative law proceedings currently or in the past?
  - 6.2 Private law proceedings or requirements?
  - 6.3 Soil decontamination order?

7 Collateral:
  - 7.1 Rate of depreciation?

8 Insurances:
  - 8.1 Environmental damage insurance present?
  - 8.2 Other relevant insurances?

9 Public opinion:
  9.1 Risk of consumer backlash?
10 Environmental information:
  10 1 General sources (newspapers, magazines, etc)?
  10.2 Legal sources?
  10.3 Government sources?
  10.4 Annual environmental report?
  10.5 External ratings?
  10.6 Internet/intranet?

Appendix V

The ICC Business Charter for Sustainable Development: Principles for Environmental Management[4]

1 Corporate priority

To recognize environmental management as among the highest corporate priorities and as a key determinant of sustainable development; to establish policies, programmes and practices for conducting operations in an environmentally sound manner.

2 Integrated management

To integrate these policies, programmes and practices fully into each business as an essential element of management in all its functions.

3 Process of improvement

To continue to improve corporate policies, programmes and environmental performance, taking into account technical developments, scientific understanding, consumer needs and community expectations, with legal regulations as a starting point; and to apply the same environmental criteria internationally.

4 Employee education

To educate, train and motivate employees to conduct their activities in an environmentally responsible manner.

5 Prior assessment

To assess environmental impact before starting a new activity or project and before decommissioning a facility or leaving a site.

6 Products and services

To develop and provide products or services that have no undue environmental impact and are safe in their intended use, that are efficient in their consumption of energy and natural resources, and that can be recycled, re-used, or disposed of safely.

7 Customer advice

To advise, and where relevant, educate customers, distributors and the public in the safe use, transportation, storage and disposal of products provided; and to apply similar considerations to the provision of services.

8 Facilities and operations

To develop, design and operate facilities and conduct activities taking into consideration the efficient use of energy and materials, the sustainable use of renewable resources, the minimization of adverse environmental impact and waste generation, and the safe and responsible disposal of residual wastes.

9 Research

To conduct or support research on the environmental impacts of raw materials, products, processes, emissions and wastes associated with the enterprise and on the means of minimizing such adverse impacts.

10 Precautionary approach

To modify the manufacture, marketing or use of products or services or the conduct of activities, consistent with scientific and technical understanding, to prevent serious or irreversible environmental degradation.

11 Contractors and suppliers

To promote the adoption of these principles by contractors acting on behalf of the enterprise, encouraging and, where appropriate, requiring improvements in their practices to make them consistent with those of the enterprise; and to encourage the wider adoption of these principles by suppliers.

12 Emergency preparedness

To develop and maintain, where significant hazards exist, emergency preparedness plans in conjunction with the emergency services, relevant authorities and the local community, recognizing potential transboundary impacts.

13 Transfer of technology

To contribute to the transfer of environmentally sound technology and management methods throughout the industrial and public sectors.

14 Contributing to the common effort

To contribute to the development of public policy and to business, governmental and intergovernmental programmes and educational initiatives that will enhance environmental awareness and protection.

15 Openness to concerns

To foster openness and dialogue with employees and the public, anticipating and responding to their concerns about the potential hazards and impacts of operations, products, wastes or services, including those of transboundary or global significance.

16 Compliance and reporting

To measure environmental performance; to conduct regular environmental audits and assessments of compliance with company requirements, legal requirements and these principles; and periodically to provide appropriate information to the board of directors, shareholders, employees, the authorities and the public.

Appendix VI

UNEP Statement by Financial Institutions on the Environment and Sustainable Development[5]

We members of the financial services industry recognize that sustainable development depends upon a positive interaction between economic and social development, and environmental protection, to balance the interests of this and future generations. We further recognize that sustainable development is the collective responsibility of government, business, and individuals. We are committed to working cooperatively with these sectors within the framework of market mechanisms toward common environmental goals.

1 Commitment to Sustainable Development

1.1 We regard sustainable development as a fundamental aspect of sound business management.
1.2 We believe that sustainable development can best be achieved by allowing markets to work within an appropriate framework of cost-efficient regulations and economic instruments. Governments in all countries have a leadership role in establishing and enforcing long-term common environmental priorities and values.
1.3 We regard the financial services sector as an important contributor towards sustainable development, in association with other economic sectors.
1.4 We recognize that sustainable development is a corporate commitment and an integral part of our pursuit of good corporate citizenship.

2 Environmental Management and Financial Institutions

2.1 We support the precautionary approach to environmental management, which strives to anticipate and prevent potential environmental degradation.
2.2 We are committed to complying with local, national, and international environmental regulations applicable to our operations and business services. We will work towards integrating environmental considerations

into our operations, asset management, and other business decisions, in all markets.

2.3 We recognize that identifying and quantifying environmental risks should be part of the normal process of risk assessment and management, both in domestic and international operations. With regard to our customers, we regard compliance with applicable environmental regulations and the use of sound environmental practices as important factors in demonstrating effective corporate management.

2.4 We will endeavour to pursue the best practice in environmental management, including energy efficiency, recycling and waste reduction. We will seek to form business relations with partners, suppliers, and subcontractors who follow similarly high environmental standards.

2.5 We intend to update our practices periodically to incorporate relevant developments in environmental management. We encourage the industry to undertake research in these and related areas.

2.6 We recognize the need to conduct internal environmental reviews on a periodic basis, and to measure our activities against our environmental goals.

2.7 We encourage the financial services sector to develop products and services which will promote environmental protection.

3 Public Awareness and Communication

3.1 We recommend that financial institutions develop and publish a statement of their environmental policy and periodically report on the steps they have taken to promote integration of environmental considerations into their operations.

3.2 We will share information with customers, as appropriate, so that they may strengthen their own capacity to reduce environmental risk and promote sustainable development.

3.3 We will foster openness and dialogue relating to environmental matters with relevant audiences, including shareholders, employees, customers, governments, and the public.

3.4 We ask the United Nations Environment Programme (UNEP) to assist the industry to further the principles and goals of this Statement by providing, within its capacity, relevant information relating to sustainable development.

3.5 We will encourage other financial institutions to support this Statement. We are committed to share with them our experiences and knowledge in order to extend best practices.

3.6 We will work with UNEP periodically to review the success in implementing this Statement and will revise it as appropriate.

We, the undersigned, endorse the principles set forth in the above statement and will endeavour to ensure that our policies and business actions promote the consideration of the environment and sustainable development.

Appendix VII

List of Signatories to the UNEP Statement by Financial Institutions on the Environment and Sustainable Development[6]

Andorra

Banca Internacional D'Andorrà – Banca Mora
Crèdit Andorrà

Angola

Banco Nacional de Angola

Argentina

Banco Frances

Australia

Westpac Banking Corporation

Austria

Bank Austria
Bank Für Tirol und Vorarlberg Aktiengesellschaft
Bankhaus Carl Spängler & Co Aktiengesellschaft
Creditanstalt-Bankverein Austria
Österreichische Investitionskredit Aktiengesellschaft
Österreichische Kommunalkredit Aktiengesellschaft
Raiffeisen Zentralbank Austria AG

Brazil

Banco do Estado de Sao Paulo SA
Banco Nacional de Desenvolvimento Economic e Social
BBV Brasil

Bulgaria

Balkanbank Ltd

Canada

Bank of Montreal
Canadian Imperial Bank of Commerce
Export Development Corporation
Royal Bank of Canada
Scotia Bank (The Bank of Nova Scotia)
Toronto-Dominion Bank

Chile

Banco BHIF

Colombia

Banco Ganadero

Cyprus

Bank of Cyprus

Denmark

Den Danske Bank, A/S
Unibank

Finland

Kansallis-Osake-Pankki

France

Banque Populaire du Haut-Rhin
Caisse des dépôts
Crèdit Local de France

Germany

Bankhaus Bauer AG
Bankhaus CL Seeliger
Bankhaus Max Flessa & Co
Bankhaus Neelmeyer AG
Bankverein Werther AG
Bayersiche Handelsbank AG
Bayerische Hypo-und Vereinsbank
Bayerische Landesbank Girozentrale

Beneficial Bank AG
Bezirkssparkasse Heidelberg
BfG Bank AG
B Metzler seel Sohn & Co KgaA
Commerzbank AG
Conrad Hinrich Donner Bank AG
DEG – German Investment and Development Company
Degussa Bank GmbH
Delbrück & Co, Privatbankiers
Deutsche Ausgleichsbank
Deutsche Bank AG
Deutsche Bank Saar
Deutsche Pfandbrief-und Hypothekenbank AG
Deutsche Postbank AG
DG Bank
Dresdner Bank AG
Eurohypo AG, Europäische Hypothekenbank der Deutschen Bank
Fürstlich Castell`sche Bank, Credit-Casse
Hamburgische Landesbank Girozentrale
Hesse Newman Co Bank (BNL Group)
HKB Hypotheken-und Kommunalkredit Bank
Investitionsbank des Landes Brandenburg
Kreditanstalt für Wiederaufbau
Kreissparkasse Düsseldorf
Kreissparkasse Göppingen
Landesbank Baden-Württemberg
Landesbank Schleswig-Holstein Girozentrale
LBS Badische Landesbausparkasse
Merck Finck & Co
MMWarburg & Co
Quelle Bank AG
Sal Oppenheim jr & Cie
SchmidtBank KGaA
Schröder Münchmeyer Hengst AG
Schwäbische Bank AG
Service Bank GmbH & Co KG
Sparkasse Leichlingen
Sparkasse Staufen
Stadtsparkasse Hannover
Stadtsparkasse München
Stadtsparkasse Wuppertal
UmweltBank AG, Germany
Vereins- und Westbank AG
Volksbank Siegen – Netphen eG

Greece

Commercial Bank of Greece

Hungary

Budapest Bank RT
National Savings and Commerical Bank Ltd

Iceland

Landsbanki Islands

India

Bank of Baroda

Ireland

Bank of Ireland Group

Italy

Banca Monte dei Paschi di Siena SpA
Istituto Nazionale di Credito Agrario SpA
Credito Italiano

Japan

Good Bankers Co Ltd
Nikko Asset Management Co Ltd
Nikko Securities Co Ltd

Jordan

Arab Bank, PLC
Export Bank of Africa Ltd

Kuwait

Kenya Commercial Bank Group
National Bank of Kuwait SAK

Madagascar

Banky Fampandrosoana ny Varotra

Mexico

BBV Probursa

Morocco

BMCE Bank

The Netherlands

Algemene Spaarbank voor Nederland (ASN Bank)
FMO
Rabobank
Triodos Bank

Norway

Den Norske Bank ASA

Peru

Banco Continental

The Philippines

Bank of Philippine Islands
Development Bank of the Philippines
Global Business Bank
Land Bank of the Philippines
Metropolitan Bank and Trust Company
Philippine Bank of Communications (PB Com)
Planters Development Bank
Rizal Commercial Banking Corporation

Poland

Bank Depozytowo-Kredytowy SA
Bank Gdanski SA, Poland
Bank Ochrony Srodowiska
Bank of Handlowy W Warszawie SA
Bank Polska Kasa Opieki SA
Bank Przemystowo-Handlowy SA
Bank Rozwoju Eksportu SA
Bank Slakski SA
Bank Zachodni SA
National Fund for Environmental Protection and Water Management
Polski Bank Inwestycyjny SA
Pomorski Bank Kredytowy SA
Powszechna Kasa Oszczednosci – Bank Panstwowy
Powszechny Bank Gospodarczy SA w todzi
Powszechny Bank Kredytowy SA

Portugal

Banco Bilbao Vizcaya (Portugal) SA
Banco Portuges do Atlantico SA

Puerto Rico

BBV Puerto Rico

Romania

Romanian Commercial Bank SA

Russia

Econatsbank

Slovenia

Kreditna banka Maribor dd

Spain

Banca Catalana SA
Banco Bilbao Vizcaya SA
Banco del Comercio SA
Banesto, Banco Espagnol de Credito
BBV Privanza Banco SA
Caixa Cataluyna
Central Hispano
Finanzia, Banca de Credito SA

Sweden

Ekobanken – Din Medlemsbank
JAK – Jord, Arbete, Kapital
Skandinaviska Enskilda Banken
Svenska Handelsbanken
Swedbank AB

Switzerland

Bank Sarasin & Cie
Banque Cantonale de Genève
Basellandschaftliche Kantonalbank
Credit Suisse Group
EPS Finance Ltd
Luzerner Kantonalbank
Sustainable Asset Management

UBS AG
Zürcher Kantonalbank

Thailand

Thai Investment and Securities Co Ltd

Uganda

Uganda Commercial Bank

United Kingdom

Abbey National Plc
Barclays Group Plc
The Co-operative Bank
Friends Provident Life Office
HSBC Holdings Plc
Lloyds TSB Bank
NatWest Group
Prudential Plc
Royal Bank of Scotland Plc
Woolwich Plc

United States of America

Citigroup
Community Capital Bank
EBI Capital Group LLP
Friends Vilas-Fischer Trust Company
Innovest Strategic Value Advisor Inc
Republic National Bank

Venezuela

Banco Provincial
Corporación Andina de Formento

Appendix VIII

Overview of Characteristics of Selected Banks

In this appendix the characteristics of the banks discussed in Chapter 9 are examined. Table VIII.1 shows the size of each of the 34 selected banks as at the end of 1999. Table VIII.2 follows with a picture of the performance and strategy of individual banks. Both tables are followed by explanations.

Table VIII.1 *Scale of selected banks: Assets, employees, offices and countries*[7]

Banks (1999 figures)	*Rank*	*Country*	*Bank assets (EUR, billion)*	*Employees worldwide*	*Offices worldwide*	*Countries worldwide*
Royal Bank of Canada	25 (55)	Canada	176.1	51,891	1509	30
CIBC	27 (59)	Canada	161.3	35,561	1250	11
Bank of Montreal	28 (61)	Canada	146.1	32,844	1198	**10**
Citigroup	2 (2)	US	714.0	**180,000**	4500e	**101**
Bank of America	5 (5)	US	630.0	155,906	4800	38
Chase Manhattan	15 (20)	US	404.4	74,801	1300	50
Bank Austria	31 (68)	Austria	140.0	19,032	655	28
Fortis Bank	19 (27)	Belgium	406.1	61,109	3000e	42
KBC Bank	30 (63)	Belgium	156.2	28,362	1470	30
MeritaNordbanken	34 (78)	Finland	**104.0**	28,220	1300	18
BNP Paribas	3 (3)	France	698.6	77,472	7900e	83
Groupe Crédit Agricole	13 (16)	France	439.5	93,244	7971	60
Société Générale	14 (19)	France	406.5	64,660	3100	75
Deutsche Bank	1 (1)	Germany	**839.9**	93,232	2374	60
HypoVereinsbank	10 (10)	Germany	503.3	46,170	1417	35
Dresdner Bank	17 (22)	Germany	396.8	50,659	1473	70
Banca Intesa	20 (29)	Italy	304.0	73,491	5089	28
UniCredito Italiano	26 (58)	Italy	171.2	62,288	3357	15
SanPaolo IMI	32 (69)	Italy	139.9	24,133	1431	**10**
ABN AMRO	11 (13)	Netherlands	457.9	104,653	3666	76
ING Group	18 (26)	Netherlands	492.8	86,040	1750e	65
Rabobank Group	22 (31)	Netherlands	281.2	53,147	1795	42
BSCH	23 (41)	Spain	256.4	95,442	**8473**	37
BBVA	24 (47)	Spain	238.2	89,235	7491	37
Svenska Handelsbanken	33 (76)	Sweden	107.8	**8520**	527	19
UBS	6 (6)	Switzerland	611.8	49,058	1200e	50
Crédit Suisse Group	12 (14)	Switzerland	450.5	63,963	1042	53
HSBC Holdings	7 (7)	UK	566.8	145,847	6000	100

Barclays Bank	16 (21)	UK	410.4	80,200	1959	60
NatWest	21 (30)	UK	299.1	72,100	2250	23
National Australia Bank	29 (62)	Australia	155.8	51,879	2339	15
Bank of Tokyo-Mitsubishi	4 (4)	Japan	710.8	33,041	742	46
Fuji Bank	8 (8)	Japan	576.6	13,976	**339**	47
Sumitomo Bank	9 (9)	Japan	505.7	29,324	688	25

Table VIII.2 *Profile of selected banks: Performance, internationalization and activities*

Banks (1999 figures)	*Rank*	*Country*	*ROaE, %*	*Cost to income, %*	*Geographic scope*[8]	*Businesses*[9]
Royal Bank of Canada	25 (55)	Canada	14.0	66.9	global (82% D)	1b, 2, 3, 4, 5
CIBC	27 (59)	Canada	9.4	**77.9**	global (83% H)	1a, 2, 3, 4 (5b)
Bank of Montreal	28 (61)	Canada	12.5	67.2	regional (94% H)	1a, 3, 4 (2, 5)
Citigroup	2 (2)	US	21.9	67.6	universal (60% D)	1b, 2, 3, 4, 5
Bank of America	5 (5)	US	17.8	57.7	global (92% D)	1a, 2, 3, 4 (5)
Chase Manhattan	15 (20)	US	**23.2**	55.0	universal (71% D)	1a, 2, 3, 4, 5
Bank Austria	31 (68)	Austria	11.9	71.7	regional + (51% D)	1b, 2, 3, 4, 5
Fortis Bank	19 (27)	Belgium	14.4	66.7	regional + (82% H)	1a, 2, 3, 4, 5
KBC Bank	30 (63)	Belgium	14.9	67.4	global (54% D)	1b, 2, 3, 4, 5
MeritaNordbanken	34 (78)	Finland	22.9	54.9	regional+ (96% He)	1b, 2, 3, 4, 5
BNP Paribas	3 (3)	France	12.7	67.8	universal (44% D)	1b, 2, 3, 4, 5
Crédit Agricole	13 (16)	France	11.6	66.8	universal (78% D)	1b, 2, 3, 4, 5
Société Générale	14 (19)	France	16.6	72.7	global (78% D)	1b, 2, 3, 4 (5)
Deutsche Bank	1 (1)	Germany	11.6	81.8	universal (41% D)	1b, 2, 3, 4 (5)
HypoVereinsbank	10 (10)	Germany	3.1	64.1	regional + (77% D)	1b, 2, 3, 4 (5)
Dresdner Bank	17 (22)	Germany	9.2	76.2	global (71% D)	1b, 2, 3, 4 (5)
Banca Intesa	20 (29)	Italy	8.5	75.9	global (70% D)	1b, 2, 3, 4 (5)
UniCredito Italiano	26 (58)	Italy	19.6	59.0	regional (88% D)	1b, 2, 4 (3, 5a)
SanPaolo IMI	32 (69)	Italy	13.9	59.9	regional (95% H)	1b, 2, 3, 4 (5a)
ABN AMRO	11 (13)	Netherlands	17.3	70.2	universal (50% D)	1b, 2, 3, 4, 5
ING Group	18 (26)	Netherlands	15.3	67.4	universal (45% D)	1b, 2, 3, 4, 5
Rabobank Group	22 (31)	Netherlands	8.8	74.2	global (74% D)	1a, 2, 3, 4, 5
BSCH	23 (41)	Spain	18.5	57.7	regional (59% H)	1b, 2, 3, 4 (5)
BBVA	24 (47)	Spain	16.1	61.5	regional + (68% D)	1b, 2, 3, 4 (5)
Sv Handelsbanken	33 (76)	Sweden	17.5	51.9	regional + (80% H)	1b, 2, 3, 5a (4)
UBS	6 (6)	Switzerland	21.2	68.0	universal (23% D)	1b, 2, 3, 4 (5a)
Crédit Suisse Group	12 (14)	Switzerland	19.2	72.9	universal (27% D)	1b, 2, 3, 4, 5
HSBC Holdings	7 (7)	UK	17.6	53.7	universal (67% H)	1b, 2, 3, 4, 5
Barclays Bank	16 (21)	UK	21.3	61.4	global (61% D)	1b, 2, 3, 4 (5a)
NatWest	21 (30)	UK	18.6	66.8	regional + (76% D)	1a, 2, 3, 4 (5)
NAB	29 (62)	Australia	16.5	56.0	regional + (54% D)	1b, 2, 3, 4 (5a)
BTM	4 (4)	Japan	4.3	45.8	global (78% D)	1a, 2, 3, 4 (5)
Fuji Bank	8 (8)	Japan	**2.8**	62.4	global (87% D)	1a, 2, 3, 4 (5)
Sumitomo Bank	9 (9)	Japan	2.9	**40.9**	global (89% D)	1a, 2, 3

The countries in the tables are first grouped into world regions (North America, Europe and Oceania), thereafter according to the country in which the headquarters is situated. The third column of Table VIII.1 shows where the bank's headquarters is situated and this is usually the country in which the bank originated. The table clearly shows the large size of the banks. In total, these 34 banks provide work for more than 2.2 million people (average 65,600 employees per bank) and possess more than 95,000 offices (average 2800 per bank). Their international reach is considerable, both in terms of the number (Table VIII.1) and the diversity of countries and world regions in which they operate (Table VIII.2). For the geographical scope in Table VIII.2, four categories have been used:

1. Regional: the bank's services are primarily offered to the surrounding countries (4 banks).
2. Regional +: ditto, but with some activities/branches in other world regions (8 banks).
3. Global: various services are offered worldwide or in many world regions (12 banks).
4. Universal: (almost) all financial services are offered in all world regions (10 banks).

Banks in the category 'universal' met at least two of the following three criteria:

1. Active in more than 50 countries (see last column of Table VIII.1).
2. Active in all business fields (see last column of Table VIII.2).
3. The percentage of international activities is greater than 50 per cent (see last but one column of Table VIII.2).

Around two-thirds of the 34 banks are worldwide in scope (global and universal). Most banks set their sights first on their neighbouring countries, then on other developed world regions and lastly on countries in transition and other developing countries. Spanish banks stand out: they are very strong in Latin America. Barclays too, stands out: it aims to be the primary player in Africa. Of the universal banks, the Swiss stand out: their domestic activities comprise only 25 per cent of their total activities (based on assets). Of the 34 banks, 66 per cent of them conduct more than 66 per cent of their activities in their domestic or home market.

In Table VIII.2 the type of financial services offered by the banks is examined. Five main business areas have been identified: retail, corporate and investment banking, asset management and insurance (with some subdivisions). The division into main and sub-areas is, however, rough. There are important variations between countries and banks in the definition of business areas. Thus some banks group 'private banking', financial services for the very wealthy, under retail banking and others under asset management. The same is true of investment banking. Nearly all banks class their activities in financial markets in this category. Only a few use the pure meaning (see Chapter 4), in which the main activity that can be distinguished is advising big multinationals during acquisitions and takeovers. The latter is primarily the domain of American, Swiss and German banks. Some other banks loan out these advisorial services, but then only in a few specific sectors. It is notable that the All Finance concept is mostly to be found in Northern European countries, in particular Benelux countries. This is a consequence of legislation and extensive domestic consolidation in traditional banking areas.

The return on average equity (ROaE) has been used as an indicator of profitability. The cost–benefit ratio is a widely used indicator of the efficiency of the bank. There are important variations between banks. Between the highest (Chase) and lowest scoring banks (Fuji) in terms of profitability is a gap of 20 percentage points; that is easily a factor of 8. In terms of efficiency the difference is 37 percentage points, ie the highest scoring bank (Sumitomo) is almost twice as efficient as the lowest scoring bank (CIBC). The non-weighted average ROaE amounts to 14.3 per cent and the non-weighted average cost–benefit ratio to 64.4 per cent. Among the Japanese banks the very low profitability paradoxically coupled with high efficiency is striking. An explanation for this is the high rate of bad loans in Japan. The provisions and actual depreciation of these loans are not counted in the cost–benefit ratio, which is therefore overstated. Alternatively, both factors influence the ROaE negatively, which is therefore relatively low.

Appendix IX

An Integral Score for Sustainable Banking[10]

In order to calculate an integral score for sustainable banking, it must first be decided which elements from Chapters 5–8 will be taken into account. Five groups have been chosen: communication, generic published data, financing, special products and social issues. In each group a maximum of 20 points can be awarded, so that in total a bank can accrue a maximum of 100 points. The points are allocated bearing in mind both current and future benefit to the environment. In addition, a key (in percentages) is established per group. This key is based on the same criteria and the relevance of the issue to the integrated score. For example, communication is important, but does not rate as highly as financing and new financial products that benefit the environment. Given that the primary source of a bank's influence is its products and services, these two groups should account for a minimum of 50 per cent.

The five groups of elements are presented below. The key for each group is shown next to the group in parentheses.[11] By each element in parentheses is the maximum number of points that a bank can score for the element. Where a bank does not have a certain element at all, it receives zero points. Moreover, where a qualitative distinction can be made, a sliding scale has been applied. Elements with sliding scales are marked with a star. The sliding scales are explained in Box IX.1. The five groups are:

1. Communication (10 per cent):
   - environmental policy (4 points)*;
   - environmental reporting (5 points)*;
   - WBCSD membership (1 point);
   - signatory to UNEP declaration (3 points);
   - signatory to ICC declaration (1 point); and
   - ISO 140001 certification (6 points)*.
2. Generic published information (25 per cent):
   - quantitative data about internal environmental care (4 points)*;
   - qualitative data about internal environmental care (1 point);
   - objectives for internal environmental care for the future (5 points);
   - quantitative data about external environmental care (4 points)*;
   - qualitative data about external environmental care (1 point); and
   - objectives for external environmental care for the future (5 points).

Box IX.1 Sliding scales

Points for environmental policy:

- comprehensive and formal policy – 4 points;
- only a few policy remarks to the environment in the financial report – 2 points.

Points for environmental reporting:

- sustainability report (environmental and social aspects) – 5 points;
- pure environmental report – 4 points;
- social report only (with environmental section) – 3 points;
- one or more sections about the environment in the financial report – 2 points;
- one or more sections in the financial report about external social aspects (but not about the environment) – 1 point.

Points for ISO certification:

- national or worldwide system certified (distinctions were difficult to assess) – 6 points;
- a few separate branches/locations certified – 4 points.

Points for the publication of quantitative data about internal environmental impact:

- absolute and relative data for most locations and subjects – 4 points;
- absolute or relative data only and for most locations and subjects – 3 points;
- data for a few sites only and for many subjects – 2 points;
- data for a few subjects only – 1 point.

Points for the publication of quantitative data about external environmental care:

- data about several products and/or years in tables – 4 points;
- data about several products but less transparent and complete – 3 points;
- data about one or a few products only – 1 point.

Points for environmental loans:

- self-financed loan – 2 points;
- loan with governmental guarantee or facilities – 1 point.

Points for investment funds:

- more than one product and/or advisorial activity – 2 points;
- one product or just advice – 1 point.

Points for insurances:

- own produced product – 2 points;
- product originates from another financial institution and is only distributed – 1 point.

Micro-credits:

- activities in developed and developing countries – 3 points;
- activities in developed countries only – 2 points.

3 Financing (15 per cent):
   - environmental risk analyses (8 points);
   - sector exclusions (4 points);
   - adherence to World Bank guidelines for financing (5 points); and

- adherence to OECD guidelines for business activities in developing countries (3 points).

4 Special products (40 per cent):
- environmental loans (2 points)*;
- sustainable investment funds or advice (2 points)*;
- environmental leasing (2 points);
- environmental savings products (1 point);
- environmental damage insurance (2 points)*;
- advisorial services impacting on the environmental care by customers (2 points);
- venture capital for environmental innovations (3 points);
- micro-credits (3 points)*;
- debt-for-nature swaps (1 point); and
- climate products (2 points).

5 Social issues and charity (10 per cent):
- credit cards/cheques for gifts benefiting nature and the environment (4 points);
- sponsoring benefiting nature and the environment (2 points);
- community involvement (8 points); and
- internal socioeconomic aspects (6 points).

The total points per group gives, after application of the sliding scales, and multiplied by its key, an integral score for sustainability per bank. The maximum integral score is 20. This integral score is then divided into four equal classes which reflect the stance of a bank towards sustainability as defined in Chapter 4. The classes, their ranges and the four stances of banks are:

- 0 to 5 integral points: defensive banking;
- 5 to 10 integral points: preventive banking;
- 10 to 15 integral points: offensive banking;
- 15 to 20 integral points: sustainable banking.

Since the group defensive banking was large and diverse, it was split into two subgroups: highly defensive (0 to 2.5 points) and lightly defensive (2.5 to 5 points).

Appendix X

Ecological Economics

A relatively new economic school, ecological economics, emphasizes physical limits. It draws from the natural sciences, specifically the First and Second Laws of Thermodynamics. The Second Law is also called the 'law of increasing entropy' (chaos). That is to say that there is no existing energy transformation which is completely efficient. Moreover, the use of energy is an irreversible process. During the transformation, a certain amount of energy is always lost and the remainder can no longer be used fully for other functions. In the absence of autonomous new energy, the Second Law of Thermodynamics implies that a closed system always uses up all of its energy. Considering that energy is a requirement for life, life ends when all energy is consumed.[12]

Neoclassical and classical economists assume a situation of relative scarcity: the scarcity of a resource compared to the scarcity of another resource or the same resource but of a lower quality is what matters. Greater scarcity results in higher prices and, as a result, resource substitution. The ecological economists however assume absolute scarcity. The amount of low entropy is limited (and there is no substitute for it). Substitution implies one low entropy resource can only be substituted by another resource of low entropy. Scarcity is therefore absolute. The fact is that prices can only reflect relative scarcity; the price mechanism cannot take account of ecological limits. That would mean that sustainable development is impossible within the current (neoclassical) economic system. The existence of physical limits is well illustrated in the historical example of Easter Island.

The solution that ecological economists propose is direct regulation by government. Because potential for substitution is not unlimited, the internalization of external effects (using regulatory levies for instance) is not sufficient. The government must intervene directly by imposing quota restrictions and command-and-control measures. This is already taking place to an extent at national level. The difference is that the entire international economic system must be restructured and that the physical limits will be much more stringent than the current ones. The transition from a classical system to an ecological one therefore also requires a paradigm shift: ecological economics is based on other worldviews (rooted in ecology and natural sciences). The neoclassical solution (internalization of external effects) is not entirely rejected; prices remain an efficient allocation mechanism (but within strict physical limits).

Appendix XI

Zero Growth and Other Solutions?

A number of economists speak of a growth obsession (or tunnel rationality; Goudzwaard, 1978): economic growth has become a goal in itself. Various research studies which show that wealth in the US, corrected for environmental quality and health among other things, has declined since the 1960s are often points of reference for this (Daly and Cobb, 1989). The drive for growth results in damage to the environment without resulting in extra wealth (this lies behind the plea for a green GNP). For these and other reasons, some economists argue for zero growth[13] while others even speak of selective contraction (Hoogendijk, 1993).

Daly goes furthest in his proposals (1991). He sees a 'steady state' economy as a necessary condition for sustainability. His solution is an institutional approach. Three institutes must be set up in order to reach the steady state, a situation in which population is stable and physical welfare reaches a certain level deemed desirable (and in which the flow of natural resources and energy through the economic system is low) in the short term. The first institute ensures a stable population level. Every individual receives a tradable right to 'produce' a maximum of one child. Second, a quota system must be put in place for the use of natural resources. Third, Daly proposes an institute that would ensure an equitable distribution of income. Minimum and maximum income levels must be determined for this (as well as a maximum level of wealth). Daly's solution requires a pervasive bureaucracy, does not bear in mind individual and national differences and exhibits totalitarian tendencies.

Some people see the existence of money or interest as the causes of unsustainable development:[14] 'Money is the root of all evil' (Russell, 1992). However, this thinking is too simple and ahistorical. The idea of abolishing money and/or interest has nevertheless been introduced to an extent. An example is LETS (local exchange trading system). A number of these systems are now in operation in a number of cities in The Netherlands, France and Spain. There is actually a group of companies in Switzerland that supplies each other with goods with no money changing hands. The assumption is that small-scale, direct contact and service orientation promote sustainable development (through, for instance, more quality and care). Of course, there are large differences between various systems (Aktie Strohalm, 1998).

Interest does not exist in an Islamic economy since this is forbidden by the Koran. Usually, alternatives for interest are sought, such as payment in kind (services or products). The ASN Bank started an experiment with interest-free banking when it introduced an interest-free fund in 1999 that works on the basis of reciprocity: those depositing money do not earn interest but points; those lending money relinquish a proportionate number of points. This concept has been operating successfully in Denmark and Sweden for a number of years. A disadvantage is that you must save (earn points) before you can do any borrowing.

Appendix XII

The Prisoner's Dilemma

The original 'prisoner's dilemma' is easiest to explain using Table XII.1.

Table XII.1 *The original prisoner's dilemma*

	Party B pleads guilty		*Party B pleads innocent*	
	A	*B*	*A*	*B*
Party A pleads guilty	9	9	0	12
Party A pleads innocent	12	0	1	1

Source: adapted from Varian (1987, p471).
Note: figures represent length of sentence in months for each prisoner.

Two individuals have committed a crime together and are interrogated in separate rooms. Each of the prisoners has the option to plead guilty. If only one of the prisoners pleads guilty of the joint act, he is released (no sentence) and the other gets full blame (12 months' sentence). If both plead innocent, they both receive a prison sentence of 1 month – for procedural reasons. If both confess, both receive a prison sentence of 9 months. Lack of trust and incomplete information of the other's choice result in both confessing. After all, confessing is optimal from *A*'s point of view since it involves a sentence of 9 months if *B* also confesses, and freedom if *B* pleads innocent. However, if *A* were to plead innocent while *B* confessed, *A* would be sentenced for 12 months. The same rationale also applies for *B*. Complete information and trust in the other's ability to keep silent could have resulted in both of them being sentenced for only 1 month! The outcome is therefore not optimal.

The 'prisoner's dilemma' has been widely examined in game theory. The dilemma assumes that the game is only a one-off event. It was shown that, in theory, the outcome is always non-cooperative, as long as the game is played once or a finite number of times (Axelrod, 1984). Only if the game is repeated an infinite number of times and both players consider the future 'returns' will a cooperative outcome (the bottom left quadrant) occur. The threat of a non-cooperative future outcome is sufficient in producing a cooperative stance in an infinitely repeated game. Theoretically, at least, there is a way out of the dilemma. The question is if this solution also exists in practice for something like the environment. After all, government intervention (stricter environmental policy, for instance) changes the nature of the game (it has become finite). It is in the individual interest of the players to obtain a maximum return by changing their strategy in the round before the policy adjustment. The same is true for the other player too, and results in a non-cooperative outcome. The simple fact that a policy change will eventually be instituted makes the game finite. This means the outcome will always be (in every round of the game) non-cooperative, at least in theory.

Glossary

absorptive capacity The capacity of the ecological system to absorb environmentally harmful substances over a certain period of time and decompose them without undermining the vitality of the system. A distinction can be made between stock pollution and fund pollution. For stock pollution (such as dioxin) the absorption capacity, in the sense of decomposition capacity, is non-existent or negligible. The aggregate fund emissions of companies are only non-sustainable once the absorption capacity is insufficient over a certain period of time.[15]

anthroposophy A worldview focusing on the awareness of the unity of life, individual and world. Anthroposophy is not a religion or belief, but a spiritual science, that is to say that man can uncover the nature of the world through insight, experience and contemplation. Rudolf Steiner (1861–1925) was the founder of this worldview.

carbon dioxide (CO_2) A colourless, odourless and non-poisonous gas that results from fossil fuel combustion or natural processes and is a normal part of the air on Earth. Excess CO_2, eg by industrial activity, can disturb the climate on Earth.

change processes, first order Incremental changes within a particular system or paradigm.

change processes, second order Changes of the system (meta-change processes), ie a paradigm shift.

corporate citizenship The idea that a company is a citizen in society, with civil rights and civil responsibilities that extend beyond the purely economic sphere.

corporate governance Involves the issues related to the way companies (listed) are managed and controlled. The emphasis is on the internal and external accountability of the company executives.

direct environmental risks (lender liability) The risk that the bank itself will be held liable for the environmental damage that has been caused by its borrower.

ecological damage A form of damage to nature that is collectively suffered (and caused). Nature is summed up here as a universal legacy to which there is no confirmed right of ownership.

ecology The branch of science that studies the relationships between organisms and their animate and inanimate environment. An ecological system is more or less wholly comprised of groups of organisms that stand in relation to each other and their environment. It is characterized by a number of natural processes that determine the development of the system. The size of the system is determined by the size, quality, nature and regenerative capacity of natural resources, the inflow of solar energy and the capacity of nature to absorb waste.

environment Environment is used as a collective term in this book, unless otherwise stated, to mean the stock of natural resources, pure nature and the pollution of the ecological system.

environmental assessment An environmental impact assessment (EIA), environmental audit, hazard assessment, risk assessment or environmental action plan (EAP), or combination thereof (terminology used by the World Bank/IFC).[16]

environmental audit An investigation of procedures and processes of a company or site regarding its compliance with applicable regulations and laws on environmental issues and its impacts on environmental conditions.

environmental damage A form of damage that is felt personally whereby the environment or nature acts as the intermediary of the damage. In broad terms, it also incorporates 'ecological damage'. In this book, however, a distinction is drawn between the individualizable form of damage – in short, 'environmental damage' – and the collective form – 'ecological damage'.

environmental due diligence Data collection and assessment of environmental conditions or impacts prior to a transaction, to identify and quantify environmental risks (commercial, legal and reputational).

environmental impact assessment (EIA) An assessment of the possible impacts on the natural or human environment of a proposed project.

ethics A component of the philosophy concerned with human morals. The prevailing moral can be contained in precepts or ethical judgements that belong to the value judgements. Ethics is the study of ethical judgements: how they originated, and their meaning and significance to society. It is equally an attempt to analyse the significance of ethical judgements. Ethical judgements are not true or untrue but must be interpreted as judgements on what is morally good, that is, what ought to be done or not done.

external effects Events that have a significant advantage or disadvantage for a particular person or group of persons not wholly involved in the decisions that led, directly or indirectly, to these events.

green financing Fiscally beneficial financing of environmental projects as approved by the government.

indirect environmental risks The materialization of the environmental risks of the borrower which leads to financial losses of the bank (eg via a decline in value of collateral or a marginalized repayment capacity of the borrower).

internalization of environmental costs A process in which the costs of preventing and restoring environmental degradation are paid directly by the parties responsible for this degradation (rather than being financed by society at large).

joint and several environmental liability The legal concept that any one of many contributors to environmental damage can be held responsible for the total damage on its own.

lender liability See *direct environmental risks.*

paradigm A paradigm reveals the underlying structure of thinking. It concerns the referential framework through which we view the world and within which we give meaning to our observations, as in a worldview, certain norms and values, a management perspective, and so on. The paradigm determines (often unconsciously) the information we absorb (and that which we ignore), the type of questions we ask and the direction in which we look for answers. If this direction does not result in answers on a regular enough basis, a crisis arises out of which new paradigms can spring: a paradigm shift.

paradigm change (or paradigm shift) A completely new way of thinking about old and new problems. This implies a change of the referential frameworks. A new paradigm does not develop gradually but is seen or experienced suddenly (possibly with a shock).

prisoner's dilemma A dilemma that shows that in the event of insufficient information, individual rational behaviour leads to irrational collective behaviour.

regenerative capacity The capacity of the ecological system to replace extracted raw materials whereby a distinction can be made between renewable and non-renewable raw materials. The regenerative capacity of a cleared wood (a renewable natural resource) whose trees were 20 years' old is, for instance, 20 years. For oil it is some 150 million years.[17]

socially responsible business Socially responsible business concerns an optimum combination of business interests and social considerations. This implies that it is not just the profit, but mainly the effect on people and the environment that is uppermost in the decision making. Internationally, the terms 'people', 'planet' and 'profit', the 'triple bottom line', form the three criteria on which businesses will be judged.

strict liability In case of environmental damage, the fault of an actor need not be established. The only thing that needs to be established is causal relation between the damage and an act or omission of an act; ie, the fact that an act or omission of an act caused the damage.

sustainable banking A modus operandi in which the internal activities meet the requirements of sustainable business and in which the external activities (such as lending and investments) are focused on valuing and stimulating sustainability among customers and other entities in society.

sustainable business An operation in which the ultimate use of raw materials and environmental pollution that can be attributed to a company does not exceed the respective regenerative or absorptive capacity of the ecological system.

sustainable development A process of change in which the pattern of investment, the focus of technological developments, patterns of consumption and institutional change are such that the consumption of raw materials they involve does not outstrip the regenerative capacity of the ecological system, nor does the environmental pollution outstrip the absorptive capacity.

sustainable society A society structured in such a way that the elements of sustainable development can be satisfied. In a broad context it consists of minimum requirements for social organization of society (such as social cohesion and justice) and the economy in addition to the minimum requirements for ecology.

tragedy of the commons A situation of inadequately diagnosed scarcity situations that results in a depletion of soils, the fostering of erosion and general disruption of ecological cycles and harm to natural reproductive systems. An individual has insufficient ownership rights to provide an incentive for action in the collective and, ultimately too, individual interest.

Notes

Chapter 1

1 Unless otherwise defined, 'environment' is used as a collective term in this book for the stock of natural resources, 'pure' nature and the pollution of the ecological system.
2 Sustainable business (or banking) is not a static concept. It is a concept whose intrinsic value will be continually adjusted in terms of the demands and issues of the time.
3 In the 1970s and 1980s, the environmental policy of the government and non-governmental organizations (NGOs) focused on the polluter. Since the 1990s, environmental issues have been considered more in terms of the chain in which financial service providers play a role. One of the first publications on banking and environmental issues was a paper by Sarokin and Schulkin (1991).
4 The conclusions must be taken with a grain of salt given that financial service providers are not keen on having any financial corpses in the cupboard. What the research did show was that the financial sector saw no possibilities of a structured and open participation. In fact, the environment was predominantly regarded as a risk, or cost item.
5 In a 1995 UNEP survey, 80 per cent of respondents adapted, to a greater or lesser extent, environmental risk analyses (see UNEP, 1995). Research in 1997 revealed that many banks had meanwhile established departments and specific products (see Ganzi and Tanner, 1997).
6 See, for instance, European Union (1993), UNEP (1997), the Dutch Ministry of Housing, Spatial Planning and the Environment (1998b), Carl Duisberg Gesellschaft (1998) and Coopers&Lybrand (1996).
7 The term 'sustainable' is susceptible to confusion. Throughout the book, this term refers to 'sustainable development' as discussed in length in Chapter 2, and is based on the definition given by the Brundtland Commission. It explicitly does not carry the narrow meaning of a 'long' duration, durable, solid or imperishable. 'Striving for sustainable economic growth' is an example of this confusion; in government circles and in business the meaning imparted here is often 'durable' and is devoid of the ecological context.
8 'Immaterial', in this context, describes qualities such as a high level of international security, social cohesion, possibilities for people to enrich themselves, minimal environmental pollution and a healthy, extensive natural world. Other, softer qualities can be added, like happiness, recognition and 'feeling at home'. The book is strongly oriented to the Western world. Further material growth remains essential for developing countries, although this also needs to be sustainable.
9 No strict division can be made between environmental pollution and the use/consumption of natural resources in view of the mutual relationship between

them. In practice, however, the Western world focuses more on pollution, while the developing world is more geared to natural resources (and the associated pollution). Such a division will not be used in the rest of the book.

10 Source: www.worldbank.org/data/databytopic/class.htm (November 2000).

11 Mexico, Korea, the Czech Republic, Hungary, Poland and Turkey are also OECD members but do not meet any of the criteria, either collectively or separately. The HDI is composed of indicators including the GDP per capita. A higher HDI than GDP per capita ranking is a positive aspect. This is the case for Sweden, whereas the opposite is true of Luxembourg.

12 Source: CIA World Fact Book 2000, www.odci.gov/cia/publications/factbook/index.html (November 2000).

CHAPTER 2

1 Property rights are comprehensive when they are universal, exclusive, transferable and enforceable. See Tietenberg, 1992, pp45–47. A fisher, for example, will indeed be stimulated to fish in a sustainable manner if the fishing grounds are his or her exclusive property. In this case, other fishers have no rights to these fishing grounds while enforcement is possible. External effects are events that have a considerable advantage or disadvantage for a certain person or group of persons who had not been sufficiently involved in the decision that led directly or indirectly to these events (Meade, 1973, p15). Other sources of environmental problems are imperfect market structures (such as monopolies) and government failures.

2 Some environmental issues are only evident in the longer term. Given that social and private discount rates differ, companies make investment decisions which, from a welfare perspective, are not optimal because they value net future return less highly than does society as a whole (ie they have a higher discount rate than that held by society). This discrepancy of values enables investment decisions to be taken in which future environmental returns play a subordinate role. Where sustainability also involves future generations, even the social discount rate of the present generation would be considered as being too high.

3 This means that there is no import into or export from the system of materials, energy, and so on. Obviously the energy and radiation from the sun are an important exception to this. This book, however, is mainly concerned with physical streams (materials). (Moreover, the import of, for example, meteorites and the export of spaceships or waste into space will be ignored.) This book will, therefore, work from the premise that together, economy and ecology form a closed system. Chapter 10 will revisit the implications of this.

4 Category 3 is a halfway category (halfway between people in the biological sense and people as initiators).

5 Note that autonomous processes within ecological systems (and processes outside the biosphere) also determine the human chances of survival, just as they make life possible (eg the natural greenhouse effect).

6 Caused mainly by the sharp increase in the use of fossil fuels since the Industrial Revolution, but also by a change in land use, including deforestation.

7 See IPCC, 2001. Over the 20th century the world temperature has already risen by 0.6 degrees Celsius and the average sea level by 0.2 metres. The projections referred to in the text are in excess of this. Whereas five years ago the IPCC (1995) was still cautious and spoke of people probably having an influence on the

climate, the most recent report (IPCC, 2001) confirms that a clear link exists and even that the greatest proportion of the warming of the Earth over the last century was caused by human activity. In the last 10,000 years (since the last great Ice Age) there has been no temperature change of greater than 1–3.5 degrees Celsius. Note that the IPPC (2001) does not make best estimates any more.

8 This book presented the world with the spectre of what would happen to humankind and other species as a consequence of the use of pesticides and insecticides, such as DDT, for the stimulation of agricultural production. A book entitled *Our Stolen Future*, by Colborn et al, did this once again in 1996. These authors write about rapidly spreading toxic substances in food intended for human consumption.

9 Boulding introduced this concept in 1966 to make it clear that the issue of environmental pollution and raw materials affects all people and nations. Nature and the environment take no heed of borders and race. This idea forms, in fact, the basis of the concept of sustainable development. The idea of a fragile 'Spaceship Earth' was given visual form by the first pictures from Apollo 8 of the blue Earth in a black universe.

10 Pezzey (1989, 1992) recorded in 1989 a total of 60 definitions of sustainable development. In 1992 he counted more than 100 different definitions in the literature. This number keeps on growing, but as a rule the definitions also refer to that given by the WCED (see below).

11 This is Ehrlich and Holdren's static formula. For a dynamic representation see Steer and Lutz, 1993.

12 Spappens (1996). The horizontal axis shows the population size (actual and anticipated) in the years 1990, 2010 and 2050. The vertical axis shows worldwide consumption of raw materials. The figure is based on the following premise: no growth in population or raw material consumption takes place in the North; in the situation 'South I' the current worldwide consumption and production pattern is maintained; in the situation 'South II' all people have the same consumption pattern as in the North. The population figures are taken from standard UN projections (United Nations, 1996). It should be noted that these figures are doubted by some because, among other things, they take no account of elevated mortality figures caused by water shortage and low birth rates caused by the increasing use of contraception (Hilderink, 2000).

13 In certain developing countries taking care of the next generation is necessary to the current generation's survival. This partly explains the high population growth (and/or high birth rate). For an elaboration of the issues facing such countries see Szirmai, 1994, for example.

14 *Financial Times* (15 March 1995), 'Nuclear Wastes Clean-up Will Cost at Least $230 bn', and the *Financieele Dagblad* (26 November 1994), 'Ex-DDR Environment Needs DM300 bn'.

15 According to some, sustainable development is by definition a subjective concept; see for example CPB, 1996, pp21–42. This is partly due to the variety in norms and values and the risk of making misdirected policy as a result of uncertainty over the capacity of the ecological system. The position this book takes is that norms and values were in fact politically defined at the Earth Summit and that the Precautionary Principle offers a way out of the uncertainty.

16 Regenerational capacity is the capacity of the ecological system to renew depleted resources. This enables the distinction to be made between renewable and non-renewable resources. The regeneration capacity of a felled forest (a renewable resource) of 20-year-old trees is, for example, 20 years. For oil the period is

around 150 million years. Absorption capacity is the capacity of the ecological system to degrade environmentally harmful substances without diminishing the vitality of the system. This enables the distinction to be made between stock pollutants and fund pollutants. For stock pollutants (such as dioxins), the absorption capacity is, by definition, zero or negligible. For fund pollutants, aggregated emissions are only unsustainable if the absorption capacity over any given period is inadequate. For sustainability, striving towards zero emissions is only necessary for stock pollutants.

17 For an analysis of the environmental problems facing for example Asia, see ADB, 2000.

18 See OECD, 1975. The OECD introduced the concept to combat subsidies, eg for soil sanitation, which may be disruptive to competition between countries. The concept has, however, acquired its own impetus and is now accepted literally (and thus more widely) as the starting point for most environmental policies.

19 The costs of environmental policy, and its advantages (estimated damage in the absence of policy), are being carefully examined. The Netherlands has a benefit-to-cost ratio of 6.5 (RIVM, 2000a).

20 This is largely in accordance with the principle of source-oriented regulation. Environmental policy must distinguish between local, regional and worldwide pollution. Seen from a cost perspective, the combating of pollution should in principle start at the source. For cross-border pollution, as worldwide pollution always is, cooperation between countries is a requirement for optimizing welfare.

21 See OECD (1994) for an evaluation of Japan's environmental policy. See USA-EPA (2000a) for the US's environmental policy and OECD (1996b) for an evaluation of it. See Commonwealth of Australia (1996) for Australia's environmental policy.

22 See also WRI (2000), OECD (1996a) and UNEP (2000).

23 The first international conference on the protection of flora and fauna (the Conference of London) took place in 1933, and in 1968 the important Biosphere Conference in Paris took place (partly as a result of Carson's 1962 book).

24 For a list of international environmental treaties, as at November 2000, see www.odci.gov/cia/publications/factbook/docs/appendix.html

25 See Keating, 1993. For the period 1993–2000 the annual costs for developing countries of adhering to Agenda 21 are estimated at €475 billion (price level 1992).

26 See www.earthsummit2002.org for details about the World Summit and www.ecouncil.ac.cr/eccharter.htm (November 2000) for details about the Earth Charter.

27 The Kyoto Protocol is relevant to six greenhouse gases. These are carbon dioxide (CO_2), methane (CH_4), nitrous oxide (N_2O), incompletely halogenated fluorocarbons (HFK), per fluorocarbons (PFK) and sulphurhexafluoride (SF_6) and other substances (substances that both damage the ozone layer and increase the greenhouse effect are CFKs, halons and chloroform). These six greenhouse gases are measured in CO_2 equivalents. Where CO_2 is discussed subsequently in the context of the Kyoto Protocol, these CO_2 equivalents are generally being referred to.

28 These are 'sinks'. 'Sinks' are still the subject of much discussion given that they offer only a temporary solution. There are a number of ways in which carbon sequestered in this way can escape (eg forest fires).

29 Measures that do not affect cost price will in any case be introduced by companies (ie when this is clear before the investment is made). Chapter 3 examines this.

CHAPTER 3

1 In many cases a win-win situation is possible in which a high degree of eco-efficiency improves a company's financial results. However, it is clear that win-lose situations exist as well in which sustainable solutions come with a higher cost price and therefore generate a lower result in a competitive market, provided that all other circumstances remain equal.
2 See The Millennium Poll on Corporate Social Responsibility (a summary is available at: www.environics.net/eil, October 2000) for the results of interviews with over 25,000 average citizens across 23 countries on 6 continents on their expectations of social corporate responsibility.
3 Yet only 17 per cent of the businesses in the same research indicate that they are well under way with the integration of sustainability into company strategy and activities. See Arthur D Little (2000).
4 Something which has become more important in Anglo-Saxon countries in which the government has had a less dominant role for a lot longer. See Reder, 1995.
5 The first three categories are based on Van Luijk, Schilder and Bannink, 1997. In the real world, it will not be easy to maintain the boundaries strictly.
6 Research has shown that share prices can come under pressure during these periods. This may also be due to government actions. See Tietenberg (1997) for an extended listing and analysis. In Canada a company lost a lawsuit related to an environmental infringement brought against it by the government. The judgement and the fine caused that company's share price to fall by 2 per cent (LaPlante and Lanoie, 1994). The announcement of litigation appears to influence share prices: Moughalu et al (1990) report an average fall of 1.2 per cent in the US. In a broad sample of 730 US companies confronted with legal action with respect to environmental infringements initiated by the government, share prices fell by an average of 0.5 per cent during the legal action (Badrinath and Bolster, 1996). Further research shows that companies adjust their environmental policy after large stock exchange and turnover losses (Konar and Cohen, 1997).
7 'As a commercial organization, our governing objective is to provide a satisfactory return on our shareholders' capital' (www.hsbc.com, October 2000).
8 See the Values Report (The Body Shop, 1998) and the Partnership Report (The Co-operative Bank, 2000) respectively.
9 See Collins and Porras (1994) or De Geus et al (1997) for instance. These are companies in categories 3 or 4. The fourth category is, in fact, more like a subcategory of the third category.
10 See Simons et al (2000) for an overview of the literature on these types of models. Companies do not necessarily go through the phases at the same tempo. Van Koppen and Hagelaar (1998) show that the preventive strategy is the maximum attainable phase for companies with large environmental risks and limited commercial opportunities. The challenge is ultimately to get these companies to become more sustainable. It should be stated explicitly that the phases and stances are defined from the point of view of the environment and sustainability.
11 The idea of this framework has evolved from Molenkamp (1995); considerable changes have been made though.
12 Under a broad definition of sustainability, that is, including the social component, this raises the question of whether a lack of personal contact can really be considered sustainable.

13 The table is largely based on Simons et al (2000), appended with Molenkamp (1995) and Shell (1998).
14 Product responsibility, phase out harmful products and processes, human rights, etc.
15 Important reasons for environmental reporting are: meeting the demands of stakeholders, improving image and identity, learning from the dialogue, and expanding awareness within the organization (Wempe and Kaptein, 2000).
16 See SustainAbility/UNEP, 1999, p9. There is little interest in pursuing environmental reporting in Japan. Transparency is seen as a 'negative virtue' (ibid, p19).
17 There are also systems of international standards for socioeconomic questions, such as the OECD Guidelines for multinational enterprises, the ILO (International Labour Organisation) declaration of fundamental principles and rights at work, the ICC (International Chamber of Commerce) code of conduct concerning corruption, the Amnesty International Business and Human Rights Guidelines and SA8000 (Social Accounting). SA8000 is a certification system based on ISO guidelines to improve working conditions around the world and has the potential to develop into an international standard for socioeconomic issues.
18 See GRI's website, www.globalreporting.org.
19 See Novo Nordisk's report, 'Putting Values Into Action' (www.novo.dk/esr99, October 2000) or Procter & Gamble 'Embracing the Future' (www.pg.com/sr, October 2000) for instance.
20 Foreign direct investment is net inflows of investment to acquire a lasting management interest (10 percent or more of voting stock) in an enterprise operating in an economy other than that of the investor. It is the sum of equity capital flows, reinvestment of earnings, other long-term capital flows and short-term capital flows as shown in the balance of payments (World Bank, 2000, pp315, 330).
21 Another possibility is that MNEs lower the standards of their activities in developing countries by recapitalizing old equipment or extending product lifecycles when this equipment and these products are no longer permitted in more regulated countries.

CHAPTER 4

1 Retail banking involves banks' business with individuals and SMEs; corporate banking involves their business with large companies. The earlier delineation by Saunders was institutional. The delineation used in this paragraph is more operational in nature.
2 See Allen and Santomero, 1998.
3 The European Union (EU) consists of 15 countries: Austria, Belgium, Denmark, Finland, France, Germany, Greece, Ireland, Italy, Luxembourg, Portugal, Spain, Sweden, The Netherlands and the United Kingdom. The Economic and Monetary Union (EMU) consists of 12 countries: this includes the above mentioned countries minus Denmark, Sweden and the UK. The difference in the degree of equity capitalization does not by definition say anything about the power of a country's public capital markets. The fact that the banking market in certain developed countries is dominant in financing the corporate world, may be a reflection of the relatively efficient banking market and tight markets caused by stiff competition for customers between the banks.

4 On top of free trade in goods, services, labour and capital (the basis of the European Union), EMU countries have fixed exchange rates through their common currency, the euro, a common monetary policy and a harmonized budgetary policy.
5 Other reasons behind consolidations are image, investment capacity (particularly with respect to information and communication technology, ICT) and pooling scarce knowledge. Consolidation concerns the institutional level and is the combination of institutions through mergers, acquisitions or strategic alliances. Concentration concerns the amount of market share and shows the cumulative market share of the largest financial institutions in any country. Measured in terms of the five largest banks as a percentage of total bank assets in the country, the UK has a concentration of 28 per cent and Sweden 90 per cent. See Berger, Demsetz and Strahan, 1999.
6 A blurring of the lines between sectors may be at the institutional level, such as the creation of bank insurers, or at the product level, such as banks offering holiday packages.
7 This trend raises the question of whether the term 'banking' is still appropriate. American Wells Fargo no longer refers to itself as a bank in its annual report, but rather an 'integrated financial service provider' (see www.wellsfargo.com/ir, June 2000). In fact, this is just another name for universal banking. The centre of gravity of certain financial conglomerates is in insurance (eg the ratio of banking to insurance at the ING Group is 40 per cent – 60 per cent with respect to its profits). The term 'banking' does little justice to these kinds of situations, which has led a number of institutions to prefer terms like all-finance service provider or integrated financial service provider. This book will not get too involved in all these distinctions. Because the emphasis will be on the depository activities of financial institutions, 'banking' will be the usual term used.
8 Although ICT developments could lead to deconsolidation and specialized financial service providers as well.
9 The table is based on the ranking of the Fortune Global 500 (March 2001); see www.fortune.com/fortune/global500 The table doesn't take into account the merger of Fuji Bank, Dai-Ichi Kangyo Bank and Industrial Bank of Japan (October 2000). The new bank generated by this merger, Mizuho Bank, will be the biggest bank in the world with assets of over US$1 trillion (cumulative US$1490 billion according to the Fortune Global 500).
10 For a good comparison between the major developed countries, see Saunders, 2000, pp437–441, 505–515.
11 An exception is the previously mentioned merger of Fuji Bank, Dai-Ichi Kangyo Bank and Industrial Bank of Japan at the end of 2000.
12 Offering banking services in supermarkets, on the internet and over mobile telephones.
13 The German VfU (1998) and the Schweizerische Bankiervereinigung (1997) refer to 'operating' and 'product' ecology. These terms are also used in the environmental reports published by UBS (2000) and CSG (2000).
14 The so-called 'environmental performance indicators'; see CSG (1999).
15 See UBS, 2000. The methodology finds business travel and paper consumption together as the second most significant environmental issues.
16 See eg NatWest (1998), ING (1999), Bank of America (2000).
17 See www.co-operativebank.co.uk/greenpeace.html (February 1998).
18 Large banks especially will need to pick the best of both sides – for example, support ecological farmers while financing resource- and pollution-intensive tradi-

tional agriculture. This is a problem which the Rabobank Group faces in The Netherlands.

19 See Van Riemsdijk (1994) for extensive research into 'consumer backlashes' against certain businesses.

20 Examples include Amnesty International, the World Nature Fund, International Chamber of Commerce (ICC) and the World Business Council for Sustainable Development (WBCSD). Incidentally, the distinction is not strict, given the fact that interest groups are increasingly and visibly prepared to do business with companies. It is sometimes also not clear where the various UN organizations should be grouped. The UNEP programmes for banks are largely supportive.

21 Source: www.worldbank.org/html/extdr/about (March 2001).

22 See www.exim.gov/menvprog.html (July 2000).

23 This depends on the perspective one has on the role and function of shareholders.

24 'Prices' is a term that is being used very broadly in this case and includes the effects of direct regulation by governments such as command-and-control approaches.

25 *Ceteris paribus* means 'with all other factors remaining equal'.

26 An exception is if the environmental measures company B takes lead to lower costs to company B relative to company A.

27 The Dutch government has also come to realize this. An 'environmental dialogue' between Dutch banks and the Dutch government started in 1999.

28 Premium differentiation is more obvious in insurance than in banking, as there is a direct relationship between the premium and the extent of insurance coverage for risks being offered. It is also more obvious to insurees that they must pay higher premiums if they are more prone to risk. This is less often the case for banks and bank customers.

29 This will be true because stricter requirements from the government and the market will make environmentally friendly companies more attractive and relatively more profitable in the future.

30 See Van der Woerd and Vellinga, 1997. The dilemma of applying premium differentiation and thereby potentially not lending credit to small companies may mean that the environmental return is negative instead, if these companies then use cheaper but older and/or more polluting technology or production methods.

31 Albeit that in a quickly internationalizing banking market environment, foreign banks can finance these healthy customers. Moreover, the initiative encourages the free-rider problem among the Swiss banks.

32 See www.naturalstep.org (March 2001).

33 This also holds true for non-banking companies. The report entitled 'Profits and Principles: Does There Have to Be a Choice?', explains how Shell (1998) is approaching this issue in an open and interesting way.

34 Once again it is true that the defensive, preventive and offensive stances are from the viewpoint of environment and sustainability.

35 Obviously, banks in the defensive stage can communicate their so-called progressive environmental stance through green marketing, for instance. The risk of consumer backlash is excessive since we live in a world in which people base their judgements on actions and not words (a 'show me' instead of a 'tell me' world).

36 See Triodos Bank (2000) and Coperative Bank (2000).

37 See Khandler et al, 1995. Moreover, see Siddiqui and Newman (2001) for an analysis of the Grameen Bank's activities towards renewable energy.

38 The Grameen Bank in Bangladesh receives 97 per cent of its outstanding credits, as a result of the personal guarantees among other things. See www.grameen-info.org (November 1997).

39 Micro-finance institutions include banks, savings banks, credit unions and NGOs. Many NGOs rely on donor funding for the majority of their lending activities (70 per cent of total NGO resources are supplied by donors). Banks and credit unions depend on customer and member deposits and commercial loans. The data is based on a World Bank survey (World Bank, 1996). The survey only looked at institutions which were founded in or before 1992, while the number of institutions has grown rapidly since then. Moreover, the figures correspond to the 206 respondents of the survey only (from the targeted 900 institutions). It's highly conceivable that the data is underestimating the total size of the micro-finance market. More recent or complete data is not available.
40 See World Bank, 2000, p143.
41 See BBA (1997) and www.deutschebank.com and www.rabobankgroup.com/community (October 2000).
42 The total debt of developing countries exceeds US$2.5 trillion (World Bank, 2000, p315).
43 See Burton (1992) for an assessment of involvement by commercial banks in debt-for-developments swaps.

Chapter 5

1 In 1995 research was carried out in England into 'ordinary' investors. Roughly 60 per cent of those questioned said they were acutely aware of a company's results. Of this group, a good 60 per cent said they took ethical behaviour into consideration when deciding whether to invest (20 per cent of all those questioned said that ethics and the environment played no part in their investment decisions) (*Source*, No 3, 1998, p14).
2 In addition to mutual funds, a large amount of money is invested sustainably by individuals; usually these individuals are private banking clients. No record is available for this and the size of such investments by wealthy individuals is therefore difficult to estimate. The same is partly true for institutional investors.
3 This would involve various evaluation methods such as contingent valuation, willingness to pay, full-cost accounting and cost–benefit analyses. For a discussion of such monitarization methods see, for example, Bowers (1993) and Braden (1991).
4 See www.kld.com. The program has data that Kinder, Lydenberg and Domini have gathered from more than 900 companies tested against more than 60 ethical criteria.
5 To evaluate a company's role in the human rights situation of a particular country, a questionnaire compiled with the assistance of Amnesty International is used.
6 Titled 'FTSE4Good indices', see www.ftse4good.com (March 2001) FTSE is the index of the *Financial Times* and the London Stock Exchange.
7 Blumberg, Korsvold and Blum, 1997. See Stigson, 2001, for work of the WBCSD to develop a framework for standardised eco-efficiency metrics which would make cross-sector comparison of eco-efficiency possible. This would enhance the possibilities of linking environmental and social performance of a company.
8 Sources: UK market – respectively Centre for Business Performance (2000, p9), www.eiris.org (June 2001) and www.uksif.org (June 2001); US market – Social Investment Forum (1999); Canada – Environmental Finance (2001); Japan – Nikko Securities/Nikko Asset Management (2000); German speaking countries (Austria, Germany and Switzerland) – www.eiris.org (June 2001); Belgium – Van Braeckel (2001).

9 The figure mentioned includes funds which only exclude tobacco, a very controversial issue in the US. Moreover, it makes plain business sense to exclude tobacco in the US because of the continuing pressure on stock prices of tobacco companies because of law suits. About 96 per cent of professionally screened SRI portfolios in the US avoid tobacco. Note that each portfolio which excludes tobacco is included in the definition of SRI in the US and therefore in the impressive volume. Furthermore, approximately 80 per cent of all assets in such portfolios are screened on environmental criteria. Sustainable shareholder advocacy (by voting on annual meetings on sustainability issues) is responsible for approximately half of the total volume of sustainable investing in the US (Social Investment Forum, 1999).

10 See www.unep.ch/etu/finserv/fin_home.htm for the report of the sixth UNEP FSI conference of November 2000.

11 In addition to the above-mentioned criteria and recognized project categories, foreign projects also have to fulfill certain social criteria which relate to development issues in the selected countries (such as child labor and ILO criteria for the right to trade union freedom). In East European countries only JI projects are eligible for green financing (see page 33). Projects abroad are difficult to get off the ground. Risks (economic and political) are, of course, higher. This is why Dutch banks have adopted the position that the government should add a guarantee mechanism to green regulation for projects abroad.

12 In parentheses is the relative size of the different categories in the accumulative financial volume of the green certificates issued up to the end of 1999 (Ministry of Housing, Spatial Planning and the Environment, 2000, p13). Within the category 'energy' windmills occupy a strong central position, followed at some distance by district heating.

13 Although the 70 per cent requirement is, from a financial viewpoint, a maximum requirement, about 90 per cent of the green invested capital in 1999 was actually invested in green projects (ibid, p11).

14 Since June 2000 ABN AMRO is offering a savings product under the green ruling as well (volume: €17m at the end of 2000). The return on the three share funds (ABN AMRO, ASN and Triodos) are average returns since the introduction of the funds (including reinvestment of dividends and based on intrinsic values) until October 1999. The difference in the rates of return is partly determined by their being introduced at different times, and factors specific to each fund: its composition, risk profile and the extent to which it is tradable. ASN and Triodos use additional, non-project-related criteria for the investments, such as an integrated environmental care system. The 'returns' on the Rabo Green Bonds (the successor of the Green Interest Fund) and the Postbank Green Interest Certificate are guaranteed and depend on the lifetime of the certificate and when it is introduced (as a result of changes in market interest rates). In fact, these are savings constructions and not investment funds.

15 The green regulation does not require funds to mention their non-financial results in their annual reports. Obviously, this information is not available to the investor in sustainable investment funds as well. The certainty that the investment contributes to sustainability is perhaps greater in the case of the fiscal green funds as a consequence of the green certification required for each project and the government's supervision of this. For example, in total under the green regulation, the following have been financed: 20,000 hectares of nature area, 14,000 hectares of biological agricultural land, 6,000 sustainable homes and 690 windmills (year end 1999; Ministry of Housing, Spatial Planning and the Environment, 2000).

16 The products mentioned can all be found on the websites of the named banks or in their environmental report (if available); see the references at the end of the book.
17 'Appealing to' is what offensive banks do, while facilitating definitely plays a role in sustainable banking.
18 The Triodos Bank has formed the 'Triodos North-South Account' in which invested capital is used to finance small-scale sustainable economic activities in developing countries (Triodos Bank, 2000).
19 ING has set up a basic service which allows clients to stipulate their own specific ethical criteria on which a personal portfolio is constructed. See Knörzer, 2001, for the approach of Bank Sarasin, a Swiss private bank.
20 See the UNEP Insurance Industry Position Statement on Climate Change of December 1997, available at www.unep.ch/etu/finserv/insmenu.htm (January 2001).
21 See www.zurich.com (March 2001). The issue of lender liability is discussed on pages 135–9.
22 In contrast to most environmental liability insurances, there is no question of 'loss occurrence' in IEDI ('loss occurrence' means that damage occurring within the term of the policy is covered; this implicitly entails that coverage is automatically extended to damage which only becomes manifest after the term). IEDI uses 'claims made' ie that only damage occurring *and* becoming manifest within the term of the policy is covered. This makes 'long tail' risks – dormant risks – such as asbestos and ground contamination more manageable. Thus, in practice the distinction between gradually and suddenly occurring damage does not exist in IEDI any more.
23 Three types of cover exist for companies with a fixed location. In increasing extent of cover, these are: basic cover (cover for environmental damage resulting from fire, explosion or lightning strike), comprehensive occurrence cover (basic, plus insurance against all environmental damage resulting from an outdoor calamity) and top cover (comprehensive, plus cover for own negligence). For insured parties temporarily working at others' premises, there is a work location cover, which has two forms: the 'by third parties' cover; and the temporary fixed policy. Environmental damage caused by criminal activity is, of course, not covered (Association of Insurers, 1997).
24 See Wohlgemuth (2001) and Hodes (2001) for an assessment of the role of banks regarding respectively energy efficiency and renewable energy.
25 The sustainable energy is generated by covering a desert area of around 800 hectares with glass. A 600-metre-high chimney in the middle of this area converts the stream of warm air created under the glass, and which can only escape via the chimney, into electricity. The covered space works as a storage space, in which energy can also be generated at night. The American energy giant Energen will take care of the building and maintenance of the power house. Investments amount to some €400 million and the production capacity is estimated at 10,000 MW (about half of the current Dutch production capacity); PolyTechnisch tijdschrift (1998) and Collaborating Elektricity Production Companies (SEP, 1998).
26 Plans in Japan are well advanced. In 2008 a project should be started that will result in the building of a floating platform measuring around 35 km^2 in the Asiatic Ocean. Sun, wind and wave power will be used to produce 30–100 million kWh per annum. The investment for one island alone amounts to around €1.8 billion (*Mens en Wetenschap*, 'Energy Park in the Asiatic Ocean', 1996, Vol 23, No 2, pp76–78).

27 Source: *Intermediair*, 16 July 1998, 'Windmills take to the sea'.
28 Sources: www.gefweb.org, www.ifc.org/enviro/epu/biodiversity/terra, www.ifc.org/enviro/epu/renewable/REEF, www.prototypecarbonfund.org and www.ifc.org/enviro/epu/renewable/photovoltaics/SDG. The list is not extensive and excludes products eg from the EBRD (for the EBRD see Pásztor and Bulkai, 2001).
29 See EIF, 1996 and Leistner, 2001. A maximum of €50 million is involved in this European programme.
30 For a complete overview of participating banks, see www.eif.org/sme/g&e/listin-term.htm (October 2000).
31 A distinction can be made between operational and financial leasing. In the former, the economic and legal property rights lie with the leasing company; in the latter, only the legal property rights lie with them. This means that only in the case of operational leasing, the lease company bears the risks of depreciation, wear and tear, loss, and any maintenance costs.
32 The leasing of cars is a good example of this. There is a risk that those interested in such an agreement would be more prone to reckless driving (adverse selection) and that the lessee will not be stimulated to operate the car decently (moral hazard).
33 Three sorts of innovation can be identified. The first type adapts the product and/or the production process with existing management techniques. The second type adapts the product and/or the production process with existing technologies. The third type is a renovation of technologies and management techniques themselves.
34 Ministry of Economic Affairs, 1996.
35 After all, a company or individual averse to risk would rather receive a guaranteed income of €1000 than a 75 per cent guaranteed income of €2000 (with an anticipated value of €1500). As a rule an investor would choose the second option.
36 In fact, wind energy and energy from biomass are also forms of solar energy. In this case study the term solar energy is used to mean picking up and converting sunlight into directly usable energy: electricity (PV energy) and warmth (thermal energy).
37 Given normal consumption, the solar panels installed can meet the total energy requirements of an energy-efficient family. The electrical grid is used as a buffer (batteries is an alternative) for periods of both under- and overproduction, bearing in mind that the quantity of sunlight is not constant. It is assumed that there is on average an underproduction of 500 kWh.
38 At current prices and with an average use (2400 kWh) the individual pays only €50 instead of €200 for electricity costs per annum.
39 A good tool for assessing CO_2 emissions is the UNEP Guidelines for a CO_2 indicator (Thomas, Tennant and Rolls, 2000).
40 The top section of the figure is partly based on Hugenschmidt et al, 2001, p51. MDBs are multilateral development banks, like the World Bank and the EBRD.
41 See www.prototypecarbonfund.org
42 Dexia and the EBRD have each allocated €20 million to the Dexia-FondElec Energy Efficiency and Emission Reduction Fund. See www.dexia.com (press release 20 February 2000).
43 See www.ing.com (January 2001).
44 There are two main methods for issuing emission rights, and each has several variants: free issuing of rights ('grandfathering') and by auction. In the first, current or past emission levels are used to determine emission rights per company.

These are re-assigned gratis (eg annually). In auctions, companies buy emission rights (eg annually). Companies do this for the portion of emissions for which the marginal costs of emission reduction are higher than the auction price. The annual revenue from the auction of CO_2 emission rights in the US is estimated at around €150 billion (at an auction price of around €100 per tonne of carbon; see Cramton and Keri, 1998). The chosen system of issuing emissions rights does, of course, influence banks. After all, in an auction scenario there is a question about whether a (borrowing) company has the financial resources to buy rights or can financially bear buying them over time. Companies with an inelastic supply (ie fixed (in terms of location) capital goods or those with a long economic lifespan) will be especially effected; examples are electricity producers and gas distribution companies. Shareholders profit in a grandfathering system: it offers attractive equity investment opportunities in certain sectors.

45 See www.cantor.com/ebs, www.ecosecurities.com and www.natsource.com/enviro-mental.

46 See www.carbontrading.com.au (March 2000).

47 See www.gerling.com and www.storebrand.com (November 2000).

48 This form is highly dependent on social preferences of economic actors and is currently the preserve of smaller niche banks. It is, however, the ultimate form of sustainable banking.

CHAPTER 6

1 In this chapter, the term 'direct liability' (lender liability) means the possibility that the bank itself will be held liable for the environmental damage that has been caused by its borrower. Indirect liability is the liability of a borrower which can then lead to financial risks for the bank.

2 The fact that little is known about the systematic materialized environmental risks run by banks is closely related to the fact that banks in general are reticent about disclosing their materialized losses. Given this, the specification of materialized environmental risks in insurance activities in Citigroup's financial annual report is interesting; it shows, for example, that up until the end of 1999 it has paid approx-imately US$1.6 billion for environment-related claims in its insurance business (Citigroup, 2000, pp25–27). A study of 32 German banks revealed that around 10 per cent of total credit losses were caused by environmental issues (Scholz et al, 1995).

3 See eg Case, 1999, pp236–249 for samples of such clauses.

4 'Lending operations' also refers to financing transactions and equity investments, throughout. The words credit risk will be used to cover these activities and all sorts of risks from Box 6.1.

5 Naturally, these costs can also result in a positive balance as a result of such factors as energy savings.

6 See Case, 1999, pp171–176 for an extensive list of environmentally sensitive branches and pp177–224 for a discussion of the environmental issues in over 20 sectors.

7 An investigation in to the environmental conduct and awareness of companies in the mid south of The Netherlands showed that 65 per cent of small businesses, 50 per cent of medium-sized businesses and 32 per cent of large businesses were in the defensive phase (Jeucken and Van Tilburg, 1999, p865). The implications of this for a bank is that its financing risks are on the whole clearly related to the size of its client.

8 Based on Tonnaer, 1994. The list is limited to the elements important for companies. Part of the legal basis for environmental policy can be traced to the Dutch Constitution: Article 21 refers to the obligation of the government to provide general environmental management. A general responsibility for environmental management for companies is included in the Environmental Management Act.
9 The first three elements of this category do not have an obligatory character but appeal to the citizen's environmental awareness (the 'internalization' of environmental policy). Pseudo-legislation means pseudo-legal power develops by means of legal precedents based on extra-legal norms such as government guidelines and policy documents.
10 ALARA stands for 'as low as reasonably achievable'. This depends on such factors as the knowledge and technology available to society.
11 However, the EU published a White Paper in 2000 which proposed the inclusion of ecological damage in an environmental liability regime (EU Commission, 2000).
12 There is no legal definition of strict liability. The judge weighs the interests based on legal precedents and requirements related to reasonableness and fairness. Strict liability can be encountered in the Convention of Lugano that was signed by many developed countries. In The Netherlands, only one party can be held liable for a certain case of environmental damage, while in the US more than one party can be held liable (joint and several liability). Another distinction is that the law relating to strict liability has no retroactive force in The Netherlands, as it does in the US.
13 Large specialized companies are considered to have more than average knowledge.
14 Both at the international and national levels, discussions are being held in regard to having the private instrument of environmental liability play more of a prominent role in environmental policy.
15 In the US, there are even tendencies to apply environmental liability in very broad terms: a subsidiary of a multinational could be held liable in the US for causing environmental damage in another country. See Millman, 1996.
16 For an overview of environmental laws in the EU, see www.eel.nl; for an overview of environmental laws in the US, see www.cec.org/pubs_info_resources/law_treat_agree (February 2001).
17 See www.ec.gc.ca/pdb/npri/npri_home_e.cfm (February 2001).
18 Sources: *Financial Times* (28 September 1994), *Nieuwsbrief Milieu en Economie* (April 1997, p5) and *Financieel Economisch Magazine* (18 April 1998, p73). The costs for soil remediation can be considerably reduced by the use of electrotechnical treatments or remediation by use of bacteria, and by not aiming to decontaminate the soil for a variety of purposes (multifunctional remediation) but to make it suitable only for its intended use (function-oriented remediation). Until 1997 in The Netherlands, the soil had to be restored to suit a multifunctional level. Actual situations, however, involve many sites that will be used only for industrial sites or parking facilities for many years into the future while the costs for multifunctional remediation (ie clean enough to grow vegetables as well) would be very high. In the case of function-oriented remediation, however, the point of departure is that neither people nor the environment should run any adverse risks on the partially decontaminated soil and is thus related to its intended use.
19 BSB stands for *Bodem-Sanering van in gebruik zijnde Bedrijfsterreinen* (soil remediation for industrial sites currently in use). These BSB organizations have been established in every province by industry and operate independently from the government. The BSB activities do not apply to agricultural operations and essentially apply only to cases of pollution that developed prior to 1987.

20 Administrative compensation will be awarded by the government to a businessperson if he or she is legally obliged to take certain measures in the interests of the environment but due to reasonableness and fairness cannot (or cannot fully) be expected to foot the bill.
21 Physical regulation (an absolute prohibition on a certain emission level) will be most commonly used (and will be best in terms of prosperity) in extremely hazardous and/or life-threatening situations. See Baumol and Oates (1988) for a discussion about the effects of various environmental policy instruments on a country's level of welfare. For many environmental problems, regulatory levies are more efficient (ie welfare benefits will be the same or higher than the costs of a certain environmental measure) than physical regulation. Although efficiency implies cost-effectiveness, cost-effective measures are not necessarily efficient (eg Tietenberg, 1992).
22 In other words, the businessperson is stimulated to reduce emissions as long as the marginal costs of emission reduction are lower than buying the emission rights.
23 The costs of soil remediation for Occidental Petroleum amounted in 1994, for example, to 107 per cent of the company's net profit before tax (and 53 per cent over the period 1992 up to and including 1994; Müller et al, 1996, p27).
24 For example, the disaster involving the Exxon Valdez in Alaska in 1989 cost Exxon at least €2 billion in compensation to third parties for environmental damage (Molenkamp, 1995).
25 In regard to risks in general, the bank was only running the risk that items in stock would gradually drop in value (unmarketability of securities). But due to environmental risks, items in stock can become worthless within a very short period of time.
26 When a company files for bankruptcy in the US, the costs that the government incurs for soil remediation occupy a position of higher priority in relationship to the collateral rights of, say, the bank. See Schmidheiny and Zorraquín, 1996.
27 See Boyer and Laffont (1997) for a study into the possibilities. Naturally, whether matters are legally possible and/or desirable should be considered (eg in conjunction with the confidentiality of personal information).
28 A separate category is the liability of banks in some countries for non-disclosure. Where, for example, a client has bought a house in an area that is frequently affected by flooding, the bank should advise the client to take out insurance against flooding. Where the bank fails to do this, and the house is affected by flooding, the bank can be held liable for failing to disclose and enabling the client to take appropriate measures. See Barannik, 2001, p253.
29 The 'security creditor's exemption' rule in CERCLA was supposed to exclude from liability those third parties who had an interest in a company for the reason of security. CERCLA resulted in a wave of legal proceedings. In some of them, the legal costs were higher than the costs for the soil remediation! In 20 years around 750 sites have been remediated under private arrangements costing around US$18 billion (USA-EPA, 2000b, p35). The average costs are estimated at US$25–40 million per site (Case, 1999, p41).
30 See Davis, 1994, p3. This fear, however, can also provide an incentive for securitizing in which the credits of such sectors are accommodated in a pool. The pool is then offered on the capital market as an investment vehicle.
31 Other laws in the US that can lead to direct liability for banks are the Toxic Substances Control Act and the Resource Conservation and Recovery Act. The above-mentioned provisions for the protection of banks are relevant to these laws too.

32 Of course, this liability applies not only to the extension of company credit, but also to equity investments and operational leases by banks and to, for example, housing finance for individuals.
33 Where the risk of bankers' liability does not exist while a liability regime for environmental pollution by companies does, the extension of credit to companies will see strong growth, principally to protect shareholders. This has been shown to be the case in empirical analyses of loans made by banks in pre-CERCLA, post-CERCLA/pre-Fleet Factors and post-CERCLA/post-Fleet Factors periods respectively, in the US. Loans to the chemical industry should, *ceteris paribus*, for example, rise by 15–20 per cent (Ulph and Valantini, 2000).
34 Examples are International Rivers Network (www.irn.org), CorporateWatch (www.corpwatch.org) and Ethical Consumer (www.ethicalconsumer.org).
35 See *het Financieele Dagblad* (3 February 1999). For the text of the adapted OECD Guidelines for Multinational Enterprises, see www.oecd.org/daf/investment/guidelines/mnetext.htm (March 2001). The OECD guidelines are, in fact, a recommendation to companies: adherence cannot be legally enforced.
36 See http://irn.org/programs/threeg (December 2000) and *Multinational Monitor* (1 March 2000)
37 For a historic analysis of the institutional development of World Bank's environmental policies, see Barannik and Goodland, 2001.
38 An EA is an Environmental Impact Assessment (EIA), Environmental Audit, Risk Assessment or Environmental Action Plan (EAP), or combination thereof.
39 An example is the IFC's *Good Practice Manual* (1998).
40 Source: www.ifc.org/enviro/enviro/Review_Procedure_Main/Review_Procedure/Annex_A (March 2001).
41 IFC defines forced labour as all work or service not voluntarily performed, that is extracted from an individual under threat of force or penalty. For IFC, harmful child labour means the employment of children which is economically exploitive, or is likely to be hazardous to, or to interfere with, the child's education, or to be harmful to the child's health, or physical, mental, spiritual, moral, or social development.
42 This does not apply to the purchase of medical equipment, quality control (measurement) equipment and any equipment where IFC considers the radioactive source to be trivial and/or adequately shielded.
43 Source: Lascelles, 1994, p32. Another example is the SERM (Safety and Environmental Risk Management) rating, which is a rating methodology similar to that of Moody and Standard&Poor's. SERM has rated 130 UK largecaps and will broaden its rating to other countries as well. See *Environmental Finance* (1999/2000, pp18–20).
44 In Deloitte Touche, IISD and SustainAbility (1993), the costs for an environmental report are estimated at between €45,000 and €180,000. Companies themselves estimate the costs of a public report at €9000 to €45,000 (see Van Dalen, 1997).
45 See Smith (1995) for sample questionnaires from the American Bankers Association (ABA) in the US. The Environmental Risk Program of the ABA was based on the guidelines of the US Federal Deposit Insurance Corporation (FDIC) from 1993 (Chapman and Cutler, 1993). Moreover, see Case (1999, pp225–235) for a UK checklist, Atkins and Pedersen (2001) for a Danish example and Barannik (2001) for a World Bank checklist.
46 See eg Case, 1999, pp236–249. Obviously, an indemnification is only as good as the financial strength of the borrower.

47 See BBA (1997) and CBA (2000).

Chapter 7

1 Research in The Netherlands has revealed that, contrary to popular assumption, business-to-business relations are a more important market force for environmental care than consumer demand (see Jeucken and Van Tilburg, 1999).
2 In The Netherlands, ABN AMRO, ING Group and Rabobank Group are together responsible for more than 80 per cent of total energy consumption within the banking sector; all three are co-signatories to the MJA. Reaching the target is not compulsory for individual banks.
3 NatWest Group, 1998, p5. The various sources of energy consumed can be converted into an integral indicator in gWh. Conversion factors such as these are, for instance, stated in VfU, 1998, p45.
4 Air traffic constitutes a huge environmental impact. ICT developments, like email and video and telephone conferencing, could reduce the number of kilometres flown (and total mobility). Another option is the planting (independently or through intermediaries) of trees and woods to compensate for CO_2 emissions by air traffic. Whenever Triodos Bank, for example, books a flight, it invests for the amount of the cost of the ticket in planting trees (Triodos Bank, 1999, p9). In 2000 Triodos opened this fund to private individuals who, along with the management and employees of the bank itself, can make deposits into this fund to help to make flying 'climate-neutral'.
5 MeritaNordbanken is another exception. In 2000 the bank started an investigation into the environmental impact of its distribution policy (press release, 30 March 2000, available at www.merita.fi/E/Merita/sijoita/uutta/20000330.STM). MeritaNordbanken is a world leader in internet banking and states to be particularly interested in the contribution electronic distribution forms can make to reducing environmental impact.
6 CSG has followed the lines of VfU in its environmental report. The Union Bank of Switzerland (UBS) also published aggregate figures according to the VfU guidelines. They do not follow the guidelines, but publish the figures to make comparison between banks possible. For most banks in other countries it is not possible to abstract the figures to standard VfU.
7 The VfU Environmental Performance Indicators are based on conventional accounting principles: completeness, authenticity, clarity, continuity, identity and prudence (VfU, 1998, p17).

Chapter 8

1 In terms of strategy this conforms to Mintzberg's contingency theory (1983), in which organization form and situation are linked together.
2 The figure is based on the wheel of sustainability by Wempe and Kaptein (2000, p47) and shows similarities with the Deming model (Walton and Deming, 1986).
3 See FORGE, 2000. As sector-specific factors it names (p7): organization structures and cultures, the importance of indirect effects (external aspects of environmental care) which are difficult for the bank itself to control, the high tempo of generic change, its worldwide reach and impact, the difficulty of balanc-

ing the long-term nature of environmental risks with the short-termism of the sector and its present (relatively low) level of awareness of environmental issues.

4 A third standard is the British BS 7750, used too in a number of other countries. International standards of certification also exist in the field of social reporting, such as SA8000 and AA1000; these relate in particular to human rights, child labour and conditions of work.

5 The following banks participated in achieving an arrangement specific to financial institutions (FEMAS): NatWest, Deutsche Bank, BBVA, Allied Irish Bank, Skandinaviska Enskilde Banken and ING. Source: www.unep.ch/etu/finserv/finserv/newslet1.htm

6 Certification cannot be regarded as a complete environmental stamp of approval because certification involves too great a degree of freedom for the certifying bodies. Bank A may have relatively low aspirations and a relatively poor environmental track record, but it can still gain a certificate by systematically focusing on the continuous improvement of its environmental performance. Bank B with a better environmental performance and greater ambitions, but no systematic EMS is, however, not eligible for certification.

7 Source: *ISO 9000 + ISO 14000 News*, April 2000, p10.

8 See ING Group (1998), Föreningssparbanken (2000) and Nikko Securities/Nikko Asset Management (2000).

9 From a sample of 140 general EMAS sites it appeared that 47 per cent also had an ISO certificate; 62 per cent of this group had first obtained ISO certification. Source: http://europa.eu.int/comm/environment/emas/faqs_en (December 2000).

10 See UNEP/ICC/FIDIC (1997) for a practical training kit.

11 Source: ISO Bulletin (January 2000), available at: www.iso.ch/index.html.

12 These institutions are: Bank Sarasin, Credit Suisse, Deutsche Bank, Gerling Konzern, HypoVereinsbank, RheinLand Versicherungen, SAM Sustainability Group, Swiss Re, UBS, Victoria Versicherungen and Züricher Kantonalbank. See EPI-Finance (2000).

13 Since consultation is very important for a fledgling environmental consciousness (and the implementation of environmental care), several committees can be set up at various levels (operation, tactical, strategic). This is, of course, partly dependent on the size of the bank.

14 Although care was taken in its construction, the table does not claim to be complete.

15 Management skills can be more important than environmental knowledge. The knowledge can always be brought in from outside.

16 The development of an EMS is usually a long and iterative process. No clear distinction can be made between a system in development and a functioning system.

17 Although in a defensive stance, little or no attention will be paid to it.

18 In the environmental annual report for 1994–1997 (ABN AMRO, 1998) are various interviews (from a member of the Board of Directors to, for example, someone from purchasing) with the interviewee's personal environmental score (for internal environmental care only).

19 Implicit in the argument of this section is that a lot of time and energy may be spent on training and education but as long as the bonus structure remains unchanged, little behavioural change will be achieved in the normal financing circuit.

20 Precisely because of this last element, this subject is dealt with in this section on external communication. Of course, a statement of intent has a definite internal effect and purpose.
21 The list of participating companies sent by the ICC in January 2001 in fact only went up to October 1997. This calls into question the true effectiviness of the declaration.
22 See Hill, Fedrigo and Marshall (1997) for a discussion on the implementation of the principles of the declaration among the signatories. In this survey by The Green Alliance it appeared that one (ethical) bank refused credit if, while government requirements were met, the activities nevertheless involved 'excessive' environmental damage. See Kelly and Huhtala (2001) for a description of the objectives and projects of UNEP's Financial Services Initiative (UNEP FSI). See UNEP (1998) for a further survey on the activities of the signatories.
23 Cowton and Thompson, 2000. The authors suggest that some banks perhaps signed the declaration purely because of political pressure and do not support or wish to/can embody the spirit of the declaration (p170).
24 The CERES principles are the follow-up to the so-called Valdez principles formulated in the US by a group of important investors after the Exxon Valdez disaster.
25 Where stakeholders are referred to, this means only those who have been identified as such in step 1 (and are involved in the EtAc process).
26 See www.unep.ch/etu/finserv/fin_home.htm for the various reports and objectives of the initiative.
27 See BBA (1997) and CBA (2000).
28 See for example Ganzi and DeVries (1998) and www.bmu.de/sachthemen/oeko/position.htm (December 2000).
29 This describes the creation of an impression that the organization is totally environmentally friendly, while this is in fact only true of a few (or no) company activities. See, for example, Ottman, 1993. In 1994 a well-known American ethical investment fund advised its clients to withdraw from The Body Shop, which presented itself as being green, because some divisions of the company were not environmentally friendly. The share price dropped by around 20 per cent. See Ryall et al (1996) and UBS International (February 1995).
30 Internationally around 40 financial institutions were challenged by the international sister NGO (Friends of the Earth) over their involvement in financing this mine.
31 See AIDEnvironment (2000). The report gives detailed lists of all projects and the amount of financing from individual banks. Banks from other countries are mentioned as well.
32 Research has shown that Brent Spar was not in fact heavily polluted, and that it would have sunk into a deep ocean rift in the Atlantic rather than being *dumped* in the North Sea, a place dear to the public's hearts. Greenpeace also acknowledged this later.
33 See Tietenberg (1997) for a broad inventory and analysis.
34 AirMiles are savings points that can be used for, among other things, flights. Shell was for many drivers an important supplier of AirMiles. The Dutch consumer is apparently happy to take action as long as it does not cost too much money.
35 Sources: *Village Voice* (29 March 2000) and *Multinational Monitor* (1 March 2000). See also www.ethicalconsumer.org (January 2001).
36 See www.nwf.org/nwf/finance/bank.html (January 2001).

CHAPTER 9

1 An exception to this rule is made for an analysis of the banks' affiliation to the activities of the UNEP, the ICC and the WBCSD, for which the following sources have been used: www.unep.ch/etu/finserv/fin_home.htm, www.iccwbo.org/index_sdcharter.asp and www.wbcsd.ch/memlist.htm respectively (February 2001). These sources are deemed to be reliable.

2 Year-end 1999 has been chosen because all banking, financial, environmental, and social annual reports for at least that reporting year were available at time of data gathering (December 2000/January 2001).

3 Note that a true comparison between a stock (assets) and a flow figure (GDP) is not possible. Taking a flow figure for the financial markets would show that a multitude of the World's annual GDP is traded every day. In short, financial flows and banks have a considerable stake in the world development (see Table 4.1 as well). The figures above shows this for the selected banks.

4 Banking is a dynamic sector. Over the period examined, many banks were engaged in merger processes. NatWest, for example, was still an independent bank but has since become an autonomous daughter of the Royal Bank of Scotland. Since 1990 NatWest has been one of the most important mainstream banks in terms of proactive sustainability. Since RBS itself has a formal environmental policy, continuity of environmental care may be guaranteed. UBS came into being in 1998 as a result of the merger of Union Bank of Switzerland and Swiss Bank Corporation; those responsible for the environment within UBS agree that the merger has helped the further professionalism towards the environment (Furrer and Hugenschmidt, 1999, p40). Bank of America arose out of a merger between BankAmerica and Nationsbank in 1998 and wants to further develop BankAmerica's sustainable approach. The consequences of such mergers for environmental policy within the new or still existing entities are not dealt with in this chapter.

5 This does not mean, however, that there are no signatories to the agreement in this region. Australia has one signatory (Westpac Banking Corporation) and Japan has two (Good Bankers and Nikko Assets & Nikko Securities). Also some other Asian countries, such as Thailand and the Philippines, have signatories (see Appendix VII).

6 Membership by ING is not shown in this figure because neither the official publications of neither WBCSD nor ING mention it. Well-informed sources have indicated ING's consideration of joining to the author.

7 This is changing slowly. A number of leading Japanese insurers now publish environmental reports and the major asset manager Nikko Securities/Nikko Assets reports on its environmental policy and activities. See Tokyo Marine (2000), Yasuda Fire and Marine Insurance (2000), Nikko Securities/Nikko Asset Management (2000). None of the three other major banks in Australia – not included in this sample – publish environmental reports as well.

8 No bank in this group devotes more than two pages of its annual financial report to the environment. The information is therefore almost entirely qualitative.

9 Of the remaining 29 per cent of the banks, most spend a few words (ie a couple of sentences) on environmental issues or (ie a couple of columns) on social issues. That is, in the following sections these banks can still score positively on certain aspects.

10 See http://about.ing.com/identity and HSBC (1999).

11 For the analysis in this chapter Fortis has been categorized as a Belgian bank, although until 2001 it had headquarters in both Belgium and The Netherlands and issues separate shares for each branch. Fortis is the fourth bank (in terms of assets) in the Dutch market.

12 By focusing on point scoring, hard and fast statements about the quality of activities is not possible. However, sustainable investment funds also investigate banks' performances and their selections of stocks concur with the conclusions in this chapter. See, for example, www.sam-group.com (December 2000). The Rabobank scores very highly in this chapter. It is not, however, a public company (it is a cooperative organization) and therefore does not appear in similar portfolios or in popular publications about investing.

13 Although other publications attempt to map bank's environmental activities, they use either interviews or banks' published environmental annual reports to gather data. The former primarily focuses on a limited group of banks within a country (eg Jasch, 2001; Barta and Éri, 2001; Schrama, 1999), while both look exclusively at banks which have long been active in the environmental arena (eg Giuseppi, 2001; Tarna, 2001). The study in this chapter deviates from this (as do its conclusions) because it focuses on the major banks in all developed countries (excluding eight smaller countries with relatively small banks, like Greece and Iceland), irrespective of the phase in which the bank operates. Moreover, the sample concerns a group in which comparisons are possible given that only big banks are involved (ie they belong to the global top 100 in terms of assets). The author is not aware of any comparable studies.

14 In 1998, American banks prioritized environmental and social issues in the following order: risk management, internal greening, community responsibility, green marketing and green/sustainable products. Source: 1998 UNEP banking roundtable (www.unep.ch/etu/finserv/fin_home.htm).

15 The Rabobank's high score was unexpected, but can be explained. Firstly, the points per element and for the keys have been examined to find out whether the high score was a fluke. This appeared not to be the case. With a very strong emphasis on internal environmental care and ISO certification, the Deutsche Bank assumed the lead from the Rabobank. (This is apparently something that the Rabobank could pay greater attention to.) A similar key was, however, not considered genuine. The Rabobank's high score is partly linked to the fact that it offers such a broad range of environmental products. Once more it is emphasized that the total score is only indicative and is simply a method of keeping a tally of what banks do. As such, the table offers no hard and fast data about the quality of any bank's activities with regard to sustainability or their relative importance in all activities of a bank. That these two large banks score sustainably is an overestimation. This is because in the methodology no elements are taken into account which explicitly adhere to the idea of sustainable banking and so are weighed zero. As such it is possible to score many points with activities which adhere to offensive banking and end up with a sustainable score. If one takes this into account and divides the integral score over three main classes (ie one abstracts from the sustainable phase), the results change to: 62 per cent defensive banking, 24 per cent preventive banking and 15 per cent offensive banking. Even so, banks which score high in the classification used in the table, will most likely have a mission or vision that sustainable banking is inevitable. A cluster analysis reveals the same five groups as presented in the table.

16 It has been investigated whether correcting for those at the extreme edge (upper or lower) of a group by omitting them would create a different picture. This was

not the case. For the degree of efficiency the cost–benefit ratio has been used: the higher the ratio, the less efficient the bank. The degree of international focus is taken by considering the share of foreign assets in the total assets of the bank. Where this information was not available, proxies have been used such as the share of profits from abroad in the total profits. Since only two banks occupy the sustainable phase, the offensive and sustainable banks have been merged (though their separate scores are also mentioned). No correlation (r) was found for profitability and the sustainability phase a bank is in ($r = 0.08$). Also, no significant correlation (ie at the 5 per cent level) was found for the sustainability phase and number of employees ($r = 0.3$) and number of countries a bank is active in ($r = 0.25$).

CHAPTER 10

1 A sociologically functional meaning of 'paradigm' is that it concerns a commonly shared standpoint. Those who do not share this standpoint (the paradigm) do not belong to the recognized experts or scientists. This implies that decision makers usually ignore what is said by those who do not accept such a paradigm. See, for example, Kastelein, 1987.
2 The dominant coalition usually 'falls' once the change has become irrevocable.
3 As an example, the recognition at the micro level of a few cases of BSE in Europe has led, by means of activism and a renewed awareness of quality, to a great decline in the beef-producing industry at the macro level. Another example is deforestation. In developing countries, at the micro level trees are chopped down primarily for small-scale agricultural purposes. At the macro level, however, such deforestation activities lead to imbalances in the ecosystem which then have serious adverse effects for people (eg loss of fertile land, a smaller buffer for the increased greenhouse effect, and air pollution). Another example is the millennium problem: at the micro level the omission, back in the 1960s, of the first two numbers of the year (19) to save on memory space in computer programs, the continued copying of parts of these programs and the following of this established standard made it necessary to invest enormous amounts to prevent all the two-digit automation systems and computers from breaking down at the turn of the millennium.
4 An economic system is an open system (although some doubt this) as is human intelligence.
5 Ferguson (1981) provides a good illustration of a paradigm shift (as experienced by an individual) and how it happens not in steps but all at once. She makes a comparison with finding camouflaged elements in a drawing. Suddenly, you see that the branches look like a pitchfork, etc. (What we had seen before was correct but incomplete.) Explaining this change in perspective to someone who has not seen it is impossible: you either see it or you don't. Seeing entirely different concrete figures within an abstract 3D drawing is another example.
6 The vertical axis (Y) has an arbitrary value. If Y = population size, then there are around 100,000 people in the year 7000BC and around 6,000,000,000 people in the year 2000 (with a forecast of 10 billion people before the year 2050; United Nations, 1996). The Y-axis can also been seen as the quantity of natural resources being used by people for production and consumption purposes (see, for example, Brown, Flavin and French, 1999; Stikker, 1992; and Harman, 1979). The Y variable can also be considered to be a measure of affluence or economic growth

(see, for example, Tietenberg, 1992). It will then be possible to distinguish certain scenarios (possible futures). Scenario A1 assumes a continued exponential economic growth (in Scenario A2, the rate of further growth is slower). Some assume that this is possible at sustainable levels. However, this scenario seems improbable (after all, there are physical – ecological – limitations). Scenario B assumes a gradually decreasing economic growth. Ultimately, a 'steady state' (see Appendix XI) will be reached in which economic growth drops to zero. Scenario E1 shows the scenario outlined by the Club of Rome. Scenario D is a variant of this one in which after a serious crisis, the level of affluence again rises to its previous level, C, that becomes this steady state. Scenarios C and E2 are scenarios that are seen as solutions for environmental problems at various places in the environmental movement; in these scenarios a steady state economy is the common element. In the case of Scenario C, this is at the current level of affluence while in the case of Scenario E2, this is set at a lower level of affluence. Scenario F assumes that no sustainability is possible as long as there is affluence. Each consumption only leads more rapidly to the destruction of the human race. The existence of renewable natural resources and a capacity of the ecological system to absorb environmental impacts, however, rules out this scenario.

7 The first part of this section provides a very brief description of the development of human philosophical thought. Only a few points are put forward and these are presented as if there had been no discussion about the conclusions drawn. Although this is far from being the case, it is the author's belief that this brief description gives a good general view of what has happened while the conclusions generally agree with current thought. For details as well as the most important sources for this description, see Bor and Petersma (1996), Stikker (1992) and Russell (1961). Furthermore, this section is limited only to *Western* thought since this is the source of current economic thought as applied, either implicitly or not, in the vast majority of the world's countries.

8 Carl Jung (1875–1961) would later (1955) speak of 'synchronicity': the idea that events occur simultaneously but have no causal connection (this idea also occurs frequently in Chinese thought).

9 The Industrial Revolution began in England as early as the 18th century (1788) and in continental Europe some years later. A second Industrial Revolution followed at around 1850 with the emergence of electricity as a source of power and chemistry as a way to create items such as plastics.

10 A description of economic thought starts here at Smith, the 'father' of current economic thought (see Smith, 1776) by means of abstracting from Greek (see Ekelund, 1990) and medieval economic thought (see Cipolla, 1981). The basis for capitalism, though, had actually been laid by the mercantilists. The predecessors of liberalism, and thinking in terms of markets, were the 'physiocrats' (17th and 18th centuries). Also see Ekelund for a description of these two streams of thought. Like the description of the development of philosophical thought, the description of economic thought is also limited to what, in this book, is taken to be the mainstream. Naturally, this does not mean that there has never been any discussion about movements or positions within economic thought (see eg Kastelein, 1987).

11 The fact that Smith's theory had an historically reactive character is important since this casts doubt on whether his ideas can be integrally applied to our own time and place.

12 Deism is the idea of God as the 'reasonable' principle, which accords with our intellectual powers (in contrast to the God of revelations).

13 This is the idea that a system will automatically go in a certain desired direction as long as all of its individual parts are given the possibility of functioning in the best possible way (eg the 'invisible hand').
14 External effects are events that provide a certain person or group of persons with a considerable advantage or disadvantage, this/these person(s) not having been sufficiently involved in the decision that had led either directly or indirectly to these events (see Meade, 1973, p15).
15 Obviously, there are differences among the Western countries in regard to their economic orders. The Anglo-Saxon model, for example, differs from the Rhineland model (in which the Dutch polder model again occupies a distinguishing role in contrast to the French regulated market economy). At a higher level of abstraction, however, the Western economies (now including a number of Asian economies) can be classed as a mixed market economy. See Albert (1992) for a discussion about the Anglo-Saxon versus the Rhineland models.
16 Even though such 'goods' (eg infrastructure) are shifting more and more towards the private domain.
17 A recurring element in ecological economics discussed in Appendix X is that placing physical limitations on the economic system is necessary to achieve sustainable development.
18 Obviously, this simplification falls far short of describing the complex development a person actually experiences.
19 'Disciples' because economic thought and the system are for some turning into a religion. Attacks on the implicit unsustainability of the system are being utterly and fundamentally disposed of.
20 If natural resources, for example, become scarcer, the price for them will rise. This will be followed by substitution (perhaps accompanied by technological innovation) or technological inventions will tap new supplies of natural resources or make them economically recoverable. Substitution and technological innovations can also be induced by the economy's demand side (eg consumers who only want to buy paper that has been recycled).
21 The 'prisoner's dilemma'. Table 10.1 lists A and B, the outcomes for both parties. See Watzlawick et al (1967) for a psychological explanation and Varian (1987) for an economic explanation.
22 This example applies to cross-country environmental pollution. If country A implements a stringent environmental policy, and country B does not do so, the environmental quality in country A can still not be improved (or may even decline) through the import of pollution from country B. This makes cooperation a necessity. See Jeucken (1998a) for a clarification. Obviously, the same goes for individual companies or for consumers. The line of thought in regard to environmentally responsible consumer behaviour, for example, is that this is useful only when others also display a similar behaviour.
23 By means of scenarios, the Club of Rome showed that, for example, if 'business as usual' continues that somewhere in the year 2020, an 'overshoot collapse' will occur. This is a situation in which the world population will have reached such a size that a large gap will exist between the availability of food and the population. The most important lesson to be learned from these scenarios (and the criticisms of them) is that although people are able to make adjustments and will do everything possible to turn the tide before disaster strikes, the time between noting the problem and the results to be gained from a reaction is too great, and that this will inevitably lead to overshooting and all of its associated problems.

24 Habermas (1981) speaks of 'the colonising of the world of the living by the world of the system'. Characteristic of many systems is that they require discipline (people must adjust themselves to the system so that they are essentially deprived of the freedom to act in their own way) and function independently from time and place. But is the first of these always desirable and/or is the second feasible considering a dynamic reality? See Van Dinten, 1996.

25 In developing countries, of course, this is a necessity and part of everyday life.

26 See Watzlawick et al (1967) for a clear, detailed discussion of both of the above-mentioned misconceptions (also see Peters and Wetzels, 1997) and the illogical character of second-order processes of change (also see the example of the arms race). An illogical solution can also be found in the adventures of Winnie the Pooh as recounted in the *Tao of Pooh* (Hoff, 1982). Pooh and his friends have become lost in a fog. They keep returning to the same hole (more of the same) until Pooh suggests that they give up trying to find their way home and try to find their way back to the hole. True to the fundamentals of Taoism, (acceptance, contemplation and harmony), they then really do find their way back home again.

27 Van Spengler (1994) gives a number of ingredients for second-order processes of change. These are: non-linearity, plurality, selection, reflection, dialogue (thinking together), creativity, imagination, humour, example figures, learning process (not just thinking but especially doing), the ability to put things into perspective, the willingness to doubt one's own values or one's own paradigm (this usually requires self-confidence), and being confronted with something that is threatening or undesirable.

28 Mechanistic involves the idea that a system automatically goes in a certain direction as long as its separate parts are given plenty of opportunity to function to the best of their ability. An organic view involves the coherence of the parts of a functional whole. Holism is the belief that a oneness exists in reality and that this is seen only when looking at the whole but cannot be found in the parts.

29 See, for example, Ballard (2000, pp53–59) for a method for 'second-order learning'.

30 According to Fukuyama (1995), trust in a society is the most important ingredient for stability and continuity. He sees society as a civil society of citizens who trust one another. This trust is based not on the reasoning of self-interest (the 'invisible hand') but on factors that are not rational such as religion, traditions, culture and public spirit.

31 And with this, a solution can also be found for the tension between economic short-term objectives and the long-term necessity to survive.

32 For more information, the reader is referred to such authors as Bor and Petersma (1996) and Stikker (1992). Interfaces between Buddhism and sustainable development are captivatingly elucidated in Tucker and Williams (1997).

33 This is expressed most strongly by Hegel (1770–1831): thesis and antithesis lead to synthesis.

34 This is the basis for Chaos Theory. This offers yet another source of learning. See, for example, Peters and Wetzels, 1997.

Chapter 11

1 Perhaps it would be more accurate to argue that sustainability is not an integral but a plural concept.

2 Regulatory institutions also encourage internalization in some situations. To be listed on the New York Stock Exchange, for example, companies need to disclose particular environmental issues (mainly risks and liabilities; past, present and future). Interestingly, other countries are picking this up as well, eg the stock exchange of Thailand.
3 An example of this is Föreningsbanken. They first give advice or information concerning how environmentally polluting companies, especially those in the agricultural sector, can make their production process less polluting. If this produces no results, they will then apply premium differentiation. They state this is separate from financial risks. Source: *Gotenborgs Posten*, 19 July 1995.
4 Obviously, however, having quantitative economic growth without taking corrective measures will lead to the further damage of the environment.
5 An example of the possible success of a change in behaviour is aviation. By using another way of ascending and descending, aeroplanes could realize a reduction in their total environmental impact of approximately 50 per cent.
6 Quieter, cleaner aeroplanes can considerably reduce the environmental impact of the aviation industry.
7 Unlike financial leasing, the lessor remains the legal and economic owner and therefore responsible for the goods with operational leasing.
8 Certainly if such developments result in a reduction in the price of the product in comparison to the traditional, relatively environmentally unfriendly products.
9 Obviously, sustainable banking does not require banks to play that much of an active role. That is, if sustainable entrepreneurship is autonomous or stimulated by consumers and/or is launched by the government, it is possible to have banks with a reactive and sustainable position at the same time.
10 This book has looked into the activities and issues of banks from developed countries. Although these banks are clearly ahead of their competitors from developing countries in general, some developing countries' banks are ahead of some banks in developed countries. An indication of this is the impressive amount of bank signatories of the UNEP-FSI declaration coming from developing countries (ie 30 per cent out of all signatories; see Appendix VII). Moreover, it would be interesting to see if differences exist between banks from developed countries to their local competitors for activities in developing countries. For Thailand operations, Zimmerman and Mayer (2001) show that no significant differences exist between foreign banks and Thai banks.

APPENDICES

1 See Verbruggen (1996) for the SEDS study (sustainable economic development scenarios). The basic path in this study is the result of the first three (of five) steps of the research method, namely: 1 Choice of environmental pressure parameters; 2 Choice of economic background scenario; 3 Estimate of technological development (including shifts within business branches). The other two are: 4 Choice of environmental objectives and other prerequisites; 5 Construction of 'sustainable economic structures'.
2 For the Weak and Strong Together scenario the GNP is also shown (the second stated value) according to a different labour market proposition that reveals the 'GNP offer' to be strongly dependent on other factors than just environmental requirements.

3 Source: www.ifc.org/enviro/enviro/Review_Procedure_Main/Review_Procedure/Annex_B (March 2001). Category FI includes eg corporate loans to banks, credit lines and private equity funds (see Box 6.5).
4 The term environment as used in this document also refers to environmentally related aspects of health, safety and product stewardship. More than 2300 companies formally support the Charter. Source: www.iccwbo.org/sdcharter/charter (September 2000).
5 As revised May 1997. Source: www.unep.ch/etu/finserv/finserv/english.htm (March 2001).
6 By country (as of May 2001). As of this date 179 banks coming from 46 countries have signed the Statement. Some of these banks have merged in recent years. Among these are Bayerische Hypotheken-und Wechselbank/Bayerische Vereinsbank AG, Föreningsbanken/Sparbanken Sverige AB, Union Bank of Switzerland/Swiss Bank Corporation, Südwestdeutsche Landesbank Girozentrale/Landesgirokasse Bank/Landeskreditbank and NatWest Group/Royal Bank of Scotland Plc. Source: www.unep.ch/etu/finserv/finserv (June 2001).
7 The second column shows the position of the bank within the group of 34 in terms of its assets (1999 year-end figures); in parentheses is the bank's position in the world rating by assets (source: *The Banker*, July 2000; see also www.cba.ca/eng/statistics/stats/bankrankings.htm, December 2000). In the last four columns the biggest and the smallest in each column are shown in bold. e = estimate (based on fragmentary information from the annual report). The figure in the column 'Countries worldwide' often offers only an indication since large banks often use the phrase 'active in more than x countries' in their annual reports. The data in the last four columns has been taken from the annual reports for 1999 for the 34 selected banks (all at group level).
8 In parentheses is shown the relative weight of the bank's activities (measured in assets) in its domestic market (D) or home market (H) in relation to its total activities. The home market for the two Canadian banks is Canada and the USA; for the two northern European banks Scandinavia and Finland; for BSCH Spain and Latin America; for Fortis Belgium and The Netherlands; for SanPaolo IMI Europe and for HSBC Europe and Hong Kong. Where no data about geographical distribution in terms of assets were available, an alternative was used: for ING and Svenska Handelsbanken total revenue; for HypoVereinsbank number of employees and for SanPaolo IMI number of offices. The data in the last two columns have been taken from annual reports in 1999 for the selected banks. The data for the ROaE and cost–benefit ratio have been taken from Bankscope.
9 This column shows the financial services on which the banks themselves report and which they themselves consider to be their most important or strategic. The following categories have been used (with, where necessary, subdivisions): 1a = domestic retail banking; 1b = domestic and international retail banking; 2 = corporate banking; 3 = investment banking; 4 = asset management; 5 = life and non-life insurance; 5a = life insurance; 5b = non-life insurance. Where a service is indeed offered, but is not considered by the bank to be a core activity, this is shown in parentheses. For retail banking a subdivision has been made between domestic and international, because this is an important distinction. For retail banking an extensive network of offices is necessary, entailing considerable investments. For other areas, such investments for an international strategy are relatively less significant.

10 This box is a methodical account of the research and the chosen parameters which aims to increase and establish the validity, reproducibility and reliability of the research (Jonker and Pennink, 2000, pp65–77).
11 Of course, the key influences the banks' final scores and countless combinations are possible. Group 5 has a relatively small weight in this paper (maximum 10 per cent) as it was chosen to look strongly at environmental issues. This group looks at the social and community activities of the bank. External environmental activities (products and services; groups 3 and 4) are given a greater weight than internal environmental care and communication (groups 1 and 2). In order to reduce the chance of the outcomes being manipulated, it has been determined at the outset that groups 1 and 2 will account for a maximum of 40 per cent and groups 3 and 4 must have a minimum of 50 per cent. In addition, each group must have a weight greater than zero and the variation in the weights must be limited to steps of 5 per cent. The results are robust. Only by attributing more weight to internal than to external activities (and so not in line with the criteria above), do real changes in the position of individual banks occur in Table 9.2. The chosen key appears, for example, to suit the research institute Oekum, which in 2000 investigated the investment services of 42 banks in Europe (the results of this investigation are however not publicly available). Environmental management and environmental data were found to account for 25 per cent and environmental products for 50 per cent of the total score. See www.oekom.de (January 2001).
12 The first law proposes that energy and material can neither be created nor destroyed. This is implicit in Figure 2.1. See for example Tietenberg, 1992, pp20–21.
13 See Ekins, 1993. 'Zero growth' is not, self-evidently, the same as a 'steady state'. Daly (1991) assumes a steady state in physical rather than monetary values. Technological innovation can thus result in economic growth without the physical flow and supply quantities changing.
14 Van Arkel and Peterse, 1993. Another idea put forward in this book is that of a tax levied on money and capital possessions. One of the problems with this idea is that the independent accumulation of an old-age pension is 'punished'.
15 For sustainability the pursuit of zero emission is therefore necessary for stock polluters and fund polluters may not exceed the absorption capacity.
16 The terminology used here is part of the environmental assessment policies (OP4.01) at IFC. For specific WB/IFC definitions, see www.ifc.org/enviro/EnvSoc/Safeguard/EA (March 2001).
17 Within sustainability, whether the regenerative capacity of each finite resource should be taken into full consideration is an ethical issue. Recycling, technological developments or substitution may offer a solution to this dilemma.

References

1 Sources comprising newspapers, news and data channels, internet and the more popular press are indicated exclusively in the footnotes in the main text.

References[1]

ABN AMRO (1989), *Recycling, een zorg minder voor het milieu*, ABN AMRO, Amsterdam

ABN AMRO (1990), *Milieu, een hele onderneming: Checklist voor ondernemers*, ABN AMRO, Amsterdam

ABN AMRO (1998), *Milieuverslag 1994–1997*, ABN AMRO, Amsterdam

ABN AMRO (2001), *Environment Report 1998–2000*, ABN AMRO, Amsterdam (available at www.abnamro.com)

Aktie Strohalm (1998), *Lokale economie wereldwijd*, Aktie Strohalm, Utrecht

Albert, M (1991), *Capitalisme contra Capitalisme,* Éditions du Seuil, Paris

Allen, F and AM Santomero (1998), 'The theory of financial intermediation', *Journal of Banking and Finance*, No 21, pp1461–1485

Arkel, H van and G Peterse (1993), 'Van grenzen aan de groei naar greep op groei', in: Biesboer, F (ed) *Greep op groei*, pp233–265, Aktie Strohalm/Jan van Arkel,Utrecht

Arthur D Little (2000), *Realising the Business Value of Sustainable Development*, Arthur D Little, Brussels

ADB (2000), *Asian Environmental Outlook 2001*, ADB, Manilla

ASN Bank (1998), *ASN Aandelenfonds Jaarverslag 1997*, ASN Bank, The Hague (available at www.asnbank.nl)

ASN Bank (1999), *Jaarverslag 1998*, ASN Bank, The Hague (available at www.asnbank.nl)

Atkins, D and C Pedersen (2001), 'Finance: Remember the environment: A Danish tool for integration of environmental aspects in credit evaluation', in: JJ Bouma, MHA Jeucken and L Klinkers, *Sustainable Banking: The Greening of Finance,* pp295–299, Greenleaf, Sheffield

Axelrod, R (1984), *The Evolution of Cooperation*, Basic Books, New York

Badrinath, SG and PJ Bolster (1996), 'The role of market forces in EPA enforcement activity', *Journal of Regulatory Economics*, pp165–181

Ballard, D (2000), 'The cultural aspects of change for sustainable development', *Eco-Management and Auditing*, No 7, pp53–59

Banco Bilbao Vizcaya Argentaria (2000), *Annual Report 1999*, BBVA, Madrid (available at www.bbva.es)

Banco Santander Central Hispano (2000), *Annual Report 1999*, BSCH, Madrid (available at www.bsch.es)

Bank Austria (2000), *Annual Report 1999*, Bank Austria, Vienna (available at www.bankaustria.com)

Bank of America (2000), *1999 Environmental Progress Report*, Bank of America, Charlotte, NC (available at www.bankofamerica.com/environment)

Bank of Montreal (2000), *Bank of Montreal Group of Companies 182nd Annual Report 1999*, Bank of Montreal, Toronto (available at www.bmo.com)

Bank of Tokyo-Mitsubishi (2000), *Annual Report 2000*, BTM, Tokyo (available at www.btm.co.jp)

Barannik, AD (2001), 'Providers of financial services and environmental risk management: Current experience', in JJ Bouma, MHA Jeucken and L Klinkers, *Sustainable Banking: The Greening of Finance*, pp247–267, Greenleaf, Sheffield

Barannik, AD and RJA Goodland (2001), 'The World Bank's environmental assessment policies: Review of institutional development', in: JJ Bouma, MHA Jeucken and L Klinkers, *Sustainable Banking: The Greening of Finance*, pp316–347, Greenleaf, Sheffield

Barclays (1999a), *1999 Environment Review*, Barclays, London (available at www.barclays.com)

Barclays (1999b), *Social Review 1999*, Barclays, London (available at www.barclays.com)

Barta, J and V Éri (2001), 'Environmental attitudes of banks and financial institutions', in: JJ Bouma, MHA Jeucken and L Klinkers, *Sustainable Banking: The Greening of Finance*, pp120–132, Greenleaf, Sheffield

Baumol, WJ and WE Oates (1988), *The Theory of Environmental Policy*, Cambridge University Press, Cambridge

BBA (1997), *The Environment: The Challenge for Business and Banking*, British Bankers' Association, London (available at www.bba.org.uk/media)

BBA (1999), *Micro Credit in the UK: An Inventory of Schemes for Businesses Supported by Banks*, British Bankers' Association, London (available at www.bba.org.uk)

Bellegem, T van (2001), 'Green fund system in The Netherlands', in: JJ Bouma, MHA Jeucken and L Klinkers, *Sustainable Banking: The Greening of Finance*, pp234–244, Greenleaf, Sheffield

Bennett, M and P James (eds) (1998), *The Green Bottom Line: Environmental Accounting for Management: Current Practice and Future Trends*, Greenleaf, Sheffield

Berger, AN, RS Demsetz and PhE Strahan (1999), 'The consolidation of the financial services industry: Causes, consequences and implications for the future', *Journal of Banking and Finance*, No 23, pp135–194

Beringer, G and E Thomas (1991), 'Lenders and environmental liability', *Practical Law for Companies*, November, pp3–11

Blair, MM (1995), *Ownership and Control: Rethinking Corporate Governance for the Twenty-first Century*, Brookings Institution, Washington, DC

Blumberg, J, A Korsvold, and G Blum (1997), *Environmental Performance and Shareholder Value,* WBCSD, MIT Press, Cambridge, MA

BNP Paribas (2000), *Annual Report 1999*, BNP Paribas, Paris (available at www.bnpparibas.com)

The Body Shop (1998), *Values Report 1997,* The Body Shop, Watersmead, Littlehampton (available at www.the-body-shop.com)

Bor, J and E Petersma (eds) (1996), *De Verbeelding van het Denken: Geïllustreerde Geschiedenis van de Westerse en Oosterse Filosofie*, Contact, Amsterdam

Boudhan, B, F Nelissen and I Vonk (1996), *Maatschappelijk Ondernemen: Dienen en Verdienen*, SMO-96-5 Informatief, The Hague

Boulding, KE (1966), 'The economics of the coming Spaceship Earth', in: H Jarret (ed), *Environmental Quality in a Growing Economy*, pp3–14, Johns Hopkins University Press, Baltimore

Bowers, J (1993), 'A conspectus on valuing the environment', *Journal of Environmental Planning and Management*, Vol 26, No 1

Boyer, M and J-J Laffont (1997), 'Environmental risk and bank liability', *European Economic Review,* No 41, pp1427–1459

Braden, JB and CD Kolstad (1991), *Measuring the Demand for Environmental Quality*, North Holland, Amsterdam

Braeckel, D van (2000), 'Ethical investment in Belgium: European certification for SRI funds', paper presented at the Triple Bottom Line Investing 2000 Conference, 2–3 November, Rotterdam (available on conference CD-Rom)
Brink, B ten (2000), *Biodiversity Indicators for the OECD Environmental Outlook and Strategy: A Feasibility Study,* RIVM Report 402001014, Bilthoven (available at www.rivm.nl)
Broekman, F (1992), *Prisma van de Algemene Economie*, Spectrum, Zwolle
Bronfenbrenner, M, W Sichel and W Gardner (1990), *Economics*, Boston, MA
Brown, LR, C Flavin and H French (1999), *State of the World 1999*, Worldwatch Institute/Norton, New York
Bryce, A (1992), 'Environmental liability: Practical issues for lenders', *Journal of International Banking Law*, Vol 7, No 4, April, pp131–137
Buiter, W, M Groenewegen and C Larsen (1995), *Communicatie bij verandering*, Kluwer, Deventer
Bulte, E and D van Soest (1997), 'Beleggen in teakhout: Geen tweesnijdend kapmes', *Economische Statistische Berichten*, ESB, 12 February 1997
Burton, E (1992), 'Debt for development: New opportunities for nonprofits, commercial banks, and developing states', *Harvard International Law Journal*, No 311, pp233–256
Business and the Environment (1996), 'UK bank launches center to offer environmental advice to clients', June
Business and the Environment (1997), 'New Swiss investment fund uses eco-efficiency criteria', August, pp6–7
Canadian Imperial Bank of Commerce (2001), *Innovation + Accountability: CIBC 2000 Annual Report*, CIBC, Toronto (available at www.cibc.com)
Carl Duisberg Gesellschaft (1998), Greening the Financial Sector: International Business Forum, 12–14 October, CDG, Berlin
Carson, R (1962), *Silent Spring*, Houghton Mifflin, Boston, MA
Case, P (1999), *Environmental Risk Management and Corporate Lending: A Global Perspective*, Woodhead, Cambridge
CPB (1996), *Economie en milieu: Op zoek naar duurzaamheid*, SDU, The Hague
CBA (2000), *Your Business, Your Bank and the Environment*, Canadian Bankers Association, Toronto (available at www.cba.ca/eng/Tools/Brochures/tools_environment.htm)
Centre for Business Performance (2000), *A Responsible Investment?,* Centre for Business Performance, London
Chapman and Cutler (1993), *Environmental Risk: A Guide to Compliance With the FDIC's Environmental Risk Program,* ABA, Washington, DC
Chase Manhattan Corporation (2000), *1999 Annual Report*, Chase Manhattan Corporation, New York (available at www.chase.com)
Cipolla, CM (1981), *Before the Industrial Revolution: European Society and Economy, 1000–1700*, Methuen, London
Citigroup (2000), *Lead by Example: 1999 Annual Report*, Citigroup, New York (available at www.citigroup.com)
Colborn, T, D Dumanoski and JP Myers (1996), *Our Stolen Future: Are We Threatening Our Fertility, Intelligence and Survival? A Scientific Detective Story,* Dutton, New York
Collins, JC and JI Porras (1994), *Built to Last: Successful Habits of Visionary Companies*, Harper Business, New York
Commonwealth of Australia (1996), *Australia: State of the Environment 1996*, CSIRO Publishing, Collingwood
Constanza, R et al (1997), 'The value of the world's ecosystem services and natural capital', *Nature*, Vol 387

The Co-operative Bank (2000), *The Partnership Report: Measuring Our Progress in 1999,* The Co-operative Bank, Manchester (available at www.co-operativebank.co.uk)
Coopers&Lybrand (1996), *De relatie tussen banken en milieu is nog niet 'natuurlijk': Inventarisatie van de opvattingen binnen de bancaire wereld inzake het Nederlandse milieubeleid en van mogelijke win–win situaties tussen de bancaire wereld en de overheid op milieugebied,* Coopers&Lybrand, Utrecht
Costaras, NE (1995), 'Environmental risk rating: A study of the development of an environmental risk rating tool for the financial sector', unpublished MBA thesis for the Netherlands Institute for MBA Studies, Utrecht
Coulson, AB (2001), 'Corporate environmental assessment by a bank lender: The reality?', in: JJ Bouma, MHA Jeucken and L Klinkers, *Sustainable Banking: The Greening of Finance*, pp300–312, Greenleaf, Sheffield
Cowton, CJ and P Thompson (2000), 'Do codes make a difference? The case of bank lending and the environment', *Journal of Business Ethics*, No 24, pp165–178
Cramton, P and S Keri (1998), *Tradable Carbon Permit Auctions: How and Why to Auction Not Grandfather*, Discussion Paper 98–34, Resources for the Future, Washington, DC
Crédit Agricole (2000), *1999 Annual Report,* Crédit Agricole, Paris (available at www.credit-agricole.fr)
CSG (1998), *Environmental Report 1997/98*, Credit Suisse Group, Zurich (available at www.credit-suisse.com/en/ecoreport98)
CSG (1999), *Environmental Performance Evaluation Switzerland 1998/99*, Credit Suisse Group, Zurich (available at www.credit-suisse.com/en/eco_performance_99)
CSG (2000), *Environmental Report 1999/00,* Credit Suisse Group, Zurich (available at www.credit-suisse.com/en/ecoreport00)
Dalen, M van (1997), *Company Environmental Reporting: Conditions for the Optimal Information Structure of Environmental Reports*, Ministry of Housing, Physical Planning and Environmental Protection, No 1997/6, The Hague
DalMaso, D, C Marini and P Perin (2001), 'A green package to promote environmental management systems among SMEs', in: JJ Bouma, MHA Jeucken and L Klinkers, *Sustainable Banking: The Greening of Finance*, pp56–65, Greenleaf, Sheffield
Daly, HE (1991), *Steady State Economics*, Island Press, Washington, DC
Daly, HE and JB Cobb Jr (1989), *For the Common Good*, Beacon Press, Boston, MA
Danthine, J-P, F Giavazzi, X Vives and E-L von Thadden (1999), *The Future of European Banking*, Centre for Economic Policy Research, London
Davis, J (1994), 'Bankers face new uncertainty over environmental liability', *Philadelphia Business Journal,* No 135, p3
Delphi/Ecologic (1997), *The Role of Financial Institutions in Achieving Sustainable Development: Report to the European Commission*, Delphi/Ecologic, Brussels
Deutsche Bank (2000), *Sustainability: Economics, Ecology and Social Responsibilty*, Deutsche Bank, Frankfurt am Main (available at www.deutsche-bank.com)
Dinten, WL van (1996), 'Organiseren in een democratie', in *Democratie, Dimensies en Divergenties: Van Descartes via Darwin naar Guéhenno,* pp79–114, VUGA/Rabobank, The Hague
Döbeli, S (2001), 'An environmental fund with the WWF label: The importance of appropriate communication tools', in: JJ Bouma, MHA Jeucken and L Klinkers, *Sustainable Banking: The Greening of Finance*, pp379–389, Greenleaf, Sheffield
Dow Jones Sustainability Group Index (1999), *Guide to the Dow Jones Sustainability Group Indexes,* Version 10, September 1999, DJSGI, Zurich
Dowell, G, S Hart and B Yeung (2000), 'Do corporate global environmental standards create or destroy market value?', *Management Science*, Vol 46, No 8, pp1059–1074

Dresdner Bank (2000), *Nachhaltiges Handeln für Umwelt und Gesellschaft: Bericht 1999 der Dresdner Bank AG*, Dresdner Bank, Frankfurt am Main (available at www.dresdnerbank.com)

DTTI, IISD and SustainAbility (1993), *Coming Clean: Corporate Environmental Reporting, Opening Up for Sustainable Development*, SustainAbility, London

Dunné, JM (1992), 'Milieu-aansprakelijkheid meer en meer gebaseerd op toerekening naar redelijkheid', *Financieel juridisch dossier*, No 10, pp88–93

Dutch Lower House (1997), *Belastingen in de 21e eeuw: Een verkenning*, vergaderjaar 1997–1998, No 25810, The Hague

EBRD (1995), *Environmental Risk Management for Financial Institutions: A Handbook*, EBRD, London

EEA (2000), *Business and the Environment: Current Trends and Developments in Corporate Reporting and Ranking*, European Environment Agency, Copenhagen (available at www.eea.eu.int)

EIF (1996), *Annual Report 1996*, EIF, Luxembourg

Ekelund, RB and RF Hébert (1990), *A History of Economic Theory and Method*, McGraw-Hill, 3rd edition, Singapore

Ekins, P (1993), 'Met groei is geen duurzaamheid mogelijk', in: F Biesboer (ed) *Greep op groei*, pp79–99, Aktie Strohalm/Jan van Arkel, Utrecht

Elkington, J (1997), *Cannibals with Forks: The Triple Bottom Line of 21st Century Business*, Capstone, Oxford

Emissions Trading Education Initiative (ETEI, 1999), *Emissions Trading Handbook*, ETEI, Milwaukee, WI

Environmental & Accounting Auditing (1997), 'Linking financial and environmental performance', May, Vol 2, No 10

Environmental Defence (1998), *Do You Know what ABN AMRO Is Doing with Your Money?,* Environmental Defence, Amsterdam

Environmental Defense et al (1999), *A Race to the Bottom: Creating Risk, Generating Debt and Guaranteeing Environmental Destruction: A Compilation of Export Credit & Investment Insurance Agency Case Studies*, Environmental Defense, New York (available at www.edf.org/programs/international/ecr/ecareportcov.html)

Environmental Finance (1999/2000), 'City welcomes new risk rating', December/January, Vol 1, No 3, pp18–20

Environmental Finance (2001), 'SRI investment in Canada hit C$50bn', February, Vol 2, No 4, p10

EPI Finance (2000), *Environmental Performance Indicators for Financial Service Providers*, EPI Finance

European Commission DG XI (1998), *Report of the Workshop: Sustainable Development; Challenge for the Financial Sector*, EU, Brussels

European Commission (2000), *White Paper on Environmental Liability*, EU, Brussels (available at www.europa.eu.int/comm/environment/liability/index.htm)

European Union (1993), *The Road to Sustainable Developmen*t, EU, Brussels

Eurostat (1995), *Enterprises in Europe*, Eurostat, Brussels

Ferderick, WC, JE Post and K Davis (1992), *Business and Society: Corporate Strategy, Public Policy, Ethics,* McGrawHill, New York

Ferguson, M (1981), *The Aquarian Conspiracy: Personal and Social Transformation in the 1980s*, JP Tarcher, Los Angeles, CA

Figge, F (2001), 'Environment-induced systematisation of economic risks', in: JJ Bouma, MHA Jeucken and L Klinkers, *Sustainable Banking: The Greening of Finance*, pp268–279, Greenleaf, Sheffield

Flatz A, L Serck-Hanssen and E Tucker-Bassin (2001), 'The Dow Jones Sustainable Group Index: The first worldwide sustainable index', in: JJ Bouma, MHA Jeucken and L Klinkers, *Sustainable Banking: The Greening of Finance*, pp222–233, Greenleaf, Sheffield

FöreningsSparbanken (2000), *Miljöredovisning 1999*, FöreningsSparbanken, Stockholm

FORGE (2000), *Guidelines on Environmental Management and Reporting for the Financial Services Sector: A Practical Toolkit*, FORGE, UK (available at www.bba.co.uk)

Fortis (2000), *Financial Report 1999*, Fortis, Brussels/Utrecht, (available at www.fortis.com)

Fransen, MAHJ, HFD Hassink and EHJ Vaassen (1997), 'Ethical accounting: Introductie en toepassing', *Maandblad voor Accountancy en Bedrijfseconomie*, June, pp312–323

Freixas, X and J-R Rochet (1998), *Microeconomics of Banking*, MIT Press, Cambridge, MA

Fuji Bank (2000), *Annual Report 2000*, Fuji Bank, Tokyo (available at www.fujibank.co.jp)

Fukuyama, F (1995), *Trust,* Hamish Hamilton, London

Furrer, B and H Hugenschmidt (1999), 'Financial services and ISO 14001: The challenge of determining indirect environmental aspects in global certification', *Greener Management International*, No 28, pp32–41

Ganzi, J and A DeVries (1998), *Corporate Environmental Performance As a Factor in Financial Industry Decisions*, Office of Cooperative Environmental Management, Chapel Hill

Ganzi, JT and J Tanner (1997), *Global Survey on Environmental Policies and Practices of the Financial Services Industry: The Private Sector*, Environment and Finance Enterprise, Washington, DC

GEMI (1996), *Environmental Reporting and Third Party Statements*, GEMI, Washington, DC

Gentry, B (1997), 'Making private investment work for the environment', in *Finance for Sustainable Development: The Road Ahead,* pp341–402, United Nations Department for Policy Coordination and Sustainable Development, New York

Geus, A de, N Stone and LH Vonk (1997), *The Living Company: Habits for Survival in a Turbulent Environment*, Nicholas Brealey, London

Giuseppi J (2001), 'Assessing the "triple bottom line": Social and environmental practices in the European banking sector', in: JJ Bouma, MHA Jeucken and L Klinkers, *Sustainable Banking: The Greening of Finance*, pp96–113, Greenleaf, Sheffield

Goudzwaard, B (1978), *Kapitalisme en vooruitgang*, Van Gorcum, Assen/Amsterdam

Green Futures (1998a), 'Money changers', January/February, pp28–29

GRI (2000), *Sustainability Reporting Guidelines on Economic, Environmental, and Social Performance*, June 2000, GRI, Boston, MA

Groene, J de (1995), *Beheersen of beïnvloeden?: De respons van bedrijven op milieuproblemen: Het belang van de omgeving*, Fanoy, Middelburg

Habermas, J (1981), *Theorie des Kommunikativen Handelns*, Suhrkamp, Frankfurt am Main

Hager, W (2000), *The Environment in European Enlargement: Report of a CEPS Working Party*, Centre for European Policy Studies, Brussels

Hardin, G (1968), 'The tragedy of the commons', *Science*, No 162, pp1243–1248

Harman, WW (1979), *An Incomplete Guide to the Future*, Norton, New York

Hawken, P (1993), *The Ecology of Commerce*, Harper Business, New York

Hilderink, HBM (2000), *World Population in Transition: An Integrated Regional Modelling Framework,* Thela Thesis, Amsterdam

Hill, J, D Fedrigo and I Marshall (1997), *Banking on the Future: A Survey of Implementation of the UNEP Statement by Banks on Environment and Sustainable Development,* The Green Alliance, London

Hodes, GS (2001), 'Sustainable finance for sustainable energy: The role of financial intermediaries', in: JJ Bouma, MHA Jeucken and L Klinkers, *Sustainable Banking: The Greening of Finance*, pp412–430, Greenleaf, Sheffield
Hoelen, H (1995), *Beschouwingen over de economische orde*, Van Gorcum, Assen
Hoff, B (1982), *The Tao of Pooh,* Dutton, New York
Hoogendijk, W (1993), 'Pleidooi voor een bevrijding', in: F Biesboer (ed) *Greep op groei*, pp187–209, Aktie Strohalm/Jan van Arkel, Utrecht
House, R (1993), 'Balance-sheet poison', *Institutional Investor,* August, pp23–28
Howe, CW (1993), 'The US environmental policy experience: A critique with suggestions for the European Community', *Environmental and Resource Economics*, Vol 3, No 4, pp359–379
HSBC (1999), *HSBC: Business Principles and Values*, brochure, HSBC, London
HSBC Holdings (1999), *Annual Report and Accounts 1999*, HSBC, London (available at www.hsbc.com)
Hubbard, RG (1994), *Money and the Financial System and the Economy*, Addison-Wesley Publishing, New York
Hugenschmidt, H, J Janssen, Y Kermode and I Schumacher (2000), 'Sustainable banking at UBS', in JJ Bouma, MHA Jeucken and L Klinkers, *Sustainable Banking: The Greening of Finance*, pp43–55, Greenleaf, Sheffield
HypoVereinsbank (2000), *Umweltbericht Im Zeichen van Nachhaltigkeit,* HypoVereinsbank, Munich (available at www.hypovereinsbank.de)
IBRD (1991), *The World Bank and the Environment: A Progress Report,* The World Bank, Washington, DC
IEA (1995), *World Energy Outlook*, IEA, Paris
IFC (1998), *Doing Better Business Through Effective Public Consultation and Disclosure: A Good Practice Manual*, IFC, Washington, DC
ING Bank (1992), *Milieumanagement: Stap voor stap naar een beter milieubeheer in uw onderneming*, ING Bank, Amsterdam
ING Group (1998), *Environmental Annual Report 1997: ING in the Netherlands*, ING Group, Amsterdam
ING Group (1999), *Environmental Annual Report 1998,* ING Group, Amsterdam (available at www.inggroup.com)
Intesa (2000), *Annual Report 1999,* Intesa, Milan (available at www.bancaintesa.it)
IPCC (1995), *Second Assessment Report: The Science of Climate Change*, Cambridge University Press, Cambridge (summary available at www.ipcc.ch/pub/reports)
IPCC (2001), *Third Assessment Report,* Working Group I, Geneva (available at www.ipcc.ch)
Jasch C (2001), 'Sustainable banking in Austria', in: JJ Bouma, MHA Jeucken and L Klinkers, *Sustainable Banking: The Greening of Finance*, pp114–119, Greenleaf, Sheffield
Jeucken, MHA (1998a), 'Milieu als comparatieve factor in de internationale handel', *Maandschrift Economie,* Vol 62, No 1, pp52–75
Jeucken, MHA (1998b), *Duurzaam Bankieren: Een visie op bankieren en duurzame ontwikkeling*, Rabobank, Utrecht
Jeucken, MHA and RA van Tilburg (1999), 'Koersen op duurzaam ondernemen', *Economische Statistische Berichten*, No 4230, pp864–866
Jonker, J and BJW Pennink (2000), *De kern van de methodologie*, Van Gorcum, Assen
Kahlenborn, W (2001), 'Transparency and the green investment market', in JJ Bouma, MHA Jeucken and L Klinkers, *Sustainable Banking: The Greening of Finance*, pp173–186, Greenleaf, Sheffield
Kaplan, RS and DP Norton (1993), 'Putting the balanced scorecard to work', *Harvard Business Review*, No 715

Kastelein, TJ (1984), *Groei naar een industriële samenleving*, Wolters-Noordhof, Groningen
Kastelein, TJ (1987), *Economie en methodologie: Een inleiding*, Wolters-Noordhof, Groningen
KBC (2000), *KBC Bank & Verzekering Activiteitenverslag 1999*, KBC, Brussels (available at www.kbc.be)
Kearins, K and G O'Malley (2001), 'International financial institutions and the Three Gorges hydroelectric power scheme', in: JJ Bouma, MHA Jeucken and L Klinkers, *Sustainable Banking: The Greening of Finance*, pp348–359, Greenleaf, Sheffield
Keating, M (1993), *The Earth Summit's Agenda for Change: A Plain Language Version of Agenda 21 and the Other Rio Agreements*, Centre for Our Common Future, Geneva
Keijzers, G (2000), 'The evolution of Dutch environmental policy: The changing ecological arena from 1970–2000 and beyond', *Journal of Cleaner Production*, No 8, pp179–200
Kelly, M and A Huhtala (2001), 'The role of the United Nations Environmental Programme and the financial services sector', in: JJ Bouma, MHA Jeucken and L Klinkers, *Sustainable Banking: The Greening of Finance*, pp390–400, Greenleaf, Sheffield
KfW (2000), *Environmental Report 2000*, KfW, Frankfurt am Main (available at www.kfw.de)
Khandler, SR, B Khalily and Z Khan (1995), *Grameen Bank: Performance and Sustainability*, World Bank, Washington, DC
Klant, JJ and C van Ewijk (1990), *Geld, banken en financiële markten,* Wolters-Noordhof, Groningen
Knoepfel, I, JE Salt, A Bode and W Jakobi (1999), *The Kyoto Protocol and Beyond: Potential Implications for the Insurance Industry*, UNEP, Geneva
Knörzer A (2001), 'The transition from environmental funds to sustainable investment: The practical application of sustainability criteria in investment products', in: JJ Bouma, MHA Jeucken and L Klinkers, *Sustainable Banking: The Greening of Finance*, pp211–221, Greenleaf, Sheffield
Konar, S and MA Cohen (1997), 'Information as regulation: The effect of community right to know laws on toxic emissions', *Journal of Environmental Economics and Management*, pp109–124
Koppen, CSA van and JLF Hagelaar (1998), 'Milieuzorg als strategische keuze: Van bedrijfsspecifieke situatie naar milieuzorgsystematiek', *Bedrijfskunde*, Vol 70, No 1
Korten, D (1995), *When Corporations Rule the World*, Berrett-Koehler Publishers, San Francisco/Earthscan Publications, London
KPMG (1999), *International Survey of Environmental Reporting 1999*, KPMG, London
KPMG/IVA (1996), *Evaluatie bedrijfsmilieuzorgsystemen 1996*, KPMG/IVA, The Hague/Tilburg
Kuhn, TS (1970), *The Structure of Scientific Revolutions*, University of Chicago Press, Chicago, IL
LaPlante, B and P Lanoie (1994), 'Market response to environmental incidence in Canada', *Southern Economic Journal*, pp657–672
Lascelles, D (1994), 'Rating environmental risk', *Professional Investor*, Vol 5, No 1, p32
Leistner, M (2001), 'The Growth and Environment Scheme: The EU, the financial sector and small and medium-sized enterprises as partners in promoting sustainability', in: JJ Bouma, MHA Jeucken and L Klinkers, *Sustainable Banking: The Greening of Finance*, pp372–378, Greenleaf, Sheffield
Lloyds TSB (2000), *Environmental Report 1999*, Lloyds TSB Group, Birmingham
Louche, C (2001), 'The corporate environmental performance: Financial performance link: Implications for ethical investments', in: JJ Bouma, MHA Jeucken and L Klinkers, *Sustainable Banking: The Greening of Finance*, pp187–200, Greenleaf, Sheffield

Luijk, H van (1997), 'Hoever strekt de maatschappelijke verantwoordelijkheid van de onderneming zich uit?', *Nijenrode Management Review*, No 5, July/August, pp56–64
Luijk, H Van, A Schilder and M Bannink (1997), *Patronen van verantwoordelijkheid: Ethiek en corporate governance*, Academic Service, Amsterdam
Maslow, AH (1970), *Motivation and Personality*, Harper & Row, New York
Matthews, E et al (2000), *The Weight of Nations: Material Outflows from Industrial Economies*, WRI, Washington, DC
Meade, JE (1973), *The Theory of Economic Externalities*, Institut Universitaire de Haute Études, Geneva
Meadows, DH, DL Meadows and J Randers (1991), *Beyond the Limits: Confronting Global Collapse: Envisioning a Sustainable Future*, Earthscan, London
Meadows, DH, DL Meadows, J Randers and W Behrens (1972), *The Limits to Growth: A Global Challenge: A Report for the Club of Rome Project on the Predicatement of Mankind*, Universe Books, New York
Merita Nordbanken (2000), *Anual Report 1999,* Nordic Balting Holding, Merita/Stockholm (available at www.nordea.com)
Milbourn, TT, AWA Boot and AV Thakor (1999), 'Megamergers and expanded scope, theories of bank size and activity diversity', *Journal of Banking and Finance*, Vol 23, Nos 2–4, pp195–214
Millman GJ (1996), 'Whose law is it anyway?', *Infrastructure Finance*, May, pp33–35
Ministry of Economic Affairs (1996), *Onbekend maar zeer bemind,* SDU, The Hague
Ministry of Economic Affairs (1998), *Energie-InvesteringsAftrek*, brochure, SDU, The Hague
Ministry of Housing, Spatial Planning and the Environment (1985), *Indicatief Meerjarenprogramma Milieubeheer 1985–1989* (*Indicative Multi-Year Programmes of 1985*), No 18602–1/2/5, SDU, The Hague
Ministry of Housing, Spatial Planning and the Environment (1991), *National Environmental Policy Plan: To Choose or to Lose (NEPP1),* VROM, The Hague
Ministry of Housing, Spatial Planning and the Environment (1994a), *National Environmental Policy Plan 2: The Environment: Today's Touchstone (NEPP2),* VROM, The Hague
Ministry of Housing, Spatial Planning and the Environment (1998a), *Policy Document on Environment and Economy*, VROM, The Hague
Ministry of Housing, Spatial Planning and the Environment (1998b), *National Environmental Policy Plan 3 (NEPP3),* VROM, The Hague
Ministry of Housing, Spatial Planning and the Environment (1999), *International Peer Review of Dutch Environmental Policy*, VROM, The Hague
Ministry of Housing, Spatial Planning and the Environment (2000), *Regeling Groenprojecten,* (revised version in Dutch), VROM, The Hague
Ministry of Public Health Care and Environmental Protection (1972), *Urgentienota 1972* (*Policy Document on the Urgency of Environmental Pollution*), No 11906–2, SDU, The Hague
Mintzberg, H (1983), *Structures in Five: Designing Effective Organizations*, Prentice-Hall, Englewood Cliffs, NJ
Molenkamp, GC (1995), *De Verzakelijking van het Milieu: Onomkeerbare Ontwikkelingen in het Bedrijfsleven*, KPMG, The Hague
Moor, APG de (1997), *Subsidies and Sustainable Development: Key Issues and Reform Strategies*, Earth Council, San Jose, Costa Rica
Moret Ernst & Young Management Consultants (1998), *Onderzoek administratieve lastendruk,* Moret Ernst & Young, Utrecht

Moughalu, MI, HD Robinson and H Glaslock (1990), 'Hazardous waste lawsuits, stockholder returns and deterrence', *Southern Economic Journal*, pp357–370
Müller, K et al (1996), *Eco-efficiency and Financial Analysis: The Financial Analyst's View*, European Federation of Financial Analysts' Societies
National Australia Bank (2001), *Annual Financial Report 2000*, NAB, Melbourne (available at www.national.com.au)
NatWest Group (1995), *Energy Management Workbook*, NatWest, London
NatWest Group (1998), *Environment Report 1997/98,* NatWest, London (available at www.natwest.com)
Negenman M (2001), 'Sustainable banking and the ASN Bank', in: JJ Bouma, MHA Jeucken and L Klinkers, *Sustainable Banking: The Greening of Finance*, pp66–71, Greenleaf, Sheffield
Nikko Securities/Nikko Asset Management (2000), *Sustainability Report 2000,* Tokyo (available at www.nikko.co.jp/sec)
Noordhoek, FK, J Abbema and R, van Ruitenbeek (1999), *Goed Geld Gids: Duurzaam beleggen, sparen, lenen en verzekeren*, Noordhoek, Utrecht
NOVEM (1996), *Elektriciteit uit zonlicht*, Nederlandse Onderneming voor Energie en Milieu, Utrecht/Sittard
OECD (1975), *The Polluter Pays Principle: Definition, Analysis and Implementation,* Organisation for Economic Co-operation and Development, Paris
OECD (1994), *Environmental Performance Reviews: Japan,* Organisation for Economic Co-operation and Development, Paris
OECD (1995), *An Assessment of Financial Reform in OECD Countries*, OECD Working Papers, Vol III, No 41 , Organisation for Economic Co-operation and Development, Paris
OECD (1996a), *Environmental Performance Reviews in OECD Countries: Progress in the 1990s*, Organisation for Economic Co-operation and Development, Paris
OECD (1996b), *Environmental Performance Reviews: United States,* Organisation for Economic Co-operation and Development, Paris
OECD (1997), *Microfinance for the Poor?,* Organisation for Economic Co-operation and Development, Paris
OECD (1998), *Improving the Environment through Reducing Subsidies, Part I and II,* Organisation for Economic Co-operation and Development, Paris
OECD (1999a), *Bank Profitability 1999: Financial Statements of Banks,* Organisation for Economic Co-operation and Development, Paris
OECD (1999b), *OECD Environmental Data: 1999 Compendium,* Organisation for Economic Co-operation and Development, Paris (tables available at www.oecd.org/env)
OECD (2000), *Foreign Direct Investment and the Environment,* Organisation for Economic Co-operation and Development, Paris
Ottman, JA (1993), *Green Marketing: Challenges and Opportunities for the New Marketing Age*, NTC Business Books, Licolnwood
Pásztor, Z and D Bulkai (2001), 'The Hungarian Environmental Credit Line', in: JJ Bouma, MHA Jeucken and L Klinkers, *Sustainable Banking: The Greening of Finance*, pp360–371, Greenleaf, Sheffield
Pearce, D (ed) (1992), *The Macmillan Dictionary of Modern Economics,* Macmillan Press, London
Peters, J and R Wetzels (1997), *Niets nieuws onder de zon: En andere toevalligheden*, Contact, Amsterdam
Pezzey, J (1989), *Definitions of Sustainability, Centre for Economic and Environmental Development,* Discussion Paper No 9, CEED, Cambridge

Pezzey, J (1992), *Sustainable Development Concepts: An Analysis*, World Bank, Washington, DC
Ponting, C (1991), *Green History of the World*, Sinclair–Stevenson, London
Projectgroep NMP4 (1999), *Duurzaamheid en kwaliteit van leven*, VROM, Den Haag
Rabobank (2000), *Annual Sustainability Report 1999*, Rabobank, Utrecht (available at www.rabobankgroep.com/sustainability)
Rabobank International (1998), *Sustainability: Choices and Challenges for Future Development*, Rabobank International, Utrecht
Reder, A (1995), *75 Best Practices for Socially Responsible Companies,* GP Putnam's Sons, New York
Renneboog, L (1999), *Corporate Governance Systems: The Role of Ownership, External Finance and Regulation,* Centre for European Policy Studies (CEPS), Working Document No 133, Brussels
Repetto, R and D Austin (2001), 'Estimating the financial effects of companies' environmental performance and exposure', in: JJ Bouma, MHA Jeucken and L Klinkers, *Sustainable Banking: The Greening of Finance*, pp280–294, Greenleaf, Sheffield
Revenga, C et al (2000), *Pilot Analysis of Global Ecosystems: Freshwater Systems,* World Resources Institute, Washington, DC (available at www.wri.org/wri/wr2000/freshwater_page.html)
Riemsdijk, MJ van (1994), *Actie of dialoog: Over de betrekkingen tussen maatschappij en onderneming* (with English summary), Eburon, Delft
RIVM (2000a), *Milieubalans 2000: Het Nederlandse milieu verklaard* (*Environmental Balance 2000*), Samson, Alphen aan den Rijn (available at www.rivm.nl/milieu/nationaal/mb2000_s)
RIVM (2000b), *Nationale Milieuverkenning 5, 2000–2030* (*National Environmental Outlook*), Samson, Alphen aan den Rijn (available at www.rivm.nl/milieu/nationaal/mv5_s)
Robeco (1999), *Duurzaam Beleggen: 'Doing Well by Doing Good',* Robeco, Rotterdam
Royal Bank of Canada (2000a), *Community Report 1999: Beyond the Bottom Line*, RBC, Toronto (available at www.royalbank.com/community/1999report)
Royal Bank of Canada (2000b), *Innovation, Opportunity, People: 1999 Annual Report*, RBC, Toronto (available at www.royalbank.com)
Russell, B (1961), *History of Western Philosophy and Its Connection with Political and Social Circumstances from the Earliest Times to the Present Day*, Allen & Unwin, London
Russell, P (1992), *The White Hole in Time: Our Future Evolution and the Meaning of Now*, London
Ryall, C et al (1996), 'Appraisal of the selection criteria used in green investment funds', *Business Strategy and the Environment*, Vol 5, pp231–241
SEP (1998), *Elektriciteit in Nederland 1997*, SEP, Arnhem
Salomon Smith Barney (1998), *Social Awareness Investment*, Salomon Smith Barney, New York
SanPaolo IMI (2000), *1999 Annual Report*, SanPaolo IMI, Turin (available at www.sanpaoloimi.com)
Sarokin, D and J Schulkin (1991), 'Environmental concerns and the business of banking', *Journal of Commercial Bank Lending*, No 745, pp6–19
Saunders, A (2000), *Financial Institutions Management: A Modern Perspective*, McGraw-Hill, Boston
SBA (1997), *Umweltmanagement in Banken*, Schweizerische Bankiervereinigung, Basel (available at www.unep.ch/eco)
Schaltegger, S and F Figge (2001), 'Sustainable development funds: Progress since the 1970s', in: JJ Bouma, MHA Jeucken and L Klinkers, *Sustainable Banking: The Greening of Finance*, pp203–210, Greenleaf, Sheffield

Schmidheiny, S and FJL Zorraquín, with the WBCSD (1996), *Financing Change: The Financial Community, Eco-Efficiency, and Sustainable Development*, MIT Press, Cambridge, MA

Schmidheiny, S, with the WBCSD (1992), *Changing Course: A Global Business Perspective on Development and the Environment*, MIT Press, Cambridge, MA

Scholtens, LJR (1997), *De ontgroening van de fiscale groenfondsen: Een tussentijdse evaluatie van de Regeling Groenprojecten,* Free University of Amsterdam, Amsterdam

Scholz, RW, O Weber, J Stünzi, W Ohlenroth and A Reuter (1995), *Umweltrisiken systematisch erfassen*, Schweizer Bank, No 95/4, pp45–47

Schrama (1999), 'Banken en milieu in Nederland', *Milieu*, No 4, pp192–203

Segerson, K (1993), 'Liability transfers: An economic analysis of buyer and lender liability', *Journal of Environmental Economics and Management*, No 25, pp46–63

Shell (1993), *Annual Report 1993*, Shell, London

Shell (1998), *Profits and Principles– Does There Have to Be a Choice?,* Shell, London

Siddiqui FA and P Newman (2001), 'Grameen Shakti: Financing renewable energy in Bangladesh', in: JJ Bouma, MHA Jeucken and L Klinkers, *Sustainable Banking: The Greening of Finance*, pp88–95, Greenleaf, Sheffield

Simons, L Ph, A Slob and H Holswilder (2000), *'The Fourth Generation': New Strategies Call for New Eco-Indicators* (available on conference CD-Rom of Euro Environment conference 'Visions, Strategies and Actions Towards Sustainable Industries', 18–20 October 2000, Aalborg; also available at www.euro-environment.dk)

Skillius, A and U Wennberg (1998), *Continuity, Credibility and Comparability: Key Challenges for Corporate Environmental Performance Measurement and Communication,* Lund University, Lund

Smith, A (1759), *The Theory of Moral Sentiments*, Liberty Classics, Indianapolis, IN, reprinted 1976

Smith, A (1776), *An Inquiry into the Nature and Causes of the Wealth of Nations*, London, Penguin Books, reprinted 1979

Smith, DR (1995), *Environmental Risk: Credit Approaches and Opportunities*, UNEP, Geneva

Social Investment Forum (1999), *1999 Report on Socially Responsible Investing Trends in the United States*, Social Investment Forum, Washington, DC

Société Générale Group (2000), *1999 Annual Report*, Société Générale, Paris (available at www.socgen.com)

Spappens, P (ed) (1996), *Nederland Duurzaam Plus: Duurzame ontwikkeling in Europees perspectief*, Vereniging Milieudefensie, Amsterdam

Spar Nord Bank (1994), *The Ethical Accounting Statement*, SNB, Copenhagen

Spar Nord Bank (2000), *Årsrapport 1999*, SNB, Copenhagen (available at www.sparnordbank.com)

Spengler, L van (1994), *Hoe waarden veranderen: Een leerproces*, Prometheus, Amsterdam

Steer, A and E Lutz (1993), 'Measuring environmentally sustainable development', *Finance and Development*, No 304, pp20–23

Stigson B (2001), 'Making the link between environmental performance and shareholder value', in: JJ Bouma, MHA Jeucken and L Klinkers, *Sustainable Banking: The Greening of Finance*, pp166–172, Greenleaf, Sheffield

Stikker, A (1992), *The Transformation Factor: Towards an Ecological Consciousness*, Element, Rockport, MA

Storebrand (2000), *Corporate Social Responsibility: Environmental Report 1998–2000 and Corporate Social Responsibility Action Plan 2000–2002*, Storebrand, Oslo

Street, P and PE Monaghan (2001), 'Assessing the sustainability of bank service channels: The case of The Co-operative Bank', in: JJ Bouma, MHA Jeucken and L Klinkers, *Sustainable Banking: The Greening of Finance,* pp72–87, Greenleaf, Sheffield

Streiff, T (2000), 'Climate Change and Proactive Strategies for Mitigation: A Swiss Re's View', paper presented at the Triple Bottom Line Investing 2000 Conference, 2–3 November, Rotterdam (available on conference CD-Rom)
Sumitomo Bank (2000), *Annual Report 2000*, Sumitomo Bank, Osaka (available at www.sumitomobank.co.jp)
SustainAbility/UNEP (1999), *Engaging Stakeholders 1999: The Social Reporting Report*, SustainAbility, London
Svenska Handelsbanken (2000), *Annual Report 1999*, Svenska Handelsbanken, Stockholm (available at www.handelsbanken.se)
Szirmai, A (1994), *Ontwikkelingslanden: Dynamiek en stagnatie*, Wolters-Noordhof, Groningen
Tarna, K (2001), 'Reporting on the environment: Current practice in the financial services sector', in: JJ Bouma, MHA Jeucken and L Klinkers, *Sustainable Banking: The Greening of Finance*, pp149–165, Greenleaf, Sheffield
The Banker (2000), 'Top 1000 world banks', July 2000, pp178–182
Thomas, C, T Tennant and J Rolls (2000), *The GHG Indicator: UNEP Guidelines for Calculating Greenhouse Gas Emissions for Businesses and Non-Commercial Organisations,* UNEP, Geneva
Tietenberg, T (1985), *Emission Trading: An Exercise in Reforming Pollution Policy,* Resources for the Future, Washington, DC
Tietenberg, T (1992), *Environmental and Natural Resource Economics*, HarperCollins, New York
Tietenberg, T (1997), *Disclosure Strategies for Pollution Control*, Keynote Address to the 8th Annual Meeting of the European Association of Environmental and Resource Economists, June, Tilburg
Tokyo Marine (2000), *Environmental Report 2000*, Tokio Marine, Tokyo
Tomorrow (1993), 'Banking on the Planet', July, pp32–34
Tomorrow (1997), 'Green' Companies Provide Better-Than-Average Investment Returns, September/October
Tonnaer, FPC (1994), *Handboek van het Nederlands milieurecht*, Tjeenk Willink, Utrecht
Triodos Bank (2000), *Milieujaarverslag 1999*, Triodos Bank, Zeist
Tucker, ME and DR Williams (1997), *Buddhism and Ecology: The Interconnection of Dharma and Deeds*, Harvard University Press, Cambridge, MA
UBS (1999), *Environmental Report 1998/99*, Union Bank of Switzerland, Zurich (available at www.ubs.com)
UBS (2000), *Environmental Report 1999/2000*, Union Bank of Switzerland, Zurich (available at www.ubs.com/umwelt)
Ulph, A and L Valentini (2000), *Environmental Liability and the Capital Structure of Firms*, paper presented at the tenth annual conference of the European Association of Environmental and Resource Economists, Crete, 30 June–2 July 2000 (paper available at www.sussex.ac.uk/Units/gec/ph4summ/ulph-mac.htm)
UNCED (1992), *Earth Summit*, Rio de Janeiro
UNDP (1998), *Debt-For-Environment Swaps for National Desertification Funds*, United Nations Development Programme, New York (available at www.undp.org/seed/unso)
UNDP (2000), *Human Development Report 2000*, Oxford University Press, New York (available at www.undp.org/hdr2000)
UNEP (1995), *Environmental Policies and Practices of the Financial Services Sector*, United Nations Environment Programme, Geneva
UNEP (1997), *Finance for Sustainable Development: The Road Ahead*, United Nations Environment Programme, New York

UNEP (1998), *UNEP Financial Institutions Initiative 1998 Survey*, United Nations Environment Programme/PWG, Geneva
UNEP (2000), *Global Environmental Outlook 2000,* Earthscan, London (available at www.unep.org/geo2000/index.htm)
UNEP/ICC/FIDIC (1997), *Environmental Management System Training Resource Kit*, United Nations Environment Programme, Geneva
UniCredito Italiano (2000), *1999 Annual Report*, UniCredito Italiano, Milan (available at www.unicredito.it/en)
United Nations (1980), *World Conservation Strategy*, United Nations, New York
United Nations (1996), *World Population Prospects: The 1996 Revision*, United Nations, New York
UNRISD and UNA (1998), *Business Responsibility for Environmental Protection in Developing Countries*, United Nations Research Institute for Social Development, Geneva
US EPA (2000a), *Strategic Plan 2000*, US Environmental Protection Agency, Washington, DC
US EPA (2000b), *Superfund: 20 Years of Protecting Human Health and the Environment*, US Environmental Protection Agency, Washington, DC (available at www.epagov/superfund)
US EPA (2000c), *Brownfields Economic Redevelopment Initiative,* US Environmental Protection Agency, Washington, DC (available at www.epagov/brownfields)
Van City (2000), *Guided by Values: The Van City Social Report 1998/99*, Van City, Vancouver (available at www.vancity.com/socialreport)
Van Melle, NV (1997), *Annual Report 1996*, Van Melle, Breda
Varian, HR (1987), *Intermediate Microeconomics: A Modern Approach*, Norton, New York
VBDO (1998), *Environmental Information for Investors, Association of Investors for Sustainable Development*, Culemborg
Veldhoen, E and B Piepers (1995), *Kantoren Bestaan Niet Meer: De Digitale Werkplek in een Vitale Organisatie*, 010 Publishers, Rotterdam
Verbond van Verzekeraars (1997), 'Duurzame onderneming straks goed verzekerd', *Dossier*, June (Dutch Association of Insurers)
Verbruggen, H (ed) (1996), *Sustainable Economic Development Scenarios (DEOS): An Analysis up to the Year 2030*, Ministry of Environment, The Hague
VfU (1998), *Time to Act; Environmental Management in Financial Institutions*, Verein für Umweltmanagement in Banken, Bonn (available at www.vfu.de)
VNO–NCW (1998), *Ontwikkelingen in milieu-aansprakelijkheid*, VNO–NCW, The Hague
Voogt, AA (1995), *Managen in een meervoudige context* (with English summary), Eburon, Delft
Wakker, E (2000), *Funding Forest Destruction: The Involvement of Dutch Banks in the Financing of Oil Palm Plantations in Indonesia*, AIDEnvironment, Amsterdam
Walton, M and WE Demming (1986), *The Deming Management Method,* Dodd, New York
Watzlawick, P, JH Beavin and DD Jackson (1967), *Pragmatics of Human Communications*, Norton, New York
Watzlawick, P, JH Weakland and R Fisch (1973), *Change: Principles of Problem Formation and Problem Resolution*, Palo Alto
WBCSD (1997), *Signals of Change: Business Progress Towards Sustainable Development*, WBCSD, Geneva
WBCSD (2000), *Measuring Eco-Efficiency: A Guide to Reporting Company Performance*, WBCSD, Geneva
WCED (1987), *Our Common Future*, Oxford University Press, Oxford

Weizsäcker, EU von, AB Lovins and LH Lovins (1997), *Factor Four: Doubling Wealth, Halving Resource Use*, Earthscan, London

Wempe, J and M Kaptein (2000), *Ondernemen met het oog op de toekomst: Integratie van economische, sociale en ecologische verantwoordelijkheden*, SMO, Den Haag

Wetenschappelijk Raad voor het Regeringsbeleid (WRR) (1992), *Milieubeleid: Strategie, instrumenten en handhaafbaarheid*, SDU, The Hague

White, AL (1999), 'Sustainability and the accountable corporation: Society's rising expectations of business', *Environment*, Vol 41, No 8, pp30–43

White, R, S Murray and M Rohweder (2000), *Pilot Analysis of Global Ecosystems: Grassland Ecosystems*, World Resources Institute, Washington, DC (available at www.wri.org/wri/wr2000/grasslands_page.html)

Woerd, KF van der and P Vellinga (1997), *Eco-rentabiliteit*, Free University of Amsterdam, Amsterdam

Wohlegemuth N (2001), 'Directing investments to cleaner energy technologies: The role of financial institutions', in: JJ Bouma, MHA Jeucken and L Klinkers, *Sustainable Banking: The Greening of Finance*, pp401–411, Greenleaf, Sheffield

World Bank (1996), *Sustainable Banking with the Poor: A Worldwide Inventory of Microfinance Institutions*, World Bank, Washington, DC

World Bank (1997), *Kan het milieu wachten?: Prioriteiten voor Oost-Azië*, World Bank, Washington, DC

World Bank (1998), *Global Development Finance 1998: Analysis and Summary Tables*, World Bank, Washington, DC

World Bank (1999), *Pollution Prevention and Abatement Handbook 1998: Toward Cleaner Production*, World Bank Group, Washington, DC

World Bank (2000), *World Development Report 2000/2001: Attacking Poverty*, Oxford University Press, New York (available at www.worldbank.org/poverty)

WRI (2000), *World Resources 2000–2001: People and Ecosystems: The Fraying Web of Life*, World Resources Institute, Washington, DC (available at www.wri.org/wri/wr2000)

Yasuda Fire and Marine Insurance (2000), *Sustainability Report 2000*, Yasuda Fire and Marine Insurance, Tokyo (available at www.yasuda.co.jp)

Zimmerman, W and B Mayer (2001), 'Banks and environmental practices in Bangkok Metropolitan Region: The need for change', in: JJ Bouma, MHA Jeucken and L Klinkers, *Sustainable Banking: The Greening of Finance*, pp133–146, Greenleaf, Sheffield

Index

Page numbers in *italics* refer to boxes, tables and figures